MYCOGENETICS

MYCOGENETICS

An Introduction to the
General Genetics of Fungi

J. H. BURNETT

Sibthorpian Professor in the University of Oxford and
Fellow of St. John's College

JOHN WILEY & SONS

London · New York · Sydney · Toronto

Library of Congress Cataloging in Publication Data:

Burnett, John Harrison, 1922–
Mycogenetics.

Bibliography: p.
1. Fungal genetics. I. Title. [DNLM: 1. Fungi.
2. Genetics. Q K602 B964m]

QH433.B87   589'.2'0415   74-13143
ISBN 0 471 12445 1

Printed in Great Britain by William Clowes & Sons Ltd.,
London, Beccles and Colchester

Preface

This book is written with the object of describing a particular and fundamental aspect of the biology of the fungi, namely, their genetics.

Many mycologists, pathologists and industrial users of fungi have a limited knowledge of genetics and they are often uncertain how far the intensive studies which have been carried out on a relatively small number of fungi are of relevance for, or applicable to, the fungi as a whole. Similarly, most geneticists do not realize the great untapped potential for genetic studies which fungi possess. Finally, biologists are often unaware of the contributions which fungi could make to general biological ideas.

In this book I have tried to provide a remedy for some of these deficiencies by describing the genetics of fungi in the broad context of their biology. I have tried to concentrate on general issues and problems rather than to report the latest findings on any particular topic. I have also attempted to indicate where knowledge is lacking, inadequate or unrepresentative.

This objective has determined the plan of the book. Formal genetics are dealt with in great depth in other texts. Here, therefore, by the use of worked examples, I have endeavoured to illustrate as simply as possible the common procedures for simple genetical analysis. More esoteric topics are held over to the last two chapters. Population genetics of fungi have not, so far, been taken up widely by geneticists nor have their consequences been fully appreciated by workers with fungi. I have, therefore, dealt with these areas in greater depth. The last section is intended for the main 'users' of fungal genetics. Their attitudes, approaches and requirements are very different. Here I have only been able to develop a limited number of representative topics of current interest in some depth.

Acknowledgements to authors for extracts, data and figures are indicated in the text. I am grateful for the assistance of my secretary, Mrs. G. Tien, for typing and my son, A. M. Burnett, for help with the Index.

v

My wife assisted in innumerable ways, not least by preparing all the illustrations. As ever, I am indebted to her for her continuing affectionate collaboration and criticism.

Oxford, April 1974   J.H.B.

Acknowledgements

Acknowledgement to individual authors, with the appropriate reference which can be found in the bibliography, for the reproduction of figures and tabular material, with or without modification, is made in the text. I am indebted to the following publishers for permission to reproduce these items and certain quotations.

Academic Press. *Advances in Genetics*, Fig. 2.4, 11.1; *Disease resistance in plants*, lines 14–18, p. 278; *The Fungi, an advanced treatise*, Fig. 5.3; *Physiological Plant Pathology*, Table 13.8.

Akademie der Wissenschaften der D.D.R. *Abhandlungen*, 1967. Fig. 12.3, Tables 12.2 and 12.3.

Annual Reviews Inc. *Annual Review of Genetics*, lines 9–14, p. 291; *Annual Review of Phytopathology*, Table 13.7.

Edward Arnold (Publications) Ltd., *Fundamentals of Mycology*, Fig. 1.1, 1.3, 1.4, 1.6, 1.7, 1.8, 1.9, 6.4, 6.6, and 11.2; Table 13.4 and 13.6.

W. A. Benjamin Inc. *Genetic Complementation*, Fig. 15.2.

Botanical Society of America. *American Journal of Botany*. Tables 8.2 and 9.1.

Blackwell Scientific Publications Ltd., *Fungal Genetics*, Fig. 2.2a.

Brookhaven National Laboratory. *Brookhaven Symposia on Biology*, Fig. 5.4.

Churchill-Livingstone Ltd. *Progress in Industrial Microbiology*, Fig. 12.4, Table 12.4.

Columbia University Press. *The Origin of Adaptations*, Fig. 8.6.

Faber and Faber. *Genetic Polymorphism*, lines 16–22, p. 211.

Friesia, Fig. 11.4.

Genetics, Figs. 3.4, 9.3, Table 6.2.

Genetics Society of Canada. *Canadian Journal of Genetics and Cytology*, Table 13.10.

Heredity. Figs. 6.2, 7.4, 9.4, Tables 5.1 and 6.1.

Long Island Biological Association Inc. *Cold Spring Harbor Symposia on Qualitative Biology*. Figs. 2.2b, 2.2c and 9.2.

National Academy of Sciences of the U.S.A. *Proceedings*, lines 1–7, p. 29.

National Research Council of Canada. *Canadian Journal of Botany*. Figs. 8.2, 9.5b, 10.1, Table 11.4; lines 14–19, p. 225.

Nature. Figures 2.2d, 4.2, 10.4; lines 13–17, p. 263.

Neurospora News Letter. Lines 13–24, p. 303.

New Phytologist Trust. *The New Phytologist*, Figure 8.4.

Phytopathology. Figures 9.5a, 13.2, Table 13.11.

Ronald Press Co. *Principles of Plant Pathology, Genetics of Sexuality in Higher Fungi*, lines 35–38, p. 167.

Royal Society of London. *Proceedings*, Figure 9.1, Table 9.2 and lines 4–11, p. 185.

Research Council of Israel. *Bulletin*, Table 7.1.

Society for Experimental Biology, Figure 15.3.

U.S. Department of Agriculture, Figure 10.2, Tables 10.2 and 13.2.

I gratefully acknowledge permission to use, in Figures 8.1 and 13.3, unpublished data presented by Drs. R. A. McIntosh, I. A. Watson and N. H. Luig at the 2nd International Congress of Plant Pathology, Minneapolis, 1973.

Contents

Section 1. Introduction

Section 2. Formal Genetics

Section 4. Applications of Fungal Genetics

Section 1

Introduction

The structural organization of fungi is almost unique. Only in one small group of algae is the basic constructional unit a fine tube which grows apically. In contrast, the fungi exploit this mode of construction in myriad ways. They also differ from most other plant and animals in their life cycles and biology. An acquaintance with these characteristic features is, therefore, a prerequisite for understanding their genetics. Yet, since fungi are eukaryotes, their genetics can be related to those organisms whose study has laid the classical foundations of genetics.

CHAPTER I

Fungi as organisms for genetic study

The study of the genetics of the fungi is of value for a variety of reasons. The biology of fungi cannot be adequately studied or assessed without a knowledge of the causes of their variability; plant pathologists, plant breeders and industrialists all require such information either to prevent, or to utilize the activities of the fungi concerned. Moreover, the fungi have provided geneticists with novel genetic systems and with the opportunity of studying particular problems with greater ease or precision than is possible in other organisms.

At present the genetics of some 20 genera and 30 species have been studied fairly intensively. How far these are truly representative of the estimated 4200 genera and 50,000+ species of fungi (Ainsworth, 1971) it is not easy to assess. It is still true that novel features continue to be revealed as new fungi are studied.

In this chapter an outline of the structure and biology of the fungi will be followed by some selected life cycles; finally, attention will be drawn to important features of genetic interest.

The main structural and biological features of fungi

Nutrition

Fungi are organisms which are heterotrophic for carbon compounds.

In many cases their nutritional requirements can be met in culture by supplying simple sugars, such as glucose or sucrose, mineral salts including a source of inorganic nitrogen, and water, e.g. *Aspergillus nidulans*. Several fungi, notably *Neurospora crassa*, greatly used in genetic experimentation, are also heterotrophic for vitamins such as thiamin, biotin or pyridoxin; of these, as in *N. crassa*, the first is most frequently required. Some fungi prefer ammonia to nitrate as a source of nitrogen, while others require organic nitrogen in forms such as asparagine, e.g. *Phycomyces blakesleeanus*, or, rarely, as amino acids. Some aquatic forms, whose genetics have not yet been studied, require novel carbon sources, e.g. *Leptomitus lacteus* cannot utilize sugars but can utilize fatty acids.

Thus the vast majority of fungi, including many that are parasitic in nature, e.g. smut fungi (Ustilaginales), can be grown readily as saprophytes on defined media between about 20–37°C. The most important exceptions are the strictly obligate parasites (obligate biotrophs, Lewis, 1973) such as the rust fungi (Uredinales) and the downy and powdery mildews (Perenosporales, Erysiphales), the vast majority of which can only be grown in conjunction with their host plants. Nevertheless, genetic studies of rust fungi and powdery mildews have been carried out.

Structure

Apart from some green algae, the fungi are almost unique in the living world in utilizing a microscopic cylinder, the *hypha*, as their basic constructional unit. Hyphae have walls and grow in length only at their tips, they are filled with cytoplasm which, apart from the growing tips may be vacuolated. In the cytoplasm mitochondria, lipid droplets, crystalline inclusions and nuclei occur. The hyphae may be partitioned along their length into cells by incomplete, or more rarely complete, transverse septa, or may lack transverse septa, i.e. are coenocytic, save where moribund or specialized hyphae are cut off. Incomplete septa are perforated either by a simple central circular pore or a complex dolipore. Cytoplasm and its inclusions can certainly pass through simple pores but there is still controversy about the passage of nuclei through dolipores. If such passage does occur it may involve the enzymatic reduction of the dolipores to the simple condition. An important property of the nuclei is that they can migrate along hyphae and through pores in certain circumstances apparently independently of cytoplasmic streaming, which is the process responsible for the transport of other inclusions. Nuclei are most frequently haploid in hyphae but in some fungi the diploid condition may persist, e.g. some yeasts, *Allomyces* spp.; or a few diploid nuclei may occur transiently amongst predominantly haploid ones, e.g. *Aspergillus niger*. The nuclei are variously distributed in the hyphae. The extreme apical region is usually entirely free of nuclei but the apical cell may regularly have one, two or more nuclei. The conditions where the cells are predominantly uni- or binucleate are described as monokaryotic or dikaryotic, respectively (see p. 8). In some fungi the apical cell is multinucleate but the distal cells are uninucleate, e.g. *Fusarium oxysporum*. Little is known concerning the processes regulating the numbers of nuclei per cell but there is a suggestion that for species, whether cellular or coenocytic, there is a characteristic nucleo-cytoplasmic ratio in active hyphae.

The hyphae are capable of branching at a regular distance, characteristic of the species, behind the apex. Secondary, tertiary and higher orders of branching can occur. When growth is confined to a single plane, as when a fungus is grown on disks of cellophane on the surface of nutrient agar, a small piece of hypha rapidly grows and branches to form a radially symmetrical, marginally expanding *mycelium*. When growth is not so confined it extends in all planes so forming a roughly spherical colony, e.g. in liquid shake culture. Such a growth pattern implies some form of internal regulation of growth between apical and lateral hyphae. Its nature is not understood. When linear growth is slow and associated with close branching the mycelium forms a discrete, spatially restricted colony, and is said to show a *colonial habit*. This habit can sometimes be induced by application of external agents, notably replacement of the normal nutrient sugar by the non-metabolized hexose, sorbose. This has valuable practical applications when it is desired to plate out several colonies of a normally, rapidly growing species on a single petri dish.

Within a mycelium the growth habit may be modified in two ways. Firstly, aggregation of hyphae, often associated with increased branching, may result in the development of complex strand-like structures which can rise up vertically as macroscopic organs—*coremia*—or form specialized horizontal strands—*rhizomorphs*—capable of very rapid growth and active translocation of metabolites. By contrast, some aggregates may form more-or-less spherical or irregular bodies called *sclerotia* which are capable of resisting desiccation and of lying dormant. At other times the aggregated hyphae form a pseudo-parenchymatous tissue called a *stroma* on, or in, which reproductive structures may later develop, e.g. *Venturia inaequalis*.

Secondly, hyphal fusions may occur between existing or induced hyphal tips within a mycelium, or between mycelia. Thus a mycelium can be transformed into a three-dimensional network and, even more remarkable, several mycelia may develop into a single biological unit as a consequence of fusions between their hyphae. This latter type of fusion provides an opportunity for genetically distinct mycelia to become associated in a single physiological unit. The differences may lie in the nuclei, cytoplasm or both and, since nuclei can migrate and cytoplasm stream, the cell contents may become intermingled. Such compounded mycelia containing genetically different nuclei are termed *heterokaryons* or, if with distinct cytoplasms or cytoplasmic factors, *heteroplasmons* (see Chapters 5 and 6). It will be realized therefore, that the factors which regulate hyphal fusion and the processes subsequent to it, migration of nuclei, of cytoplasm or both, are of great importance for the biology and genetics of fungi. In general, such fusions are confined to cellular fungi and, even here, a variety of environmental and inherited factors regulate and restrict fusions and their consequences. In coenocytes fusions are rare (the genus *Mortierella* is possibly an exception) but heterokaryons or heteroplasmons can arise in time by the accumulation of mutant nuclei or mutant cytoplasmic factors, respectively, within a single mycelium.

Attention has been directed, so far, to hyphal fungi, but several, notably the yeasts, *Sporobolomyces* and some aquatic forms either lack hyphae or only develop them under certain conditions. Yeast cells are unicellular and usually uninucleate and grow either by fission, e.g. *Schizosaccharomyces* spp., or by budding, as in baker's or brewer's yeast, *Saccharomyces* spp. (Figure 1.5).

The cells often stick loosely together, forming little clones of cells which repeat the growth process. Details of their structure, ultrastructure and composition are similar to those of hyphal forms. Indeed, many hyphal forms, e.g. *Mucor* spp., can be induced to develop as unicellular, yeast-like cells—the so-called *Torula* condition—by appropriate treatments, e.g. liquid culture with high CO_2 or sugar contents. In uninucleate, unicellular fungi, cell fusions are usually associated with sexual reproduction so that heterokaryons and heteroplasmons do not arise so frequently but may arise, as in coenocytes, by accumulated mutations.

Nuclear cytology

The nuclei of fungi are minute and exhibit a number of technical problems in relation to the usual stains and procedures employed for chromosomes. There is no doubt that mitosis occurs in the sense that a process of division occurs which results in two daughter nuclei each possessing a complete chromosome complement. The mechanics of the process which brings this about vary considerably. There seem to be some fungi in which the process resembles that of a classical mitosis, e.g. *Macrophomina phaseoli*, but even here the processes probably occur within the nuclear membrane, until the final separation, and there is no distinct metaphase-plate arrangement of chromosomes. In most fungi there are even greater differences: mitosis is nearly always intranuclear, the spindle of microtubules, visible by electron microscopy, arises in various unusual ways and the mechanics of chromosome alignment and chromatid separation on it are very different from the classical situations. Nothing is known of how nuclei fuse somatically or of how somatic segregation occurs.

Meiosis is best documented for the Ascomycetes, partly because the ascus is a large, readily accessible cell and partly because the chromosomes are larger than at mitosis. Broadly, meiosis in fungi resembles that in higher plants, most of the differences being in the prophase. In *Neurospora crassa*, for example, the chromosomes are highly contracted when pairing commences in zygotene; leptotene—in the usual sense—is lacking. Maximum chromosome extension is reached at pachytene.

Fungal cytology is in an active state at present and many of its problems should be soluble by a combination of light and electron microscopy. So far as the genetics of fungi are concerned, the consequences of both mitosis and meiosis are the same as for other organisms.

Sporulation

Most fungi multiply and are disseminated by means of spores. Those which do not are termed *Mycelia sterila* and have either lost the ability to develop spores or have been prevented by intrinsic or extrinsic factors from so doing. Spores can be divided into three types, sporangiospores, conidiospores and chlamydospores, although cases exist where the spore-type is in some respects intermediate in character.

(i) *Sporangiospores* arise within a special structure, a *sporangium*, and are eventually liberated from this enclosing structure. They may be uni- or multinucleate, motile, with 1 or 2 flagella, i.e. zoospores, as in *Phytophthora* spp., or lacking flagella, and non-motile, e.g. *Phycomyces* spp. (Figures 1.2 and 1.3). In certain cases spores develop in a specialized sporangium in which nuclear fusion and meiosis, i.e. sexual reproduction, has occurred. Each of these special sporangia is termed an *ascus* or a *basidium*; the spores they produce are called *ascospores* and *basidiospores* respectively (Figures 1.6–1.9).

(ii) *Conidiospores* or, more usually, *conidia* arise externally to a hypha as a result of various types of abstriction processes. The hypha giving rise to conidia is often specialized and is called a *conidiophore*. The kinds of conidia and conidiophore are legion and are used as generic characteristics. Several hundred genera and thousands of species are known which lack sexual reproduction and are multiplied and dispersed only through conidia. These are the *Fungi Imperfecti*. Apart from the many branched and unbranched forms of conidiophore, there are three other structures on, or in, which conidia develop. These are synnemata (or coremia), acervuli and pycnidia. A *synnema* or *coremium* is a vertical aggregation of hyphae in parallel alignment, often joined together laterally through hyphal anastomoses. Conidia are borne either at the tip or all over the surface of the structure. The technical distinction between the two forms is that a synnema is more compact than a coremium.

An *acervulus* is an aggregation of short, much-branched conidiophores to form a saucer or disk-like clump of conidia-bearing structures. A *pycnidium* is a hollow spherical or pyriform structure, usually with an apical aperture or *ostiole*. The conidia, called here pycnidiospores, are borne on the surface lining the structure and liberated into its cavity, from whence they emerge either through the pore or by rupture of the wall.

In the toadstools and their allies, conidia are usually developed only on monokaryotic hyphae (see p. 9). They are often termed *oidia*. Conidia may be unicellular or multicellular and each cell is uni- or multinucleate.

(iii) *Chlamydospores* are intercalary structures developed within hyphae. They usually acquire thick walls and their cytoplasm accumulates lipid material and glycogen. They are generally multinucleate.

Spores derived from a cell in which meiosis has just preceded spore formation are sometimes called *mitospores*. The majority of spores arise by a mitotic process and can be described as *mitospores*.

Spores are liberated by a variety of devices and are then disseminated most frequently by the wind, their dispersal following the same pattern as that of smoke particles or pollen grains (see Gregory, 1973). In some cases they are dispersed through being ingested by a variety of animals and deposited in their droppings. Such fungi usually develop on dung and are termed *coprophilous*. A number of species used for genetic study belong to this biological group, e.g. *Ascobolus immersus*, *Sordaria* spp., but can often be grown also in plate culture on simply defined media. In other cases, the spores adhere to the surface of the animal and are eventually brushed off. Spores dispersed by animals are often sticky, being developed in association with glutinous materials or in sporangial, or conidial drops, e.g. *Verticillium* spp.; those dispersed by the wind are usually dry, e.g. *Penicillium* spp.

Sexual reproduction

Apart from the Fungi Imperfecti and Mycelia sterila, fungi reproduce sexually. It is convenient in fungi to recognize three stages in sexual reproduction: *plasmogamy*, *karyogamy* and *meiosis*, i.e. the fusion of the cytoplasm of conjugant cells, however specialized or unspecialized; the fusion of conjugant nuclei; and the reduction divisions of such fusion nuclei.

Plasmogamy may occur between cells of four kinds, (a) between motile gametes, (b) between a motile gamete and a gametangium, (c) between gametangia or (d) between unspecialized hyphae. The first two are restricted to aquatic fungi. When gametes are involved they are isogamous, heterogamous or even oogamous, e.g. *Allomyces* spp. Plasmogamy between gametangia covers a wide range of morphological manifestations. In some the gametangia are isogamous and little more than swollen ends of hyphae, or hyphae somewhat thicker than usual, e.g. *Mucor* spp. (Figure 1.3); in others they may be similarly differentiated morphologically but differ in size, e.g. *Zygorrhynchus heterogamus*; whereas in others there is a clearly differentiated oogonium and antheridium, e.g. *Phytophthora* spp. (Figure 1.2). In many Ascomycetes plasmogamy occurs between a conidium, which can act as a fertilizing agent, and a modified, often spirally wound hypha, which may or may not possess a fine, branched *trichogyne* (Figure 1.6). Such structures locate the site of future reproductive structures or *ascocarps*, e.g. *Neurospora*, *Ascobolus*. In some Ascomycetes, although such structures exist, plasmogamy occurs by fusion of unspecialized vegetative hyphae, e.g. *Humaria granulata*. In many Basidiomycetes this is the normal pattern of fusion but the location of the reproductive structures (basidiocarps) is not predetermined. When sexual reproduction is initiated by fusions of normal vegetative hyphae it can be called *somatogamy*, as opposed to *gametogamy*. An inevitable consequence of somatogamy is that potentially conjugant nuclei have to migrate through hyphae to the site(s) of karyogamy. In many Basidiomycetes the potentially conjugant nuclei become associated soon after plasmogamy and thereafter multiply in synchrony to give the binucleate cells of the *dikaryon*; hyphae with uninucleated cells existing before plasmogamy are then termed *monokaryons*.

Karyogamy is usually confined to specialized cells, even in fungi with somatogamous fusion, which become the specialized sporangia termed the *ascus* in Ascomycetes and the *basidium* in Basidiomycetes. An ascus is, initially, a swollen subterminal cell which usually elongates to form a cylinder. Two conjugant nuclei enter the base of the cell, fuse and immediately undergo meiosis. The four nuclei so formed usually divide once more, mitotically, to give eight nuclei and these become surrounded by cytoplasm and then isolated as eight, endogenous ascospores (Figure 1.6). These are actively or passively extruded from the ascus tip.

A *basidium* is a swollen terminal cell which shows a development similar to that of an ascus until meiosis is over. Basidia are usually club shaped and towards their distal ends, narrow, pointed sterigmata are extended. Their tips enlarge to form basidiospores and the nuclei, derived from meiosis, migrate from the basidium via the sterigmata, becoming extremely thin and elongated en route, into the basidiospores (Figure 1.9).

Initially each basidiospore is typically uninucleate but a further mitosis often occurs, giving rise to binucleate basidiospores; other variants are also known. Basidiospores are violently discharged except when the basidium is totally enclosed, as in the Gasteromycetes. There are also different forms of the basidium which may become multicellular, as in the rust and smut fungi (Figures 1.7 and 1.8).

There is thus a great range in the morphological events preceding and immediately following karyogamy. There is an equally great range of physiological and genetical controls which can affect both plasmogamy and karyogamy, or either process alone. Organized processes which ultimately control the kinds of nuclei that fuse in sexual reproduction are termed *mating systems*, of which many different kinds are known in the fungi (see Chapter 8).

Parasexuality

Some fungi, in addition to or in place of, normal sexual reproduction, show *parasexual reproduction*. This term, first coined by Pontecorvo (1949) to describe the situation which he found in *Aspergillus nidulans*, involves the sequence of events (a) hyphal fusion and plasmogamy, (b) karyogamy in vegetative hyphae (*not* in specialized cells), (c) somatic recombination which may, or may not, be followed by non-meiotic reduction of the diploid nucleus to the haploid condition. In general, this is thought to be a rare series of events but data as to the frequency of parasexual events are still very sparse. Variations in this sequence are also possible.

Life cycles

The occurrence, sequence of processes and types of products involved in asexual and sexual reproduction, as well as the occurrence of hyphal fusions, have profound effects upon the life cycles and hence the whole biology of the fungi. Five basic life cycles, reduced from the seven described by Raper (1954), can be recognized (Figure 1.1). They are:

1. *Asexual*, in which sexual reproduction is apparently lacking entirely. Since this definition is based upon the absence of a phase, it is somewhat

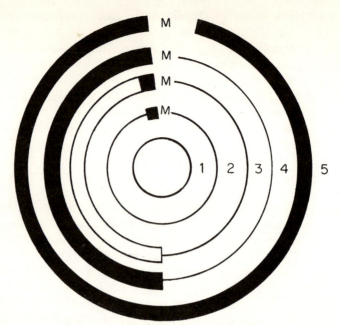

FIGURE 1.1 A diagram to illustrate the five basic life cycles in fungi. Each circle represents a life cycle and should be followed clockwise; M represents meiosis; ——— a haploid phase, ═══ a dikaryotic phase and ▬▬ a diploid phase. The life cycles shown are 1. Asexual; 2. Haploid; 3. Haploid dikaryotic; 4. Haploid diploid; 5. Diploid. (From Burnett, 1968.)

artificial but it has great convenience and accurately describes the situation in many fungi, viz, Fungi Imperfecti.

2. *Haploid* (or *haploid monokaryotic*), in which meiosis immediately follows nuclear fusion and the meiotic products are then dispersed. The diploid phase is, therefore, of minimal duration. This cycle is shown by many Phycomycetes and some Ascomycetes.

3. *Haploid dikaryotic*, similar to the preceding cycle save that paired, potentially conjugant kinds of nuclei persist in close physical association in the same hyphal segment (hence dikaryon) and divide synchronously for a greater or lesser period. At one extreme the nuclear association may be for a few cell generations only, e.g. in many Ascomycetes, binucleate ascogenous hyphae are developed just prior to ascus development, and such a dikaryon cannot apparently exist independently of the haploid phase (Figure 1.4). At the other extreme is the condition where the meiospores fuse immediately to re-form a dikaryon so that the fungus is dikaryotic throughout its life cycle, save for the moment of fertilization and immediately after meiosis. This can occur in yeasts (Saccharomycetales) but is more frequent in certain smut fungi (Ustilaginales). An intermediate and highly characteristic type of life cycle is shown by the

majority of Basidiomycetes. Here the mycelium derived from germination of a meiospore may persist in the haploid condition, as a monokaryon, but once a dikaryon is formed, through hyphal fusion for example, it shows potentially unrestricted and independent growth so that it may well comprise the most long-lived phase of the life cycle. For example, the dikaryotic phase of various fairy-ring fungi has probably persisted for several centuries.

4. *Haploid diploid*, in which these phases alternate regularly, or can be induced so to do. Regular alternation is unusual in fungi and is probably restricted to a few species of aquatic Chytridiomycetes. Irregular alternation with potentially persistent haploid or diploid phases occurs in the yeasts.

5. *Diploid*, in which the haploid phase is restricted to the gametes or gametangial phase. It seems likely that the majority of Oomycetes may well conform to this pattern.

Classification

It is necessary to give a very brief outline of fungal classification. Excellent texts exist giving fuller details (see General References), so here, only the barest skeleton is given in the form of a key, but one which utilizes various features of the whole fungal organism, or *thallus*, and its biology.

1   Thallus unicellular or composed of hyphae; spores, if present, motile, or non-motile; incapable of sexual reproduction.

1.2 Thallus as above but *lacking either spores or sexual reproduction*.

MYCELIA STERILA

1.3 Thallus as above; spores motile or non-motile; *but sexual reproduction lacking*.

DEUTEROMYCETES (FUNGI IMPERFECTI)

2   Thallus unicellular but, if so, *multiplying by budding* or *simple fission*, or a mycelium of *septate* hyphae; *spores non-motile;* capable of sexual reproduction.

2.1 Thallus if unicellular, not multiplying by budding or simple fission, or a mycelium of hyphae which are *aseptate*, at least when young; spores, motile or non-motile; capable of sexual reproduction.

PHYCOMYCETES
(Lower Fungi)

2.2 Thallus rarely unicellular, multiplying by budding or simple fission, or more usually a mycelium of septate hyphae; spores either *conida* or produced endogeneously in an *ascus* (ascospores); usually capable of sexual reproduction.

ASCOMYCETES

2.3 Thallus rarely unicellular, but if so multiplying *only by budding*, or predominantly mycelial of septate hyphae; spores either conidia, absent or produced *on a basidium* (basidiospores); capable of sexual reproduction.

BASIDIOMYCETES

12

A classification of those fungi whose genetics have received appreciable study is provided in the appendix (p. 327).

It is now convenient to consider the life cycles of a number of fungi used in genetical work in order to see how the general processes described apply in particular instances. These are illustrated in Figures 1.2–1.9.

PHYTOPHTHORA

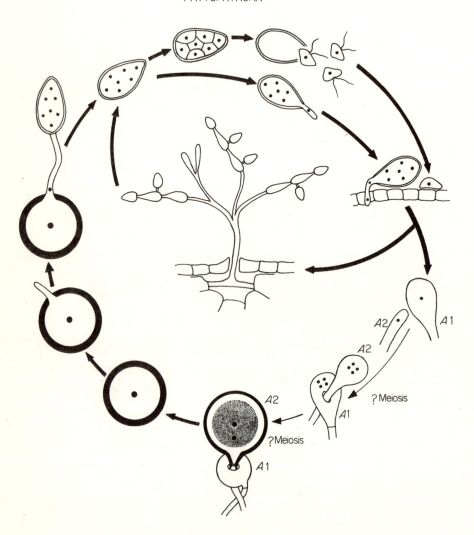

FIGURE 1.2 Life cycle of a dimictic *Phytophthora* (Phycomycetes–Oomycetes). In this diagram meiosis is shown to occur at the gametangial stage, i.e. the mycelium is diploid. There is still some controversy on this. The necessity for two mating types *A1* and *A2* is indicated. Interaction between mycelia can be complex. (See Figure 8.2, p. 159.)

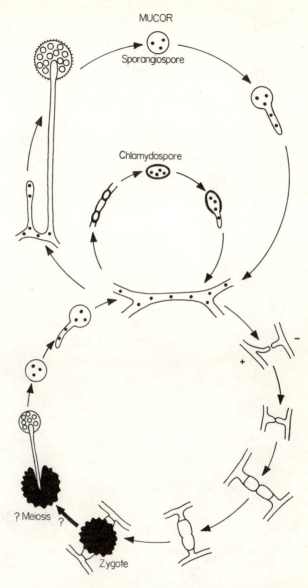

FIGURE 1.3 Life cycle of a dimictic *Mucor* (Phycomycetes–Zygomycetes). In fact, unequivocal cytological and genetical evidence for meiosis in the zygote is not shown but there is circumstantial evidence for supposing it occurs there. *Phycomyces* has a similar life cycle.

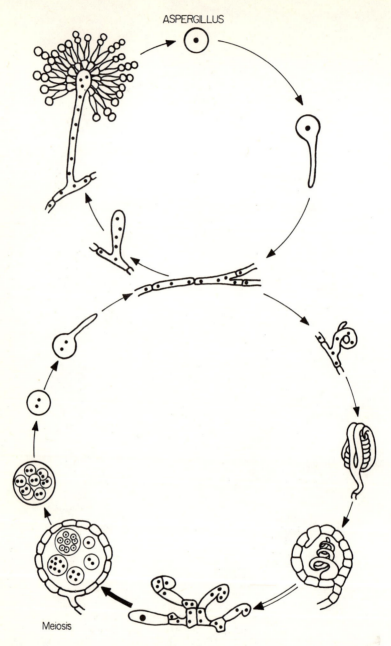

ASPERGILLUS

Meiosis

FIGURE 1.4 Life cycle of a homomictic *Aspergillus* (Ascomycetes–Plectomycetes). Note the brief dikaryotic phase in the ascogenous hyphae within the developing cleistothecium which eventually ruptures irregularly, as do the asci, releasing the binucleate ascospores. Meiosis is followed by a mitotic division in the ascospore.

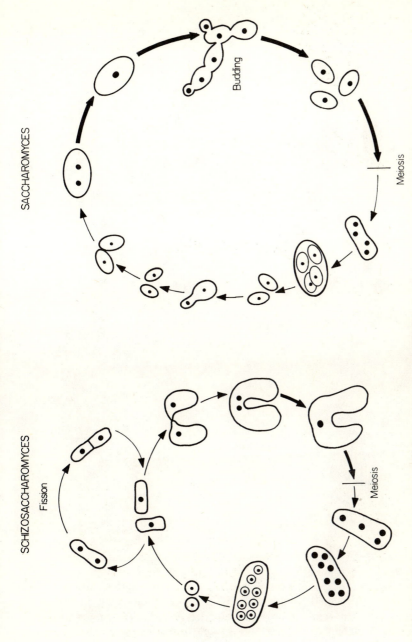

Figure 1.5 Life cycles of a dimictic budding yeast *Saccharomyces* and a fission yeast *Schizosaccharomyces* (Asco-mycetes–Hemiascomycetes). The principal difference lies in the mode of division of both haploid and diploid cells, budding or fission, and the number of ascospores.

16

NEUROSPORA

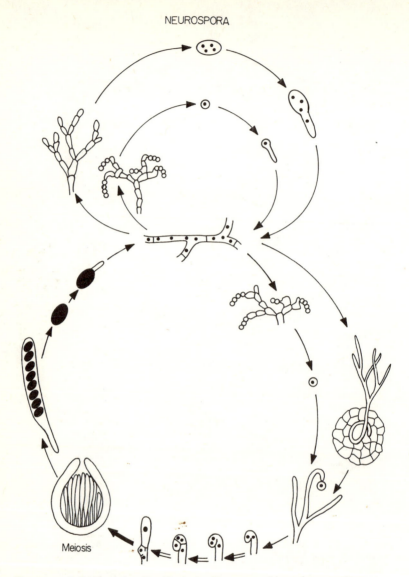

FIGURE 1.6 Life cycle of a dimictic *Neurospora* (Ascomycetes–Pyreno-mycetes). Note that both uninucleate and multinucleate conidia occur. The former act as gametes fusing with the fine hyphae of the branched trichogyne developed by the protoperithecium. Microconidium and trichogyne must be of opposite mating type, *A* and *a*. Ascospores are discharged violently. *Sordaria* and *Venturia* show similar life cycles, *Podospora anserina* is homodimictic and its asci develop differently (see Figure 8.5b, p. 169). *Ascobolus* is similar but develops an open cup-shaped apothecium, not a perithecium.

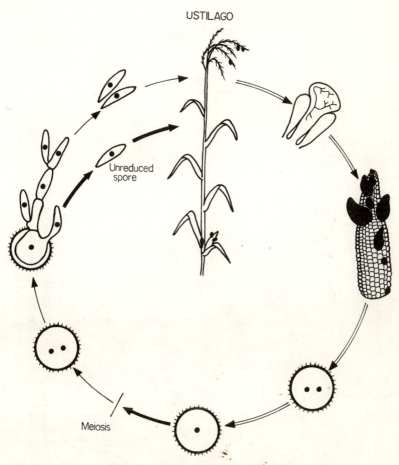

FIGURE 1.7 Life cycle of *Ustilago maydis* (Basidiomycetes–Ustilaginales) This is one of the more complex life cycles. The dikaryotic phase is not readily maintained in culture and the mating system is unifactorial diaphoromictic; most species can be cultured readily and have dimictic systems. Unreduced diploid teleutospores can infect a host and give rise to solopathogenic lines.

18

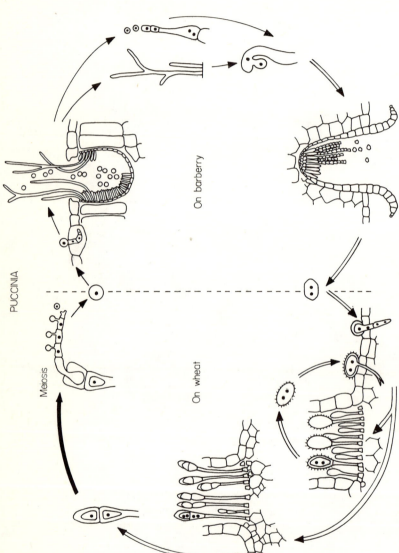

PUCCINIA

Meiosis

On barberry

On wheat

FIGURE 1.8 Life cycle of *Puccinia graminis tritici* (Basidiomycetes–Uredinales) another obligate pathogen, although successful culture of a few strains and species has now been achieved. On barberry the flask-shaped pycnidia produce pycnidiospores which are only capable of fertilization of the fine flexuous hyphae (cf. a trichogyne). Aecidiospores can only infect wheat plants. In the absence of barberries and in a favourable climate, e.g. Australia, Mexico, the uredospores can maintain the fungus indefinitely. Teleutospores usually overwinter before producing basidiospores which only infect barberry.

COPRINUS

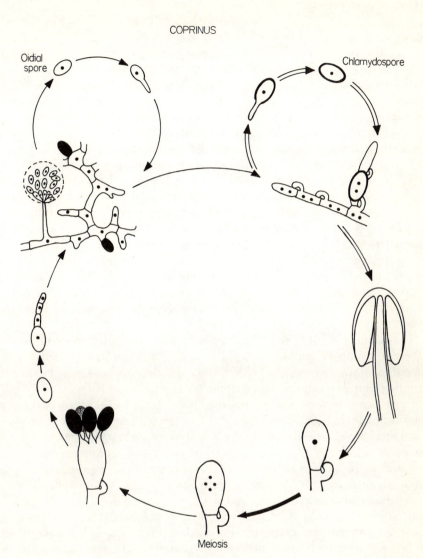

FIGURE 1.9 Life cycle of *Coprinus* (Basidiomycetes–Agaricales). This is typical also of bracket fungi (Aphyllophorales), e.g. *Polyporus*, and Gasteromycetes, e.g. *Cyathus*, but, of course, the form of the basidiocarp differs. In outbreeding species diaphoromixis is the rule. Not all species produce asexual oidiospores while others produce them on both the monokaryotic and dikaryotic mycelium, e.g. *Collybia*.

Features of genetical interest

It only remains in this introductory chapter to consider those features of the fungi which are of especial interest to geneticists.

The ability of many fungi to produce large numbers of identical, predominantly haploid, uninucleate conidia enables such units to be used for the investigation of spontaneous and natural mutation. Because of the simple nutritional requirements of many fungi such mutants can often be circumscribed in fairly precise biochemical terms. Such 'biochemical mutants' have also proved of value to biochemists in the elucidation of specific synthetic pathways. Since conidia can be readily exposed to a wide variety of mutagens they are of particular value in such studies (Chapter 2).

The range of environments in which many fungi can be grown, varying in both physical and chemical factors, is very great. The conditions can often be easily controlled with great precision and reproduced readily. This permits replication of particular inocula to be achieved fairly easily and so characters which vary quantitatively can be studied more readily than with other organisms where the control of the environment poses great problems. Moreoever, the effective control of the environment can be turned to the purpose of investigating genotype-environment interactions both as a means of studying quantitative variation and of gene expression in general (Chapter 7).

The occurrence of hyphal fusions leading to heteroplasmon and heterokaryon formation enables cytoplasmic and nuclear interactions to be investigated under controlled conditions. Mycelia can be sampled either by isolating spores or from hyphal fragments. The properties of such genetically heterogeneous systems are of particular importance in the study of fungi in relation to their natural variability (Chapter 5). The occurrence of parasexuality in synthetic or spontaneous heterokaryons enables an analysis to be made more readily of somatic (non-meiotic) recombination than is possible in diploid organisms such as *Drosophila melanogaster*, where the phenomenon was first detected (Stern, 1936). The significance of parasexuality as a recombination-producing mechanism for Fungi Imperfecti and as an alternative to sexuality is also capable of study (Chapter 4). Heteroplasmons have permitted far-reaching analyses to be made of the behaviour of extrachromosomal elements. Indeed, the fungi are especially valuable in the study of cytoplasmic inheritance (Chapter 6).

The control and regulation of in- and out-breeding, as determined by mating systems is a further field of study. It is important in its own right because of the numerous systems possessed by fungi and for comparative study with other genetic systems (cf. Darlington, 1958; Chapter 8). The facility with which the immediate products of meiosis can be obtained and the large numbers in which such products are produced is of particular value to the study of the mechanism of crossing over. Moreover, Ascomycetes and Basidiomycetes produce tetrads of spores derived by meiosis and this enables reciprocal and non-reciprocal exchanges to be detected, although usually somewhat laboriously. The linear, ordered tetrad of Ascomycetes has the further advantage that the order of its spores enables first-division segregation to be distinguished from second-division segregation, thus facilitating studies of interference, for example (Chapters 3 and 14).

The ease with which biochemical mutants can usually be obtained, the variety of ways their interactions can be studied, as homokaryons, heterokaryons, in haploid or diploid nuclei, at somatic or meiotic crossing over and in different cytoplasms, indicates the value of fungal material for the study of gene action and interaction from the molecular to the morphological level (Chapter 15).

Valuable as fungi are to geneticists, their use is not free from difficulties. They cannot be manipulated as readily as bacteria or bacteriophages, for example, nor can they be cultured quite so readily in bulk. Particular attention has to be paid to technique and sterile procedures to avoid contamination with its consequence of biased interpretations. In this respect, conidia are a particular hazard. The small size of most fungi is also disadvantageous in some circumstances. This can render manipulation, as of single spores, for example, either difficult or extremely time consuming, e.g. a skilled operator can dissect free hand the linear tetrads of the asci of *Neurospora crassa* at the rate of 30/40 an hour; an unskilled operator finds the task almost impossible for some days! The smallness of most fungal chromosomes is a further disadvantage wherever a cytogenetic approach is desirable. Indeed, the uncertainty which exists concerning somatic division is a genuine handicap to the complete analysis of the parasexual cycle. Because of their simple morphology the action of genes in development can only be studied in certain limited situations. Nevertheless, this is not such a great handicap as might appear.

At present the fungi have not contributed greatly to the understanding of problems in population genetics, speciation or evolution. There is no doubt that such topics can be studied in fungi and that they may provide some rather unusual features (Chapters 8–11). The present dearth of knowledge reflects, on the one hand, the neglect of such topics by geneticists in favour of the more immediately rewarding studies of recombination and gene action in fungi and, on the other hand, the neglect of mycologists who, for too long, have failed to use genetical methods to analyse the variability of their organisms.

GENERAL REFERENCES

Mycology
Ainsworth, G. C. (1971). *Ainsworth and Bisby's Dictionary of the Fungi*. 6th edn. 631 pp. Commonwealth Mycological Institute.
An invaluable and succinct compendium of all matters mycological.

Alexopoulos, C. J. (1962). *Introductory Mycology*. 2nd edn. 613 pp. Wiley.
  A standard taxonomic text: well referenced.
Burnett, J. H. (1968). *Fundamentals of Mycology*. 546 pp. Arnold.
  Deals with the general biology of the fungi.
Ingold, C. T. (1973). *The Biology of Fungi*. 3rd Edn. 176 pp. Hutchinson
  Educational.
  The best short introduction to the structure and classification of the fungi.
  Elementary.
Webster, J. (1970). *Introduction to Fungi*. 424 pp. Cambridge University Press.
  A good taxonomic and biological introduction. Well illustrated.

Genetics and fungal genetics

Catcheside, D. G. (1951). *The Genetics of Micro-organisms*. 223 pp. Pitman.
  Old and fairly elementary but still useful.
Day, P. R. (1974). *Genetics of host-parasite interaction*. 238 pp. Freeman.
  A most stimulating treatment of this specialized topic.
Esser, K. and Kuenen R. (1968). *Genetics of fungi*. English edn. 499 pp. Springer,
  Berlin.
  An advanced text particularly valuable for references.
Fincham, J. R. S. and Day, P. R. (1971). *Fungal Genetics*. 3rd edn. 402 pp.
  Blackwell Scientific.
  An advanced text particularly good on recombination and gene action.
Sermonti, G. (1969). *Genetics of antibiotic-producing microorganisms*. 389 pp. Wiley.
  A readable and comprehensive text on this broad topic.
Srb, A. M., Owen, R. D. and Edgar, R. S. (1965). *General Genetics*. 2nd edn. 557
  pp. Freeman.
  A most readable general text.
Strickberger, M. W. (1968). *Genetics*. 868 pp. Macmillan.
  Comprehensive and readable general text.

Section 2

Formal genetics

Formal genetics seeks to devise methods to locate the genetic determinants in space by a study of the breeding behaviour of individuals and their progeny. It is based upon the consequences of segregation and recombination at meiosis and fertilization. These methods are applicable to fungi, allowance being made for most of them being haploid organisms. In addition, however, because of their characteristic biology, genetic analysis can be extended to mitotic recombination and the recombination and segregation of whole nuclei as well as to extrachromosomal elements. Finally, application of these principles and techniques to apparently continuous variation is as relevant to fungi as to other eukaryotes although, as yet, understanding of this extension of formal genetics has only received limited attention.

Formal genetics provides the basic framework upon which population genetics can be built.

CHAPTER 2

Genetic markers

Mutant genes are the markers whereby the genetics of an organism can be investigated and mapped; an adequate supply of such markers is, therefore, of great importance to distinguish them from the standard form of the organism or 'wild-type'.

The first mutants employed in a genetical experiment with a fungus were the spontaneous morphological mutant, *piloboloides* type v. normal sporangiophore and mating types, mt^+/mt^- in *Phycomyces blakesleeanus* (Burgeff, 1912, 1914). Spontaneously occurring characters of this type plus colour mutants and pathogenicity reactions were virtually the only kinds employed until the advent of induced biochemical mutants in the classic experiments of Beadle and Tatum (1941). Since then their numbers, diversity, relative ease of production, as well as the insight they may give into cellular and gene function, has resulted in biochemical markers being the preferred mutant types. In a sense, of course, all mutant genes are potentially biochemical markers but frequently the biochemistry involved is not known.

CHROMOSOMAL MARKERS

Gene mutants
Morphological mutants

These affect either the vegetative or the reproductive features of the fungus. Such mutants presumably have a biochemical basis but this is rarely known or understood. Table 2.1 sets out details of the basis of a few morphological mutants from the 80 or so known in *N. crassa*. Commonly the morphology of the whole colony is affected and most frequently its component hyphae grow more slowly and show either more laterals per unit length or reduced lateral production. The result is either a more compact colony showing so-called colonial growth or else one showing a more diffuse growth pattern to which a variety of names may be applied—*diffuse, streaky, thin*, etc. One problem, indeed, with such mutants is to so name them

25

TABLE 2.1 Morphological mutants of *Neurospora crassa* which also possess a biochemical defect. (Data of Brody and Tatum, 1967; Mishra and Tatum, 1970; Scott and Tatum, 1970.)

Mutant	Phenotype	Biochemical lesion
col-2	Colonial growth	
bal	Balloon, hemispherical colony, slow growth $\Big\}$	Altered glucose-6-P-dehydrogenase
frost	Frost	
col-3 *col-10*	Colonial growth	6-Phosphogluconic acid dehydrogenase
rg-1 *rg-2*	ragged, colonial and poorly conidiating	Deficient in different phosphogluco-mutases

that a reader has some clear notion of their phenotype. Thus, for example, in *N. crassa, bu—button*, a tight colonial form, is understandable, but it is more difficult to visualize *ro—ropy*, with curled hyphae, and impossible to get a clear notion of *sc—scumbo* with 'scummy-like' surface growth and no conidia. A practical difficulty with such mutants is their isolation or recognition in the presence of wild-type mycelia which can readily overgrow them. This can be avoided either by examining plates with few, widely dispersed propagules on them, by reducing the wild-type growth with an inhibitor like sodium desoxycholate, or by the replacement of some of the readily utilized carbohydrate in the medium by sorbose—as with *Neurospora* spp., or mucoraceous forms (Tatum and others, 1949). Moreover, if more than one locus carries a morphological mutant there may be complementation in a heterokaryotic propagule so that the mycelium cannot be differentiated phenotypically from the true wild type.

Those morphological mutants which affect reproductive structures have the disadvantage, as it usually is, that they may be difficult to maintain or propagate. The appearance of the colony as a whole may be altered in some mutants, e.g. *pi—pile*—dense aconidial growth in *N. crassa*. The spore-producing apparatus may be impaired, altered or totally lacking, e.g. *stu—stunted* conidiophore; *aba—abacus*, swellings replace conidia; and *bri—bristle*, no vesicle or conidia; all in *A. nidulans* (Clutterbuck, 1969); and, in mutants affecting sexual reproduction, various degrees of partial or complete sterility may occur. The principal uses of such mutants are either to prevent self-fertilization in cases where conidia, for instance, act as the fertilizing agents, or in the genetic analysis of morphogenesis. An admirable example of the latter is to be found in the many mutants known to affect different stages in the development of perithecia in Pyrenomycetes (Table 2.2).

Colour mutants

Colour mutants usually affect the pigment in the spores although albino forms of species with coloured hyphae are known, e.g. spontaneous hyaline

TABLE 2.2 Occurrence of morphological mutants in some Pyrenomycetes which affect perithecium development. (Data from Wheeler and McGahen, 1952; Esser, 1956; Esser and Straub, 1958; Carr and Olive, 1959; Horowitz and others, 1960; Rothschild and Suskind, 1966.)

Effect	Sordaria macrospora	S. fimicola	Species Glomerella cingulata	Podospora anserina	Neurospora crassa
No reproductive organs	+	+	+	+	+
Ascogonia but no protoperithecia	+	−	+	+	+
Protoperithecia but no perithecia	+	+	−	+	−
Sterile perithecia	+	+	+	−	+
Fertile perithecia but no spore discharge	+	+	−	−	−

variants from the black hyphal form of *Verticillium albo-atrum*. In this particular case, however, the albino phenotype was not determined by a single mutant gene (Hastie, 1962). In heavily pigmented species mutants with differing spore colours often show epistasis. For instance, in *A. niger* the conidia are normally black and the three mutants *white w1*, *fawn a* and *olive o* show the relationships *w1* epistatic to both *a* and *o*, and *a* epistatic to *o*. The value of spore-colour mutants is also related to whether or not the expression of spore colour in uninucleated conidia is autonomous, i.e. the pigment being determined by the nucleus in the spore, or non-autonomous, where it is determined by the genotype of the mycelium as a whole. In the former case different spores or spore-chains may show different colours, in the latter all conidia are identical in colour (e.g. the case of *A. heterothallicus*, p. 94 and Figure 2.1, p. 28).

A particularly valuable type of colour mutant is that which affects the ascospores of linear asci. A few are known in *N. crassa*, e.g. *asco*—pale ascospore, *ts*—*tan* colour, *ws*—*white*, but they have been exploited most fully to investigate the mechanism of genetic exchange at meiosis in species of *Sordaria*, where several ascospore-colour mutants are known (e.g. Olive, 1956; Kitani and Olive, 1967, 1969; see also pp. 294–297, Chapter 14).

Biochemical mutants

The commonest mutants of this type result in a requirement for a specific nutrient additional to those needed by the wild-type strain for normal growth. Such mutants are termed *auxotrophs* in contrast to the wild type which is termed a *protroph*. For example in the classic, first example, Beadle and Tatum (1941) grew *N. crassa* on a medium containing inorganic salts, sucrose, biotin and agar—this they termed the *minimal medium*. They argued as follows:

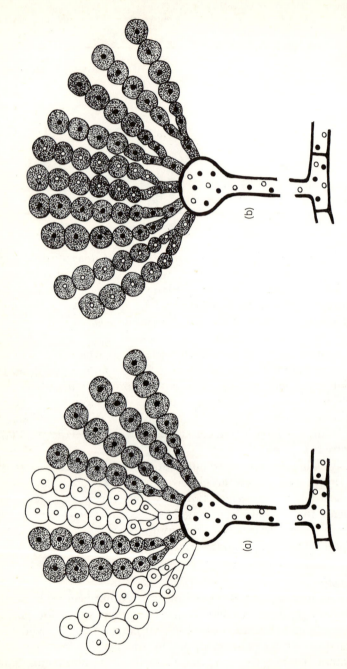

FIGURE 2.1 (a) Autonomous and, (b) non-autonomous gene expression exhibited by uninucleate conidia of an aspergilloid fungus. In both cases the mycelium is a two-member heterokaryon. In (a) the genotype of each uninucleate conidium determines the colour of the conidium but in (b) the colour of all the conidia is determined by the interaction of the nuclei in the heterokaryon.

'The procedure is based on the assumption that X-ray treatment will induce mutations in genes concerned with the control of known specific chemical reactions. If the organism must be able to carry out a certain chemical reaction to survive on a given medium, a mutant unable to do this will obviously be lethal on this medium. Such a mutant can be maintained and studied, however, if it will grow on a medium to which has been added the essential product of the genetically blocked reaction'.

Accordingly they plated their irradiated spores on a *complete medium* of agar, inorganic salts, malt extract, yeast extract and glucose in the hope that this would include the 'essential product' of the blocked reaction. Once they had found a treated strain that grew on the complete medium but not on the minimal medium they tested it systematically to see what substance it needed for growth by inoculating it on minimal medium plus (a) vitamins, (b) amino acids, (c) glucose. If, for example, it grew on (a) then it was further tested on minimal medium supplemented by individual vitamins. In this way they demonstrated the induction of three single-gene mutants which required pyridoxin, thiazole and *p*-aminobenzoic acid respectively. In essence, this method has been employed for the production of all further auxotrophic mutants. It has been modified in various ways with the objects either of producing mutants with a particular phenotype, e.g. exclusively histidine-requiring kinds, or to improve the speed and efficiency of mutant selection and identification. Some of these methods will be described later in this chapter.

Apart from the use of auxotrophs and, in back-mutation studies, the production from them of protrophs, other kinds of biochemical mutant have been employed. The two most common ones are fermentation mutants and resistance mutants. Fermentation mutants have been extensively isolated in yeasts. The vast majority are spontaneous and occur naturally. Initially Winge's laboratory studies, commencing in 1948, demonstrated that enzymes responsible for the fermentation of different sugars are determined by specific genes, many of them polymeric (see p. 125). Other mutant genes have been obtained which result in the loss of ability to ferment particular sugars; in particular, attention has been paid to galactose. In some cases, e.g. *Gal-2*, a so-called permease mutant, the yeast cell is no longer able to transport the sugar into the cell, in others, either no enzyme is produced or only a defective one, e.g. *Gal-1* and *Gal-5*, respectively.

A class of biochemical mutants which has not often been analysed genetically but is of considerable potential economic importance, is that which increases the yield of some metabolite. Such mutants have been obtained for increased penicillin production by *P. chrysogenum*, citric acid by *A. niger* and itaconic acid by *A. terreus* (see Chapter 12). Fermentation mutants are similar in that they also affect the yield from fermentable products.

Mutants which result in the loss of ability to transport a compound into a cell, i.e. due to the loss or impairment of a permease, can also render a cell resistant to anti-metabolites. For example a toxic metabolic analogue of methionine is ethionine and ethionine-resistant mutants are known in several fungi. In yeast the resistance appears to be due to a reduced ability to accumulate the compound, i.e. defective permease activity, but in *Coprinus lagopus* the ability to use ethionine is blocked after its entry into the cell (Lewis, 1963; Gits and Grenson, 1967). Regardless of the mechanism of resistance, the principal employed to obtained such mutants is to grow the fungus, or its propagules, in concentrations of the anti-metabolite or drug sufficient to inhibit normal growth. It is of some historical interest that possibly the first fungal mutant recorded was so obtained by Schiemann (1912) who grew *A. niger* on toxic levels of zinc sulphate. An important practical consequence of this type of mutant is the development of fungicide-resistant strains, e.g. benomyl-resistant strains of *Erysiphe cichoracearum*, the powdery mildew of Cucurbitaceae.

An entirely different type of resistance mutant is that resistant to irradiation by ionizing radiations or ultraviolet light. Related to these are mutant strains showing enhanced sensitivity to such radiations.

Suppressor mutants

Such mutants are best regarded as a special class of biochemical or physiological mutants. They are usually detected as apparent reversions of a mutant to a wild-type phenotype, but the term can be applied to any gene which unmasks the expression of another. So the phenotypes of the various combinations of two alleles A/a and its suppressors $+/su$ are:

A +   normal, i.e. non-mutant condition
a +   mutant, i.e. mutant condition
A *su*   normal, i.e. non-mutant condition, suppressor has no action
a *su*   normal, i.e. mutant condition suppressed by mutant suppressor.

An early example studied was the *td-24* mutant of *N. crassa*. This produces a highly zinc-sensitive tryptophan synthetase which is not usually detectable at 25°C so that the mutant is auxotrophic for tryptophan. A suppressor gene *su-24* is known which restores the protrophic phenotype. The synthetase is still highly zinc sensitive but the suppressor apparently acts by preventing the entry of Zn^{2+} ions into the cell (Suskind and Kurek, 1959).

An entirely different type of suppressor mutant is known as a super-suppressor and was first demonstrated in fungi in yeast by Hawthorne and Mortimer (1963). Super-suppressors act on many different mutant genes at several different loci simultaneously. For example, *SUP-1* suppresses mutant genes causing requirements for adenine, uracil, arginine, histidine, leucine, lysine, tryptophan and tyrosine, so rendering the yeast cells protrophic. Although acting on several genes the super-suppressors show specificity to particular mutant alleles at any one locus.

Physiological mutants

This class is distinguished, somewhat arbitrarily from biochemical mutants, by the fact that their response to some physiological situation has altered without any concomitant morphological or known biochemical change. It includes mutants with altered responses to physical factors such as temperature, light and gravity. A not uncommon class of induced mutants is the temperature-sensitive one. These show a normal phenotype and/or nutrient requirement at one temperature but a mutant phenotype at a higher or lower temperature; some mutant phenotypes are lethal (see Lethal mutants below). In respect of light, several mutants are known with a reduced or lost phototropic sensitivity affecting the orientation of some structure or morphogenetic development, e.g. *mad* mutants of *Phycomyces blakesleeanus*, with normal morphology but lost or impaired phototropic reactivity of their sporangiophores.

In this rather general class may also be included mutants with altered permeability to ionic solutions or water, e.g. *osm*—osmotic, where permeability properties of the wall have changed.

Two other groups are of importance in particular fungi, namely mutations in mating-type specificity and pathogenicity. So far as can be ascertained the mt^+/mt^- genes of most dimictic fungi are exceedingly immutable, itself a remarkable fact, but in the yeasts mutant mating-type factors, both spontaneous and induced, do not seem uncommon. In diaphoromictic fungi, with multiple series of mating-type factors, large numbers of allelomorphs occur naturally but spontaneous mutation is rare. Induced mutant factors have now been obtained in *Schizophyllum commune* and *C. lagopus* but they do not seem to resemble those which occur naturally (see p. 167, Chapter 8).

In a few pathogenic fungi several genes are known which affect the host-parasite response and, indeed, on Flor's gene-for-gene hypothesis there will be as many genes determining virulence as there are resistance genes in the host. Spontaneous mutations to virulence have been recorded in a number of rusts and mutant genes, both for virulence and avirulence, have been induced in several fungi, e.g. *P. coronata avenae* (Griffiths and Carr, 1961), *Venturia inaequalis* (Keitt and Boone, 1954). Obviously a mutation to avirulence in an obligate pathogen would, in effect, be lethal.

Lethal mutants

This is a class about which very little is known, although a technique exists for detecting recessive lethals in *Neurospora* (Atwood and Mukai, 1953). In essence this method depends upon obtaining the mutant gene in a heterokaryon where it can be maintained indefinitely. So this method is, in principle, applicable to any fungi which can form stable heterokaryons and where complementation can occur. Recently an important and interesting class of so-called conditional lethal mutants has been isolated from yeast. Phenotypically they are temperature sensitive and grow with a normal phenotype at 23°C but die at 36°C. They are, in essence, irreparable auxotrophs whose synthetic deficiencies cannot be made good by exogenous replacements. The genetic blocks appear to be in essential and fundamental processes such as the synthesis of protein, RNA, DNA and cell walls (Hartwell, 1967).

Another class of lethals arises from chromosome damage, notably deletions following irradiation by X-rays for example. This class too has not yet received much study but they have been used in an ingenious way to define the limits of genetic loci by complementation mapping (de Serres, 1969).

Mutagenic agents

Spontaneous mutants occur in most cases with low frequency. Auerbach (1959) has shown that dry macroconidia of *N. crassa* stored at 32°C and 4°C showed weekly increases of recessive lethals of only 0_r3 and not more than 0.1%. Tatum and others (1950) tested *Neurospora* for a wide range of auxotrophic mutants and found only 1 in 3000 cultures tested, i.e. only one mutant gene from all the thousands tested. In tests for mutation at specific loci the frequency is equally low, e.g. $0/10^6$ conidia to either *ad-3A* or *ad-3B*—purple, adenine-requiring in *N. crassa* (de Serres and Kølmark, 1958); $3/10^6$ oidia from methione-requiring *met-8* to wild-type, *met -8+* in *C. lagopus* (Moore, 1969).

In most genetic studies, therefore, it is necessary either to induce or to select the genetic markers and fungi have been employed for many years in both types of study. The principal inducing agents have been those which affect DNA, namely ionizing and ultraviolet radiations and chemical mutagens. Work over the last two decades, with prokaryotes in particular, has greatly increased understanding of how chemical mutagens operate and so has led to their more rational use.

Mutation-inducing agents employed with fungi are illustrated in Table 2.3.

TABLE 2.3 Some common mutagens which have been used with fungi.

Mutagen	Typical genera tested with indication of effectiveness					
	Phycomyces	*Aspergillus*	Yeast	*Neurospora*	*Ustilago*	*Coprinus*
u.v.-irradiation						
(254 nm)	+	+ +	+ +	+ +	+ +	+ +
γ or X-irradiation	±	+	+	+		+
'Mustard gas'		+ +		+ +		+
2-AP				+		
DES		+ + +	+	+		
EMS		+	+ + +	+ +		+ +
NA		+ + +	+ +	+ + +		
NG	+ +					+ +
DEB		+ +		+ +		

Key: Mustard gas is used to cover all the nitrogen mustards and related compounds: 2-AP, 2-aminopurine; DES, diethyl sulphate; EMS, ethylmethane sulphonate; NA, nitrous acid; NG, N-methyl-N-nitrosoguanidine; DEB, diepoxybutane.

Relative efficacy indicated from ±, least, to + + +, most.

It should be realized that no two fungi necessarily respond alike to the same mutagenic agencies, nor indeed in the same way as do other organisms. For example, NG is a most potent chemical mutagen for *Escherichia coli* and quite effective with *Saccharomyces cerevisiae* but it is no more effective with *C. lagopus* than u.v.-irradiation (Moore, 1969); by contrast no *A* mating-type factor mutants were induced in *S. commune* by u.v. or X-rays but were readily produced by EMS treatment (Raper, Boyd and Raper, 1965). It is necessary, therefore, to experiment with different mutagenic agents if work with any fungus not hitherto studied is contemplated. The following notes, therefore, are indicative only and do not apply necessarily to all fungi; general references only are given.

It is usually found with all mutagens, however, that an effective yield of mutants is only achieved when the proportion of survivors from the material exposed is very low, say 5–1% or less. Large numbers of spores or nuclei, therefore, must be exposed to obtain a worthwhile yield.

Ionizing radiations

In general the yield of mutants is usually proportional to the dose and such mutants are normally single 'point' mutations, i.e. changes in individual genes. In general, yields of mutants are high if irradiated material (e.g. spores) is wet and well oxygenated. Ionizing radiations also induce breaks in chromosomes and these can result in chromosomal changes such as translocations, inversions and deletions; the latter are frequently lethal (Figure 2.2a). Not much is known about the frequency of chromosomal mutants in fungi and this reflects the fact that their small chromosomes have not proved cytologically rewarding.

(General references: Pomper and Atwood, 1955; Giles, de Serres and Partridge, 1955; de Serres and Webber, 1967.)

Ultraviolet irradiation

The classic studies of Emmons and Hollaender (1939) with monochromatic u.v. light using the dermatophyte *Trichophyton mentagrophytes* implicated the nucleic acids as the mutable material since 253·7–265 nm were the most effective wavelengths (Figure 2.2b). Because of the wavelength-dependency effect a mercury-resonance lamp is usually employed as a source of u.v. light; such lamps have some 95% of their output confined to the resonance line at 253·7 nm. The relationship between dose and mutation frequency is most usually non-linear, often rising sharply with increasing dose to an optimum and then falling (Figure 2.2c). In many cases, e.g. in *Schizosaccharomyces pombe*, ultraviolet sensitivity varies with the growth cycle (Figure 2.2d).

Ultraviolet irradiation is frequently a potent mutagenic agent for fungi with hyaline spores but not with heavily pigmented or thick-walled spores because of absorption losses. It is also susceptible to a variety of effects from

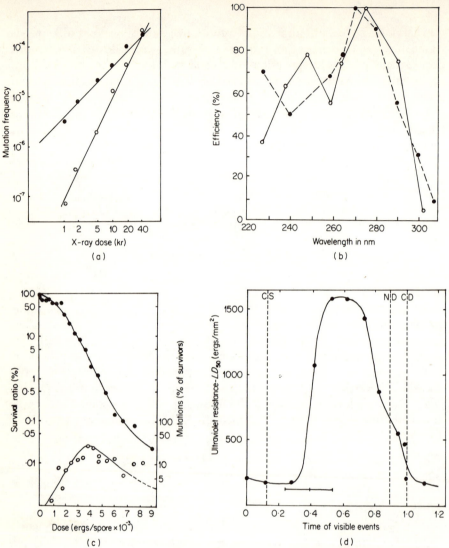

FIGURE 2.2 Effects of radiation on fungal cells. (a) Logarithmic plot of the frequencies of *ade-3* (adenine-requiring) mutations in *Neurospora crassa* as a function of X-ray dose on conidia. Point mutations (●) and deletion mutations (○), viable only in heterokaryons were scored separately. (Based on Webber and de Serres, 1965, from Fincham and Day, 1971.) (b) Relative effectiveness in inducing mutations in conidia of *Trichophyton mentagrophytes* of different wavelengths of u.v.-irradiation (− ○ −) and relative absorption spectrum of sodium thymonucleate, taking absorption of 260 nm as 100% (− − ● − −). (Based on Hollaender and Emmons, 1941.) (c) Logarithmic plot of survival (%) and mutations (as % of surviving cells) of conidia of *Trichophyton mentagrophytes* exposed to increasing doses of monochromatic, 265 nm, u.v.-irradiation. (Data from Hollaender and Emmons, 1941.) (d) Resistance to killing by far u.v.-irradiation (200–300 nm) of cells of the fission yeast, *Schizosaccharomyces pombe*. The start of the cycle coincides with the appearance of the cross-cell plate, signifying the completion of cell division and is followed by cell separation (CS), nuclear division (ND) and the next cell division (CD). (Based on Swann, 1962.)

other agencies, both physical and chemical. For example, its effectiveness is reduced by pre-treatment with infrared radiation or nitrogen mustards or by post-treatment exposure to visible radiation (photoreactivation). This last effect has been studied intensively in bacteria. For most experiments of about 10 minutes duration carried out in normally lit rooms it is of little practical significance. The genetic and biochemical control of photo-reactivation has begun to be studied in *Neurospora* (Terry and others, 1967).

(General references: Pomper and Atwood, 1955; Setlow, 1964; Zetterberg, 1964; Witkin, 1966.)

Other radiations

High temperature appears to be mutagenic but there has been neither careful nor analytical study of the phenomenon. Some of the earliest fungal mutants were obtained by exposure to a temperature of 60°C or more for brief intervals (Barnes, 1928). Elevated temperature has also been employed to obtain extrachromosomal variants (see p. 46) but in none of these cases is it clear whether mutations have been induced or selected.

In bacteria, intense visible light, its effects enhanced photodynamically by the presence of certain vital dyes such as methylene blue, has been shown to be mutagenic (Webb, Malina and Benson, 1967). No such effects have yet been recorded in fungi.

Chemical mutagens

As Auerbach (1967) wrote 'We now take it for granted that mutations are chemical changes in DNA (or RNA), and we no longer ask *whether* a mutagen reacts with DNA but *how* it reacts with it.' This attitude has led to attempts to classify chemical mutagens in terms of their action on DNA but this has not been entirely satisfactory.

It is ironic that one of the most effective mutagens, nitrous acid, which deaminates bases, substituting hydroxyl groups for the amino groups, was originally tested on Aspergilli on the hypothesis that the genetic material was proteinaceous (Steinberg and Thom, 1940, 1942). It is a relatively unstable substance obtained by dissolving sodium nitrite at pH 4·6. A consequence of its deaminating ability is that bases are changed chemically so their pairing relations may also change:

adenine $\longrightarrow$ hypoxanthine which now pairs with cytosine (A-T $\longrightarrow$ H-C).

guanine $\longrightarrow$ xanthine which shows no change in
$\qquad\qquad\qquad\qquad$ base pairing (G-C $\longrightarrow$ X-C).

cytosine $\longrightarrow$ uracil which now pairs with adenine (C-G $\longrightarrow$ U-A).

Hydroxylamine also brings about a base change in cytosine and the modified molecule then pairs more readily with adenine than with its normal partner, guanine. Another important group of mutagens is formed by the alkylating agents which include ethylmethane sulphonate (EMS) and

N-methyl-N′-nitro-N-nitrosoguanidine (NG), which presumably ethylate and methylate guanine to give 7-ethyl- or methylguanine respectively. Either of these compounds can easily be lost by hydrolysis from the DNA and could then be replaced by almost any other base. Nitrogen mustard was not only one of the earliest compounds to be demonstrated, unequivocally, to be a mutagen but it was also used with fungi in early investigations. It is typical of a group of compounds, the tri-β-chloroethyl amines or sulphides, which are also alkylating agents; another such group is that of the epoxides. However, the precise mode of action of these compounds is not known. Finally, there is a class of compounds used extensively with prokaryotes but not often with fungi, the base analogues, e.g. 5-bromouracil (BU), 2-amino-purine (2-AP), which can be incorporated in place of thymine and adenine respectively when new DNA is being synthesized.

The general technique employed is to expose spores in a dilute solution of the mutagen and after an appropriate treatment, usually resulting in very high lethality, to filter the spores off and, after appropriate washing out or blocking of the mutagen, to plate out the survivors.

(General references: Westergaard, 1957; Heslot, 1962; Auerbach and Ramsey, 1968.)

Selective agents

The most commonly employed selective agents are toxic compounds such as heavy metals, Cu, Zn; fungicide and fungistatic compounds such as nystatin and cycloheximide; metabolite analogues such as canavanine or ethionine, which are analogues of arginine and methionine respectively, and cellular poisons such as sodium azide or sulphonamide. The principal of application is the same as with mutagens, namely that cells or spores are exposed to a concentration such that most of the population are killed. In some cases it proves necessary to 'train' the fungus, i.e. carry out selection at a relatively low concentration of the selective agent, then transfer survivors to a higher concentration and so on. A very simple device for applying an increasing concentration of a selective agent is the gradient plate designed by Sybalski (1952; Figure 2.3). The inoculum is placed at the side of the petri plate with the lowest concentration and either spores or hyphae are subcultured from the mycelium which extends furthest across the plate. Usually it is the hyphae which are subcultured since aerial parts of the mycelium, including spores, can develop without internal translocation of the toxic agent to them and thus they are never exposed to selection.

(General references: Howe and Terry, 1962; Ashida, 1965; Ahmad and Woods, 1967; Georgopoulos and Zaracovitis, 1967.)

Isolation and selection of mutants

Techniques for obtaining mutants will depend partly upon the biology of the organism employed and partly on the specific purpose for which the mutant is required.

(a)

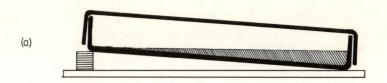

(b)

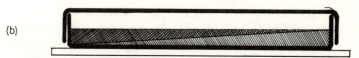

FIGURE 2.3 Gradient plate technique of Szybalski (1952)
for selecting resistance mutants. (a) The plate at an angle
is partly filled with agar medium and then (b) agar +
toxic agent is added after placing the plate horizontal. The
drug is diluted by the underlying agar and so is distributed
in a decreasing gradient from left to right.

An idealized procedure is set out below and thereafter the details and
modification of each stage are discussed. The procedure is:

(i)   Employ thin-walled, permeable, hyaline uninucleate conidia which
      have high viability in aqueous solution, buffered if necessary.
(ii) Treat with mutagen, wash to remove traces of agent and aim for
      95–97% lethality.
(iii) Enrich yield of auxotrophs by eliminating, mechanically or chemically,
      surviving protrophs.
(iv) Plate on complete medium at low density. Incubate at two tem-
      ·peratures.
(v)   Replica-plate, if possible, to minimal medium with or without appro-
      priate supplements at same temperatures as parent plates.
(vi) Isolate individual mutants and identify requirements of auxotrophs,
      do check tests on temperature-sensitive mutants, etc.

Mutable material

Many fungi do not produce spores which meet the criteria set out above,
the principle variants being wall permeability, pigmentation and nuclear
content or the absence of viable spores. In general, wall variants are not
serious problems. Impermeability occurs in many ascospores but this can
usually be overcome by some treatment, e.g. 60°C for 1 hour in *N. crassa*,
furfural treatment, or by employing radiations as the mutagenic treatment.
In a similar way, chemical mutagens can be used with pigmented spores
where u.v. light might be attenuated by undesirable absorption. When the
spores are multinucleate there are two possible procedures. One is the

classical method employed by Beadle and Tatum (1941) where irradiated macroconidia were used to fertilize protoperithecia of opposite mating type, and the ascospores isolated. This is based on the assumption that each perithecium arises from fertilization by a single nucleus—an assumption usually but not completely valid (e.g. Nakamura and Egashira, 1961)—so that 50% of the ascospores will be of mutant phenotype. Alternatively, the mutagenic agent can be used at a level where the proportion of survivors is less than 1% so that in any multinucleate spore the probability is increased that only one nucleus is viable.

Some fungi appear to produce spores ideally suitable as mutable material but unfortunately they do not germinate and are only effective as fertilizing agents, e.g. microconidia of *Podospora anserina* or the pycnidiospores of rust fungi. In some species it is possible to treat the conidia and then use them immediately for effecting fertilization, e.g. *P. anserina* (Esser, 1969), but other species have to be treated in the same way as sterile mycelia. In such cases the mycelium is macerated in a blender to give fragments of, preferably, 2–4 cells in length and these are then treated with the mutagenic agent, e.g. *S. commune* and *Agaricus bisporus* (Raper and Miles, 1958; Raper, Raper and Miller, 1972).

Mutagen treatment

Examples of mutagens have been give in Table 2.3 (p. 32). With chemical mutagens it is especially important both to test a range of conditions, especially a range of pH, to find the most efficacious and to devise an effective way of stopping the treatment when required. Frequently the mutagen can be removed by centrifuging the treated material and resuspending it in water; the process can be repeated several times if need be. In other cases, for example with EMS, the treatment can be terminated by adding non-mutagenic sodium thiosulphate which combines with EMS and renders it ineffective. It is absolutely essential to have carried out trial runs to determine a suitably high percentage mortality.

Enrichment techniques

Fries (1947, 1948a, b) developed two simple techniques with *Ophiostoma multiannulatum* for increasing the yield of auxotrophic mutants. The principle is to suspend the treated material in minimal medium so allowing protrophs to germinate and grow; they can then be filtered out readily through sterile cotton wool or muslin gauze. The suspension is agitated or oxygenated briskly to reduce the possibility of fusions between auxotrophs and protrophs and growth must not continue for too long or filtration becomes impossible and fusions common. An effective schedule has been published by Woodward and others (1954) for *Neurospora* but the technique has been applied to *Coprinus* (Day and Anderson, 1961) and even to flocculating yeasts where the loosely joined flocculating habit has the same consequences on filtration as does mycelial growth (Takahashi, 1959). A different technique for eliminating protrophs is that employing an agent which is differentially toxic to growing and non-growing cells. This has been achieved with the fungistatic nystatin which kills only growing cells of yeast (Snow, 1966) or growing hyphae of *P. chrysogenum* (MacDonald, 1968).

Fries's second method is only suitable for isolating double mutants. He showed that for *O. multiannulatum* doubly auxotrophic mutant spores survived longer than did single auxotrophs when incubated in minimal

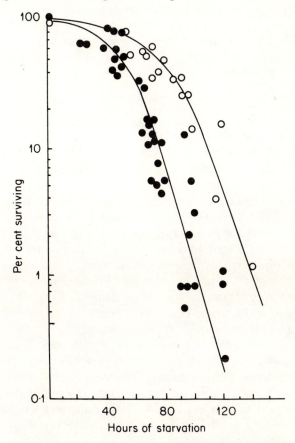

FIGURE 2.4 Survival rates of single and double auxotrophic mutant conidia of *Aspergillus nidulans* on minimal media. At various intervals complete medium was added and the number of viable colonies surviving estimated. Key: − ● − *bio-1* or *y bio-1* conidia, i.e. green or yellow biotin-requiring; − − ○ − *y bio-1 thi* or *w ade-1 bio-1* conidia, i.e. yellow biotin- and thiamine-requiring or white adenine- and biotin-requiring. Note the doubly auxotrophic conidia survive starvation better (Redrawn from Pontecorvo and others, 1953b.)

medium. MacDonald and Pontecorvo (in Pontecorvo and others, 1953b) applied the technique to *A. nidulans* with considerable success, treating a biotin-requiring strain with either X-rays or u.v. light (Figure 2.4). They give a very full and useful discussion of the whole technique which has also been applied to *Neurospora* (Lester and Gross, 1959) and to yeast (Roman, 1955).

Both techniques have limitations. Filtration enrichment is neither effective with mutants that do not survive long in minimal medium nor with those auxotrophs than can grow with trace amounts of their required supplement, since these may leak from the growing protrophs. The starvation technique is limited by the choice of the initial mutant strain used. This must have a high rate of dying off under starvation conditions if the technique is to be effective.

Plating techniques

It is obvious that if total mutant recovery is the aim then the plating density must be low enough for clear separation to be achieved between adjacent, treated propagules. However, there are further additional reasons for this to be desirable practice. The first is that some auxotrophs may grow as a result of leakage of appropriate adjuvants from adjacent protrophs—the so-called 'breast-feeding' or syntrophic effect. A more remarkable effect was discovered in *Neurospora* by Grigg (1952), namely competitive inhibition. He showed, by reconstruction experiments, that at conidial densities of 10^6– 3×10^7 per plate, protrophs were inhibited by auxotrophs on sorbose-complete medium. In this case the effect was sorbose mediated, for on glucose (sorbose free)-complete medium the inhibitory effect was lost (Kølmark and Westergaard, 1952). Nevertheless, effects of this sort, due to density or plating medium, have been found in *A. nidulans* (Roper, in Pontecorvo and others, 1953b), *O. multiannulatum* (Zetterberg and others, 1969), *Saccharomyces* (Meuris and others, 1967) and *Schizosaccharomyces* (Pourquié, 1968). The consequence of this type of effect, which may well be more widespread than is usually realized, is that certain classes of mutants never appear to arise, e.g. *arg*, *lys* and *met* were the only amino acid requiring mutants known until 1968 in *O. multiannulatum*, when *trp* was obtained by plating on minimal + tryptophan rather than on complete medium containing tryptophan.

If the fungus studied has a spreading habit this may be overcome either by using for treatment a mutant with colonial growth or by incorporating sorbose or sodium desoxycholate in the plating medium. Sorbose may, however, affect auxotrophic growth as just discussed.

The object of plating at two temperatures, say 24°C and 34°C, is an attempt to detect temperature-sensitive mutants.

Replica plating and subculturing

The simplest way to test survivors on complete medium for auxotrophy is to replicate the colonies on minimal media. A circular block which fits into the petri plate is covered with velveteen possessing a short pile. It is first pressed into sterile agar to moisten the pile and to render the hairs slightly tacky, then it is placed lightly on the complete plate, removed and pressed on to the minimal plate, so transferring a 'replica' of adherent cells or spores (Lederberg and Lederberg, 1952; Mackintosh and Pritchard, 1963). If replica plating is not possible then the colonies or the complete plate have to be transferred individually to minimal test plates or slopes.

Mutant identification

The procedures employed for identifying auxotrophs are essentially a series of successive tests in which the number of possible requirements is progressively reduced. Once a requirement of some sort is established by demonstrating no growth on minimal medium but growth on complete medium, the auxotroph is tested on supplemented minimal media. The classic supplements are minimal medium plus (a) a vitamin mixture, (b) a mixture of amino acids, usually in the form of casein hydrolysate, (c) purines and pyrimidines, usually as nucleic acid hydrolysate. Once growth has been achieved on one of these supplemented media the mutant is tested on a range of minimal media to which each component of the mixture has been added singly.

Alternatively, the mutant can be grown on complete medium and spores then spread on or in minimal medium on which minute amounts of various supplements, e.g. amino acids, have been placed in known and well-separated positions. This is the so-called auxanographic method developed originally by Pontecorvo for *A. nidulans* (1949). Provided that competitive inhibition is not likely to occur or is not regarded as of importance, a less time-consuming technique is to employ a series of supplemented minimal media such that the growth on the different media indicates the requirements of the mutants. The 12-plate system used by Holliday (1960) for identifying auxotrophs of *Ustilago maydis* was:

Plate No.	1	2	3	4	5	6
7	adenine	biotin	phenylalanine	alanine	arginine	leucine
8	hypoxan-thine	folic acid	serine	cysteine	ornithine	glycine
9	cystine	pantothenic acid	tryptophan	threonine	aspartic acid	isoleucine
10	guanine	pyridoxin	tyrosine	thiosulphate	proline	histidine
11	thymine	thiamin	*p*-aminobenzoic acid	methione	glutamine	lysine
12	uracil	riboflavin	nicotinic acid	choline	inositol	valine

Thus, for example, an auxotroph requiring leucine would only grow on plates 6 and 7, for methionine on plates 4 and 11, and for both methionine and cysteine on plates 4, 8 and 11.

Another procedure is the layered-plate technique of Reaume and Tatum (1949). This involves reversing steps (iv) and (v) of the procedure (p. 37).

The treated propagules are plated at low density on minimal medium and incubated for a day or so to allow protrophic growth. These colonies are marked and then cooled, complete or supplemented minimal medium is gently poured all over the plates. The auxotrophic mutants can now grow and are identifiable as unmarked colonies.

A variant of this method is simply to examine incubated plates of minimal medium, on which the treated propagules are spread, either for slightly growing mycelia or for non-growing spores. These are then removed, transferred to complete medium and, if they grow, are further identified. This method has been used by Boone and others (1956) for *Venturia inaequalis*, the apple-scab fungus.

Temperature-sensitive mutants can, of course, be identified by transferring them from the temperature at which they were first isolated.

The procedures outlined here are shown diagrammatically in Figure 2.5.

Many comparative lists have been published of the types of mutants produced by particular mutagens. Similarities and differences in kinds of mutant produced have been related either to the genetic architecture of the species or to the specific action of the mutagen. It is, of course, possible that either or both of these explanations apply in any particular case but it seems at least as likely that similarities and differences in mutant production reflect the selective procedures adopted after treatment with a mutagen. Selection can occur at any stage: plating, transfer, by inhibition or enrichment, or even in the identification procedures. As yet, clear evidence of mutagen specificity does not exist.

Chromosomal mutants

Segmental mutants

This is a useful term to cover those mutations in fungi which involve structural changes within a chromosome but for which there is not much supporting cytological evidence because of the small size of fungal chromosomes and the technical difficulties in their study. As a consequence the detection of segmental mutants is usually confined to aberrant tetrad and linkage data. McClintock (1945) has provided in *N. crassa* one of the rare genetical and cytological demonstrations of a translocation in fungi.

Genetically detectable interchanges of various types, especially translocations, frequently arise after X-, γ- and u.v.-irradiation and these have been studied especially in *A. nidulans* (Käfer and Chen, 1964; Käfer, 1965). A high proportion of aborted asci or of aborted ascospores after treatment with ionizing radiations suggests the occurrence of interchanges (McClintock, 1945; Heslot, 1958).

Inversions are known in *Neurospora* and if involved in meiotic crossing over may give rise to progeny with duplications and deletions (Newmeyer and Taylor, 1967). Deletions are probably the most studied and best understood segmental mutants in the fungi. They can be recognized as loss

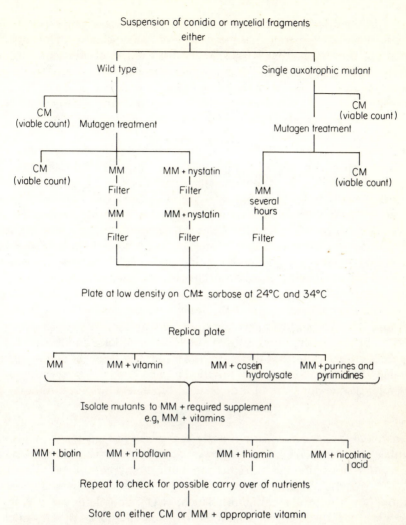

Suspension of conidia or mycelial fragments
either

Wild type — Single auxotrophic mutant

CM (viable count) — Mutagen treatment

CM (viable count)

CM (viable count) — Mutagen treatment

CM (viable count)

MM — MM + nystatin
Filter — Filter
MM — MM + nystatin
Filter — Filter

MM several hours
Filter

CM (viable count)

Plate at low density on CM± sorbose at 24°C and 34°C

Replica plate

MM — MM + vitamin — MM + casein hydrolysate — MM + purines and pyrimidines

Isolate mutants to MM + required supplement
e.g, MM + vitamins

MM + biotin — MM + riboflavin — MM + thiamin — MM + nicotinic acid

Repeat to check for possible carry over of nutrients

Store on either CM or MM + appropriate vitamin

FIGURE 2.5 Diagram to illustrate procedure for mutant isolation and identification. CM, MM; complete and minimal medium, respectively.

mutations which never revert and which are associated with distorted linkage relationships. They were found to be not uncommon after X-irradiation of *N. crassa* (Giles, 1955) and have been intensively studied at the *ade-3* locus in the same organism (de Serres, 1964; de Serres and Webber, 1967).

The most remarkable studies of duplications are those of Roper and his colleagues in *A. nidulans*. Haploids with duplications show reduced rates of growth and their colonies exhibit a characteristic crinkled appearance. They are unstable and sectors either recover normal growth rates or, rarely, become more degenerate. Recovery appears to be associated with the loss of the duplication in part or in whole; the cause of degeneration is not known. The phenomenon has been termed 'mitotic non-conformity'. If the loss of the duplication is associated with the loss of adjacent chromosomal material then a variegated position effect is exhibited. The phenomenon is not fully understood and so far does not seem to have been encountered outside *A. nidulans* (Nga and Roper, 1968, 1969; Roper and Nga, 1969; Cooke and others, 1970; Azevedo and Roper, 1970).

Aneuploidy

Aneuploids occur whenever diploid nuclei undergo haploidization; the condition is usually a transient one. Aneuploids are usually not very vigorous compared with euploids so that they are not often easily detected. Nevertheless, by providing conditions where they were neither suppressed nor overgrown, it proved possible to isolate and maintain 0·6–2·5% aneuploids from diploid strains of *A. nidulans* (Käfer, 1961). Since haploidization can be induced quite specifically by *p*-fluorophenylalanine this provides a means, if coupled with an appropriate selective technique, for inducing aneuploids at will (Lhoas, 1961). Aneuploids of constitution $n + 1$ (disomics) have arisen spontaneously in both *Neurospora* and yeast (Pittenger, 1954; Cox and Bevan, 1962) but here too they are transient conditions.

It can be seen both from the observations on aneuploids and those referred to earlier on deletions that there appears to be very heavy selection against any departure from the euploid condition in such fungi as have been studied. Aneuploids, therefore, are of little general biological interest in fungi.

Polyploidy

Polyploids appear to occur naturally as diploids or higher polyploids in a number of fungi such as *Allomyces arbuscula* and *S. cerevisiae* where $2n$ stages occur quite normally and can alternate with the haploid phase. The most effective polyploidogenic agent for fungi, excluding Phycomycetes, is without doubt natural *d*-camphor (Bauch, 1941). Polyploids can usually be recognized by their 'gigas' characters—increased spore volume, nuclear volume and, of course, DNA content. If a diploid nucleus arises from the fusion of two nuclei a further technique can be employed since, if the two haploid nuclei carry non-allelic auxotrophic markers, the diploid nucleus will usually be protrophic as a consequence of intranuclear complementation. This is the basis of Roper's (1952) technique for obtaining heterozygous diploid strains of *A. nidulans*, in which a balanced heterokaryon was exposed to *d*-camphor vapour and its conidia plated on minimal medium. The mode of action of camphor is not understood. It may act as a spindle inhibitor since it acts like colchicine on onion root-tip nuclei (Östergren and Levan, 1943) or it may act as a selective agent for spontaneously formed $2n$ polyploid nuclei. There is a slightly higher probability that the latter explanation is correct since other spindle inhibitors, such as colchicine and acenaphthene, either have no polyploidogenic effect or only a doubtful one on most fungi

(Sansome and Bannan, 1946). The only unequivocal exception is the action of colchicine on *Allomyces arbuscula* where Sost (1955) was able to produce polyploid plants. This is a lower phycomycete and there is some possibility that its mitotic mechanics may differ from those of higher fungi (Burnett, 1968).

Ultraviolet light increases the frequency of diploid nuclei in *U. violacea* (Clements and others, 1969).

Haploidy and haploidization

In fungi where the 'normal' condition is haploid, diploids can be induced. A proportion of such nuclei spontaneously break down, by non-disjunction, through aneuploidy to the haploid state. In many Aspergilli and related species the frequency of haploidization can be increased greatly by treatment with low concentrations of DL-*p*-fluorophenylalanine (*p*-FPA), e.g. 10^{-5} (w/v) for *A. niger* (Lhoas, 1967). The action appears to be a direct one on the nucleus. With a synthetic diploid strain of *A. niger* the hyphal tip cell showed a reduction in mean size of the nuclei and an increased number of micronuclei; the latter are thought to be formed by nuclear fragments. Even in haploid strains some micronuclei are induced. Evidently *p*-FPA, an amino acid analogue, favours chromosome loss (Lhoas, 1968).

While *p*-FPA will induce haploidization in a number of Ascomycetes and Fungi Imperfecti, including the normally diploid *Verticillium dahliae* var. *longisporum* (Ingram, 1968), it has proved ineffective against yeasts where a diploid stage is a normal phase and against Basidiomycetes such as *Ustilago maydis* and *Coprinus lagopus*. Although non-disjunction to form aneuploids does occur in *S. cerevisiae* (Emeis, 1966; Gutz, 1966), there appear to be compensating non-disjunctional events which result in the restoration of the diploid state (Strömnaes, 1968). Conceivably this also occurs in Basidiomycetes such as *U. maydis*, for in *U. violacea* *p*-FPA is effective in inducing haploidization (Day and Jones, 1968).

EXTRACHROMOSOMAL MARKERS

Extrachromosomal variants

These differ in showing non-mendelian inheritance rather than by their phenotypic attributes. They parallel chromosomal markers and include variants of many morphological types, colour variants, variants showing diverse types of defect especially for respiratory enzymes, and variants resistant to toxic agents. Very few auxotrophic variants are known but this may reflect their mode of selection and isolation rather than any inherent inability for them to occur.

In so far that the physical basis of several extrachromosomal variants appears to be mitochondrial DNA (M-DNA), it might have been supposed that such variants should be inducible by mutagens for DNA. There is some similarity. For example, production of some M-DNA based variants in *S.*

cerevisiae shows an action spectrum typical for DNA, while irradiation with u.v. light is effective both in *N. crassa* (Pateman, 1960) and *A. nidulans* (Arlett, 1957). However, chemical mutagens do not seem to be particularly effective and the most effective agents are a number of toxic substances like acriflavine, ethidium bromide (Ephrussi, 1953; Slonimski and others, 1968) and a number of antibiotics such as erythromycin and others which probably inhibit mitochondrial protein synthesis (Linnane and others, 1968a, b). M-DNA differs from nuclear DNA in a number of ways, in particular the A-T groups may account for almost 80% of the bases. Whether this is a sufficient cause for the differences in reaction to mutagens or selective agencies is not clear. Since M-DNA appears to resemble prokaryotic DNA more than it does eukaryotic nuclear DNA, it might be expected that mutagens effective with prokaryotes would be particularly effective in inducing M-DNA based extrachromosomal variants in fungi. As yet insufficient data is available to test this expectation and even if it were the case there could, of course, be other mutable sites which might determine extrachromosomal variant characters.

High-temperature treatments have been found to be quite effective and produce both transient and permanent changes in extrachromosomal elements (Barnes, 1928; Arlett, 1960). Such treatments could have a selective effect on the cytoplasm comparable with that which operates in the heat treatment of virus-infected plants.

CHAPTER 3

Recombination, segregation and linkage I. Meiotic systems

Sexual and parasexual cycles

Most fungi exist in the haploid state most of the time, whereas the diploid phase when it occurs is usually of relatively short duration (see Figures 1.2–1.9, pp. 12–19). Diploid nuclei occur either in specialized reproducive structures—zygotes, asci or basidia—or, under certain circumstances in some fungi, in vegetative hyphae. The mechanics of mitosis and, to a lesser extent, of meiosis differ in fungi from those of green plants but the genetical consequences of these divisions do not. Thus, techniques of genetical analysis originally devised for diploid organisms can be applied to fungi with appropriate modifications. So far as meiotic recombination is concerned there are two differences from diploid organisms, namely:

(i) Segregants are usually haploid and their phenotypes can, therefore, be assessed directly. Moreover, since the segregants frequently occur in enormous numbers, it is only feasible to study a random sample of them.

(ii) The four products of meiosis can often be isolated together, as a *tetrad* or its equivalent. In the commonest cases—*unordered tetrads*—the sequence of segregation, whether at the 1st or 2nd division of meiosis, cannot always be ascertained with certainty but, in favourable cases—*ordered tetrads*—it can, since the segregants are disposed in a manner which reflects the spatial orientation of the two meiotic divisions.

Mitotic recombination is a rare event in most sexually reproducing fungi, as in higher organisms, so its detection is difficult. However, in those fungi where its frequency is increased, e.g. *Aspergillus niger*, or, if it is also associated with haploidization, detection is readily achieved. In practice, therefore, genetic analysis employing mitotic recombination is largely confined to those fungi which exhibit the parasexual cycle. The cycle was first

47

described in 1952 by Pontecorvo and Roper in *Aspergillus nidulans*; it is the sequence:

Fusion of dissimilar nuclei: mitotic recombination: haploidization.

A detailed comparison of the sexual and parasexual cycles and of the consequences of meiotic and mitotic recombination shows that the essential difference can be summarized as:

(i)   The sexual cycle and meiosis are highly regulated procedures; para-sexuality seems to be a sequence of more fortuitous events.

(ii) In the sexual cycle, nuclear fusion is frequently regulated either by specific mating-type factors or through morphological differentiation. Thereafter, a high proportion of compatible haploid nuclei appear to fuse once they are isolated in pairs in a specialized zygote cell. In contrast, nuclear fusion appears to be a rare, unlocalized and probably random event in the parasexual cycle.

(iii) Crossing-over is a regular attribute of every chromosome pair of the nuclear complement and multiple exchanges are frequent in meiosis but, in mitotic recombination, crossing-over is an event of variable frequency either a single or a few reciprocal exchanges confined to one arm of one or a few chromosome pairs of the complement, never with a frequency of exchanges approaching that in meiosis.

(iv) Segregation is effected through the ordered processes of two successive nuclear divisions in the sexual cycle but, in the parasexual cycle, by several successive mitoses from a hyperdiploid $(2n - 1)$ nucleus, probably by the loss of chromosomes derived through accidental non-disjunction of the diploid nucleus.

It would be extremely convenient if either the whole sequence, or component events of the cycles, both sexual and parasexual, could be induced at will by investigators. In some cases such possibilities exist, but, broadly speaking, at present, investigators can but provide reasonable conditions for normal growth and then leave a fungus to carry through the sequence. There is a vast literature on the induction of the reproductive phase (e.g. Hawker, 1957; Turian, 1969) from which few generalizations can be made. Suitable media for promoting the development of reproductive structures have been developed for fungi whose genetics have been studied intensively, e.g. the synthetic crossing medium for *Neurospora* (Westergaard and Mitchell, 1947). It should also be noted that in those fungi where a carpophore is normally developed, enclosing the meiocytic cells, the sexual cycle can sometimes be completed without normal development. For example, *Polyporus betulinus* normally produces a large bracket-like basidiocarp bearing basidia in pores on the lower surface, but a mycelial, dikaryotic culture inverted and kept in the dark will develop a pore surface with fully functional basidia in which meiosis occurs (Burnett, unpublished).

The induction of phases of the parasexual cycle has been attended with more success than those of the sexual cycle. Although there is little evidence to support the view that *d*-camphor vapour induces the development of diploid nuclei in hyphae, it certainly appears to select for them in some fungi, e.g. *Aspergillus nidulans, Penicillium chrysogenum*. For fungi with uninucleate condia a more generally applicable technique has been to force a heterokaryon between two auxotrophic mutant strains and plate the conidia on to minimal medium; only protrophic diploid conidia are then capable of growth, e.g. *Verticillium albo-atrum, Ustilago violacea, Aspergillus niger*. A further technical advance was the discovery that several agents such as X-irradiation, u.v. light and various chemicals (mitomycin C, nitrous acid) increase the frequency of mitotic segregation, while haploidization was shown to be effectively promoted by treatment with low concentrations of *p*-fluorophenylalanine (Murpurgo, 1961).

The rest of this chapter will be confined to examples of the application of formal genetic analysis subsequent to meiotic recombination in particular fungi to illustrate the techniques and procedures employed. In a few fungi, both meiotic and mitotic, analyses have been applied, e.g. *Aspergillus nidulans, Saccharomyces cerevisiae, U. maydis* and *S. commune*.

RECIPROCAL MEIOTIC RECOMBINATION

The use of randomly isolated products of meiosis

This method of analysis may be far more rapid than tetrad analysis when all the products of meiosis, usually spores, can be collected easily and their genotypes classified readily. Analysis of a large, random sample of segregants provides information concerning the number, location and linkage relationships of the segregating genes. In all situations, the larger the numbers of progeny scored, the more accurate is the information derived.

Numbers of genes segregating

Provided that the genes are unlinked and the genotypes of equal viability, the number of genes segregating can be determined from the number of genotypes produced. So,

(i) *For n segregating genes 2^n possible genotypes should occur with equal frequency;*
(ii) *Departures from equality are due either to differential viability or to linkage.*

Differential viability must be guarded against constantly. When the cause is genetical it can sometimes be allowed for if comparisons are made between the inviable, or semi-lethal mutant type and the standard wild, or non-mutant type in a model situation. Corrections can then be applied to the experimental segregations observed in the light of the model experiments. Differential viability as a consequence of environmental selection sometimes operates in an unexpected way. For example, although an auxotrophic mutant obviously will not grow on minimal medium, it may not grow on a complete medium either. Thus, some histidine-requiring mutants of *Neurospora crassa* are inhibited by the presence of other amino acids, notably arginine, or mixtures of methionine and lysine, so that such mutants will not grow on complete media containing casein hydrolysate. Another possible source of error arises from plating spores at too high a density. Thus, Grigg (1958) showed that non-growing mutant spores could inhibit the growth of non-mutant spores on the same plate, at total concentrations of 10^6 to 3×10^7 per petri plate, as a result of competition for respirable substrates between the two types of spore. The effects of competitive suppression on rare mutant protrophs can be particularly important when they arise through recombination between genes showing very close linkage. A deficit of recombinants in these circumstances makes the linkage appear tighter than it really is (Jinks, 1952).

Linkage and the location of segregating genes

(i) If linkage is present then the proportion of recombinants R can be determined in crosses in either the presumed coupling or repulsion phases. Suppose the loci are designated a^+/a and b^+/b then:

	In coupling	*In repulsion*
Cross	$ab \times a^+b^+$	$ab^+ \times a^+b$
Progeny	$(1 - R)\,ab$	$(1 - R)\,ab^+$
	$(1 - R)\,a^+b^+$	$(1 - R)\,a^+b$
	$R\,ab^+$	$R\,ab$
	$R\,a^+b$	$R\,a^+b^+$

and *linkage value is*

$$\frac{100\,R}{(1-R)+R}$$

Note, *neither double cross-overs nor centromere position can be detected from such data.*

(ii) If differential viability is suspected its consequences can be ameliorated, even eliminated, by making the cross in both phases and estimating R. Suppose the viability of a^+b compared with a^+b^+ is as $V:1$ then,

	In coupling	*In repulsion*
Cross	$ab \times a^+b^+$	$ab^+ \times a^+b$
Progeny	$(1 - R)\,ab$	$(1 - R)\,ab^+$
	$(1 - R)\,a^+b^+$	$V(1 - R)\,a^+b$
	$R\,ab^+$	$R\,ab$
	$VR\,a^+b$	$R\,a^+b^+$

so, computing R and eliminating differential viabilities:

$$R = \sqrt{\frac{VR}{(1-R)} \times \frac{R}{V(1-R)}}$$

only the positive root being significant.

(iii) When biochemical mutants are employed a useful adjunct to this technique is selective plating. Two auxotrophic mutant strains are crossed and the spores formed subsequent to meiosis are plated on minimal medium. Here, only recombinant protrophs can grow. If the loci involved are unlinked 25% of the spores will be protrophic; if linked, then less than 25%. Since a number of undetected double mutants *equal to the number of* protrophs must also have arisen through reciprocal recombination, *the actual linkage distance is twice the frequency of the protrophs.*

RANDOM MEIOTIC PRODUCTS: EXAMPLES

Example 3.1   Independent segregation

Two strains of *N. crassa* were crossed. One was normal in growth and produced macroconidia (wild-type); the other exhibited colonial growth and produced microconidia. Seventy-seven ascospores were isolated as follows:

Normal with macroconidia	20
Normal with macro- and microconidia	23
Colonial with no conidia	15
Colonial with microconidia	19

Note

(i)   Four phenotypic classes produced.
(ii) Numbers in each class approximately equal. Tested by χ^2, this was found to be the case ($\chi^2 = 1 \cdot 698$; $P = 0 \cdot 78$).

Deduction

(i)   Four genes are segregating and there is no linkage, i.e. the four phenotypes arise from genotypes $col\text{-}1^+\ m^+ : col\text{-}1^+\ m : col\text{-}1\ m^+ : col\text{-}1\ m$.
(ii) The mutant *col-1* not only impairs normal growth but prevents macroconidium formation.
(iii) The mutant *m* determines microconidium production.

(Data of Barratt and Garnjobst, 1949.)

Example 3.2   Linkage

The doubly mutant strain (*col-1 m*) was crossed

(a) with two stocks having normal growth and conidiation but requiring pathothenate (*pan*) and pyridoxin (*pdx*) respectively,
(b) with two other stocks having normal growth and conidiation but possessing peach-coloured (*pe*) and white (*al-2*) condia, respectively.

It is known that *pan* and *pdx* are both on the right arm of linkage group IV. Similarly *pe* is in linkage group II and *al-2* in linkage group I. Ascospores were isolated from the crosses with the following results.

(a) Considering only *col-1*, *pan* and *pdx*

	Parental	Recombinant
col-1 × *pan*	388	0
col-1 × *pdx*	176	48

(b) Considering only *m*, *pe* and *al-2*

m × *pe*	400	0
m × *al-2*	496	588

Note

(i) In each case there is a departure from equality in the classes but viability, when tested, was shown to be comparable.

Deduction

(a) (i) Since *col-1* and *pan* show no recombinants they are probably closely linked.
   (ii) Since parentals exceed recombinants for *col-1* and *pdx* they are probably linked. Estimate recombination as

$$\frac{100 \times 48}{176 + 48} = 21 \cdot 4\%$$

(b) (i) Since *m* and *pe* show no recombinants they are probably closely linked.
   (ii) As recombinants slightly exceed parentals but are only just significantly different from equality ($\chi^2 = 4 \cdot 05$; $P < 0 \cdot 050$), *m* and *al-2* are either far apart on the same chromosome or unlinked. Since *al-2* is in group I and *pe* in group II, the latter hypothesis is the more likely.

Thus *col-1* is in group IV close to *pan* but separate from *pdx*, while *m* is in group II close to *pe*.
(Data of Barratt and Garnjobst, 1949.)

Example 3.3   The selection of protrophic recombinants by selective plating

A cross was made between two strains of *N. crassa*, one auxotrophic for pantothenate and one for pyridoxin. The progeny were plated on minimal medium, i.e. lacking both required nutrients for growth. From this cross, $pan^+ pdx \times pan\, pdx^+$, some 392 ascospores were plated, 44 grew on minimal medium.

Deduction

(i)   Only ascospores of genotypes $pan^+ pdx^+$ could grow on minimal medium. These must have been recombinant or reassorted types.
(ii) The frequency of protrophs 44/392 is significantly less than 98, the number expected if the two genes are segregating independently, i.e. 25%. Therefore linkage is suspected.
(iii) Since the $pan^+ pdx^+$ ascospores have arisen by crossing-over, there must have been an equal number of undetected double mutants, *pan pdx*. Thus the total number of recombinants is twice the number of protrophs and the recombination frequency is

$$\frac{2 \times 44 \times 100}{392} = 22 \cdot 5\%$$

(Data of Houlahan, Beadle and Calhoun, 1949.)

Tetrad analysis

The use of tetrads

In virtually all Ascomycetes and Basidiomycetes the immediate products of meiosis are retained as they mature either within or upon the terminal cell, i.e. the ascus or basidium in which the diploid nucleus, established by nuclear fusion, has undergone meiosis. Not infrequently the tetrad of meiotic products is either discharged forcibly in a group, or the products can be removed by mechanical means. So the isolation of a tetrad is often relatively easy. For example, with large basidia and basidiospores as in some species of *Coprinus*, the 4 spores will adhere electrostatically to a fine dry needle brought towards them (Hanna, 1924). With smaller basidia or non-discharging asci, however, a micromanipulator more often has to be employed; that of Kemp (Kemp and Bevan, 1959; Kemp, 1960) is simple, effective and readily constructed. Tetrads of species such as *Ascobolus immersus* or *Neurospora crassa*, which discharge their ascospores violently, can be collected readily on a slide held at an appropriate height above the ascocarp (Lissouba and Rizet, 1960; Strickland, 1960). Other special techniques exist; for example, it is possible to isolate tetrads of *Saccharomyces cerevisiae* either by first separating an ascus cell and then dissolving away its wall with snail-gut enzyme (Johnston and Mortimer, 1959) or by pre-incubation on nutrient medium followed by mechanical disruption (Cox and Bevan, 1961).

In many Ascomycetes with linear asci, ascospores can be dissected in the same order in which they are arranged in the ascus. This order is related to the orientation of the meiotic and subsequent mitotic spindles, so that such that asci are described as *ordered tetrads*. The classical example where such analysis has been employed is in *Neurospora*, whose asci contain 8 ascospores, each genotype being duplicated as a result of the further post-meiotic mitosis. In practice, therefore, either only one of each of the four pairs of spores needs to be isolated or they can be isolated as adjacent pairs, provided that errors due to spindle disorientation, transposition of nuclei or spores, or non-reciprocal exchanges are either negligible or non-existent. In some species, e.g. baker's yeast, some 4-spored asci are linear but usually the majority are oval so that certainty of spore order is not high. The 4-spored basidia of rusts and smuts are linear tetrads.

In cases where neither the spore order nor their position provides reliable information concerning the sequence of meiotic events the tetrads are described as *unordered tetrads*.

Whether or not the tetrads are ordered or unordered, three classes can be recognized in respect of the patterns of distribution of the genotypes when two pairs of allelomorphic genes are segregating.

For example, consider the cross $a^+b^+ \times ab$. Three kinds of tetrads can be derived from it, viz.

$$
\text{PD} \quad
\begin{matrix}
a^+b^+ \\
a^+b^+ \\
ab \\
ab
\end{matrix}
\qquad
\text{NPD} \quad
\begin{matrix}
a^+b \\
a^+b \\
ab^+ \\
ab^+
\end{matrix}
\qquad
\text{T} \quad
\begin{matrix}
a^+b^+ \\
a^+b \\
ab^+ \\
ab
\end{matrix}
$$

(1) Parental ditype (PD) tetrads. Two of the meiotic products have the genotype of one parent, the other two that of the other parent. Note that each pair of alleles segregates in a one-to-one fashion (2:2— $a^+b^+:ab$).

(2) Non-parental ditype (NPD) tetrads. Here again each gene pair shows a one-to-one segregation but the two genotypes differ from the parental genotypes (2:2— $a^+b:ab^+$).

(3) Tetratype (T) tetrads. Four different genotypes are found, two identical with the parental genotypes, the other two identical with the non-parental genotypes (1:1:1:1— $a^+b^+:a^+b:ab^+:ab$).

Note, however, that there can be great differences in the proportions of the three types of tetrad. In Figure 3.1a the number of PD and NPD tetrads would be approximately equal, while in 3.1b their numbers would be very unequal. These differences arise from the fact that in 3.1a the loci are on different chromosomes which, therefore, show independent segregation whereas in the latter case the loci are linked. This is set out diagrammatically in Figure 3.1, which also illustrates the origin of each tetrad type.

When the loci are unlinked, PD and NPD tetrads arise from the independent alignment of chromosomes at the first meiotic division, whereas the T tetrads arise as a result of segregation at the second meiotic division consequent upon a cross-over in at least one of the paired chromosomes at the first meiotic division. When the loci are linked, however, PD tetrads reflect those cases where the linkage is unbroken, but both NPD and T tetrads arise through recombination. T tetrads reflect in the simplest case, the occurrence of a single cross-over between the loci, whereas NPD tetrads can only arise most simply as a consequence of a double cross-over involving all four chromatids of the paired chromosomes. So information concerning independent segregation, linkage and the position of the centromere can be derived from the study of unordered tetrads.

Ordered tetrads have the advantage that the positions of spores of particular genotypes indicates directly whether the allelomorphs have segregated at the first or second division of meiosis. If a single pair of alleles a^+/a, is considered six different segregation patterns can be recognized (Table 3.1).

(a)

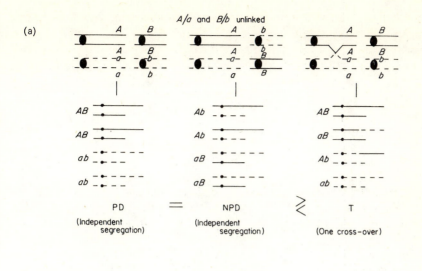

(b)

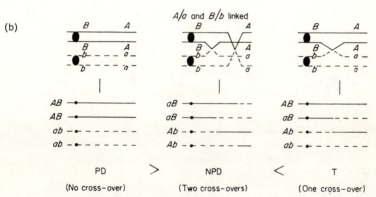

FIGURE 3.1 Diagrams to illustrate the segregation of a pair
of alleles *A/a* and *B/b* in a tetrad when they are on separate
chromosomes or on the same chromosome. Tetrad types are,
PD—parental ditype; NPD—non-parental ditype; and T—
tetratype.

TABLE 3.1 The six patterns of arrangement of a pair of alleles, *A/a*, in an
ordered tetrad such as that of *Neurospora crassa*.

Segregation type	Pattern	Spore number (apex–base) and genotype			
		1 and 2	3 and 4	5 and 6	7 and 8
1st-division segregation	I	*A*	*A*	*a*	*a*
	II	*a*	*a*	*A*	*A*
2nd-division segregation	III	*A*	*a*	*A*	*a*
	IV	*a*	*A*	*a*	*A*
	V	*A*	*a*	*a*	*A*
	VI	*a*	*A*	*A*	*a*

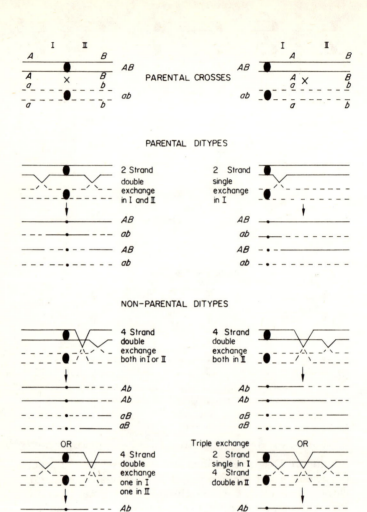

FIGURE 3.2 The origins of the three classes of tetrad, PD, NPD and T derived from the cross $AB \times ab$ where the A/a and B/b loci are linked but either on opposite sides of the

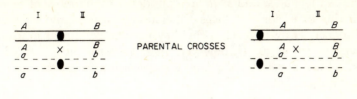

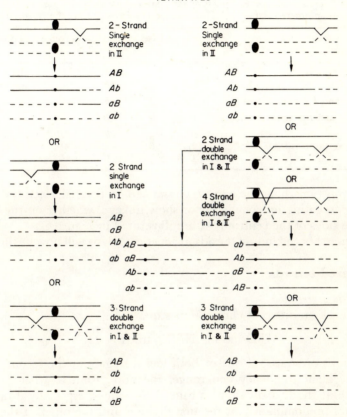

centromere or on the same side of the centromere. For crosses in 'repulsion' ($Ab \times aB$), the cross-overs accounting for classes 1 and 2, 5 and 6 are reversed. For further explanation see p. 58.

Patterns I and II arise from separation of the alleles at the first division, III–VI from second division segregation.

When two loci a^+/a and b^+/b are considered the number of segregation patterns rises to 36, for each locus can exhibit 6 patterns and since they can occur independently for each locus they may be combined in all possible ways, viz. $6 \times 6 = 36$. If, however, these patterns are classified *only in respect of whether first or second division segregation is exhibited they are reduced to seven basic classes* (Table 3.2).

TABLE 3.2 The seven classes resulting from the joint segregation of two pairs of genes, A/a and B/b, in an ordered tetrad such as that of *Neurospora crassa* from the cross $AB \times ab$.

Segregation type	Ascus type	Class	Spore arrangement and genotype			
Both loci at	PD	1	ab	ab	AB	AB
1st division	NPD	2	aB	aB	Ab	Ab
A/a at 1st and B/b at 2nd division	T	3	ab	aB	Ab	AB
A/a at 2nd and B/b at 1st division	T	4	aB	AB	ab	Ab
	PD	5	ab	AB	ab	AB
Both loci at 2nd division	NPD	6	aB	Ab	aB	Ab
	T	7	ab	AB	aB	Ab

The ways in which these seven classes can be derived when the loci are either segregating independently or show linkage, whether on the same or opposite sides of the centromere, are illustrated in Figure 3.2.

The use of both kinds of tetrads in genetic analysis will now be illustrated.

Unordered tetrads

Independent segregation

If loci are on different chromosomes then:

$$\text{PD tetrads}:\text{NPD tetrads} = 1:1$$

T tetrads only arise if one or both loci segregate at the second meiotic division. Their frequency can range, therefore, from 0, when both loci segregate at the first division, to quite high values, especially if one locus segregates predominantly at the first division and the other at the second division of meiosis. The frequency of T tetrads is given, in fact, by the equation (Perkins, 1949)

$$\text{T frequency} = x + y - \frac{3xy}{2}$$

where x and y are the frequencies of the second division segregations of the two loci respectively.

On the whole, therefore, *good evidence for independent segregation is given by*: PD = NPD and low values of T tetrads. A χ^2 test may be used to test equality.

If the ratio PD/NPD is one but the T tetrad frequency high the loci may be far apart on the same chromosome, so appearing to segregate independently (see below).

Linkage

The more closely two loci are linked the less likely they are to be separated and so the greater the frequency of PD tetrads and the fewer the NPD tetrads. *So, in well-defined cases, for linkage*:

$$\text{PD} \neq \text{NPD and PD} > \text{NPD}$$

Estimates can also be made of the recombination percentage. Each NPD tetrad has four recombinant strands and each T tetrad only two (cf. Figure 3.1). The total number of recombinant tetrads, r, is, therefore:

$$r = \sum \text{NPD} + \tfrac{1}{2}\text{T}$$

or, since $\text{T} = 1 - \text{NPD} - \text{PD}$,

$$r = \tfrac{1}{2}(1 + \text{NPD} - \text{PD})$$

So the *percentage recombination is*:

$$\frac{100(\tfrac{1}{2}\text{T} + \text{NPD})}{\text{total tetrads}}$$

This is, of course, an estimate uncorrected for cross-overs; with experience of the behaviour of particular species it is possible to modify the relationship so that it approximates more closely to the usual situation. For example, Perkins (1949) has shown that if the sample of tetrads includes only those with 0, 1 or 2 cross-overs between the linked loci and if NPD tetrads only account for $\tfrac{1}{4}$ of the double cross-overs, then the relationship

$$\frac{100\ (\text{T} + 6\text{NPD})}{2 \times \text{total tetrads}}$$

provides a more accurate estimate. This has been employed by Mortimer and Hawthorne (1969) for yeasts.

If the linked loci are far apart on the chromosome the PD:NPD ratio may approximate to unity, suggesting independent assortment. It may then be possible to utilize the T tetrad value to distinguish this situation from that where the loci are, indeed, on different chromosomes.

So long as loci are on the same chromosome most of the T tetrads will arise from single cross-overs while most NPD tetrads arise from double cross-overs. For n cross-overs it can be shown that the frequencies are

$$\text{NPD} = \tfrac{1}{6} + \tfrac{1}{3}(-\tfrac{1}{2})^n \quad \text{and} \quad \text{T} = \tfrac{2}{3} - \tfrac{2}{3}(-\tfrac{1}{2})^n$$

so in the limiting case where n is large

$$\text{NPD} \longrightarrow \tfrac{1}{6} \quad \text{and} \quad \text{T} \longrightarrow \tfrac{2}{3} \quad \text{or} \quad \text{NPD:T} = 1:4$$

It will be recalled that the frequency of T tetrads in the case of two un-linked loci can vary from o to quite high values. Nevertheless, if (a) the PN:NPD ratio approximates to equality, (b) the frequency of T tetrads approximates to 66·7% and (c) the T:NPD ratio never falls below 4:1, then distant linkage may be suspected. It is, of course, preferable to estimate recombination from more closely linked loci.

Centromere mapping

Two methods are generally available: a further method is applicable if tetraploid tetrads can be obtained.

(i) The first depends on the fortunate occurrence of a marker gene so close to the centromere that it segregates with it regularly. Thus 1st and 2nd division segregations can be distinguished with high reliability. The most extensive use of this method has been, to date, with yeast where tetrads involving several centromere-linked genes have often been used (e.g. Hawthorne and Mortimer, 1960). An example of such a gene is *trp-1* (tryptophan-requiring) which shows less than 1% recombination with the centromere of chromosome IV in yeast.

(ii) The second method, devised by Whitehouse (1950, 1957), has more general usefulness. It utilizes three unlinked loci. It stems from the fact that, with independently segregating loci, the frequency of T tetrads depends on the frequencies of 2nd division segregation of the loci concerned as set out earlier. Thus, for three loci, a, b and c with 2nd division segregation frequencies of x, y and z respectively, T tetrads for loci a and b, b and c, and a and c are, respectively:

$$T_{ab} = x + y - \frac{3xy}{2}; \; T_{bc} = y + z - \frac{3yz}{2} \quad \text{and} \quad T_{ac} = x + z - \frac{3xy}{2}$$

The solution of these three simultaneous equations for x, y and z are

$$x = \frac{2}{3}\left[1 \pm \sqrt{\frac{4 - 6T_{ab} - 6T_{ac} + 9T_{ab}T_{ac}}{4 - 6T_{bc}}} \right]$$

$$y = \frac{2}{3}\left[1 \pm \sqrt{\frac{4 - 6T_{ab} - 6T_{bc} + 9T_{ab}T_{bc}}{4 - 6T_{ac}}} \right]$$

and

$$z = \frac{2}{3}\left[1 \pm \sqrt{\frac{4 - 6T_{bc} - 6T_{ac} + 9T_{bc}T_{ac}}{4 - 6T_{ab}}} \right]$$

Each of these equations can have two solutions; when both are real, one is greater than $\frac{2}{3}$, the other less. In this case the latter value is the more likely, since 2nd division segregations greater than 66·7% are likely to be rare (Mather, 1938).

It will be clear that this method is only effective if at least two of the three loci are fairly near their respective centromeres. If this is not so then the T tetrad frequencies will each approximate to $\frac{2}{3}$ so that little precision will be obtained from the calculated values. This method has been used especially with *Coprinus lagopus* (e.g. Moore, 1967).

(iii) The third method, depending on the availability of tetraploid asci, has only been used for yeast. Roman and others (1955) have shown that three types of segregant asci arise when classified by their phenotypes. Suppose a^+ is dominant to a, then the three types are:

4:0, i.e. all ascospores a^+/a; phenotype a^+
3:1, i.e. $1a^+/a^+ : 2a^+/a : 1a/a$; phenotypes $3a^+ : 1a$
2:2, i.e. $2a^+/a^+ : 2a/a$; phenotypes $2a^+ : 2a$

Providing that pairing is by bivalent formation or by quadrivalents showing specific types of exchange (Leupold, 1956a, b), the frequencies of these three types will be 4:4:1 for a gene which recombines freely with the centromere of its chromosome and 2:1:1 if immediately adjacent to the centromere. So the percentage of asci segregating phenotypically 3:1, a^+/a provides an estimate of the frequency of 2nd-division segregation. Hilger, (1973) has shown how tetraploid yeast sets may readily be synthesized by a controlled crossing procedure and subsequently analysed following one or two successive meioses. These sets show regular tetrasomic inheritance.

This method has very limited usefulness for fungi in general. In part this derives from the few polyploid fungi, natural or induced, which have been studied but, of far more importance, from the assumptions which have to be made concerning the kinds, frequencies and orientation of bivalents and multivalents at meiosis. These cannot be substantiated cytologically because of the small size and technical difficulties associated with fungal chromosomes. In species with larger chromosomes than those of yeast where polyploidy is suspected, e.g. *Cyathus stercoreus* (Lu, 1964) it might be possible.

Unordered Tetrads: Examples

In the genetic analysis of *Coprinus lagopus* four characters were employed: the mating types, *A* and *B*, and two auxotrophs for adenine (*ade-5*) and choline (*chol-1*).

Crosses were made, tetrads of basidiospores isolated and the genotypes of each spore ascertained. The tetrads were then scored as either parental, non-parental or tetratype. Data obtained are set out in the form PD:NPD:T below:

	B	ade-5	chol-1
A	239:237:253	39:35:38	46:63:97
B		77: 2:33	102: 7:47

(Data of Moore, 1967.)

Example 3.4   Independent assortment
Note

(i)   In the crosses involving A/B, A/ade-5 and $A/chol$-1 the PD:NPD ratio approximates to 1:1 and a χ^2 test indicates no departure from this.
(ii)   The values of the T tetrads are high but are less than 66·7%.
(iii)   The T:NPD ratio is below 4:1 in each case.

Deduction

(i)   The 1:1 PD:NPD ratio suggests independent assortment, although distant linkage could be responsible.
(ii)   The facts that in no case does the value of T approach 66·7% nor does the T:NPD ratio approach 4:1 confirm that independent assortment is probable.

Example 3.5   Linkage
Note

(i)   In the crosses involving B/ade-5 and $B/chol$-1 the PD:NPD ratio is $\neq 1$ and PD > NPD.

Deduction

(i)   Since PD:NPD is $\neq 1$ and PD > NPD, it is probable that the genes are linked and NPD tetrads represent recombinants.

To estimate recombination the simplest formula can be applied so that:

$$\text{for } B/ade \quad \frac{100(\tfrac{1}{2}(33) + 2)}{112}, \quad \text{for } B/chol\text{-}1 \quad \frac{100(\tfrac{1}{2}(97) + 7))}{206}$$

$$= 16\cdot5\% \qquad\qquad = 26\cdot9\%$$

or, employing Perkins formula to allow for double cross-overs,

$$\text{for } B/ade \quad \frac{100(33 + (6)(2))}{(2)(112)}, \quad \text{for } \quad B/chol\text{-}1 \quad \frac{100(97 + (6)(7))}{(2)(206)}$$

$$= 20\cdot1\% \qquad\qquad = 33\cdot7\%$$

Note that the latter estimates are higher than the former as is to be expected if double cross-overs had occurred, and this is not improbable in distances of the order of magnitude of those obtained. In both cases it would be desirable to find another marker between the B factor locus and the mutant locus and so reduce the uncertainty over the recombination data.

Example 3.6   The use of unlinked loci to determine second division segregation frequencies and hence gene–centromere distances

The A, B and nic-4 (nicotinamide-requiring) loci are unlinked in $C.$ $lagopus$ as tested by the method illustrated in Example 3.4. Frequencies of the tetrad types involving them (PD:NPD:T) are:

$$
\begin{array}{llll}
A\text{—}B & 239{:}237{:}253 & \text{i.e. T as \%} & 34\cdot71 \\
A\text{—}nic\text{-}4 & 92{:}\ 99{:}145 & \text{i.e. T as \%} & 43\cdot15 \\
B\text{—}nic\text{-}4 & 136{:}132{:}\ 68 & \text{i.e. T as \%} & 20\cdot24 \\
\end{array}
$$

Employing Whitehouse's formula

$$x = \frac{2}{3}\left[1 \pm \sqrt{\frac{4 - 6(0{\cdot}347) - 6(0{\cdot}432) - 9(0{\cdot}347)(0{\cdot}432)}{4 - 6(0{\cdot}202)}}\right]$$

or $$y = \frac{2}{3}\left[1 \pm \sqrt{\frac{4 - 6(0{\cdot}347) - 6(0{\cdot}202) - 9(0{\cdot}347)(0{\cdot}202)}{4 - 6(0{\cdot}432)}}\right]$$

or $$z = \frac{2}{3}\left[1 \pm \sqrt{\frac{4 - 6(0{\cdot}202) - 6(0{\cdot}432) + 9(0{\cdot}202)(0{\cdot}432)}{4 - 6(0{\cdot}347)}}\right]$$

Solving for x

$$x = 1{\cdot}49 \text{ or } 0{\cdot}339$$

Evidently the former is unreal hence A shows 33·9% second-division segregation. The equations for y and z may be solved independently or the relationship

$$T_{AB} = x + y - \frac{3xy}{2}$$

may be used. Thus

$$0{\cdot}347 = 0{\cdot}339 + y - \frac{3(0{\cdot}339)y}{2}$$

$$0{\cdot}008 = 0{\cdot}491y$$

$$y = 0{\cdot}163$$

hence B shows 1·6% second-division segregation.

Similarly nic-4 shows 18·9 second-division segregation. Hence centromere distances are

$$A, 17{\cdot}0; \quad B, 0{\cdot}8; \quad nic\text{-}4, 9{\cdot}5.$$

Example 3.7 The use of centromere-linked genes in yeast to determine centromere–gene distances

Auxotrophs for the following nutrients are known in yeast: tryptophan (*trp-1*) uracil (*ura-3*), histidine (*his-6*) and adenine (*ade-2*) and the mating-type factors a/α are also employed. Twelve asci derived from a cross were scored in respect of mating type and each auxotroph and gave the following results (PD+NPD:T)

a/α	*trp-1/trp-1*$^+$	12:0	hence no 2nd-division segregation
a/α	*ura-3/ura-3*$^+$	12:0	hence no 2nd-division segregation
a/α	*his-6/his-6*$^+$	7:4	hence 36·8% (4/11) 2nd-division segregation
a/α	*ade-2/ade-2*$^+$	3:9	hence 75% (9/12) 2nd-division segregation

Therefore centromere–gene distances are

$$trp\text{-}1 \text{ and } ura\text{-}3, 0; \quad his\text{-}6, 18{\cdot}4; \quad ade\text{-}2, 37{\cdot}5.$$

(Data of Mortimer and Hawthorne, 1966.)

Ordered tetrads

As was explained earlier (p. 58 and Figure 3.2, pp. 56, 57), first and second division segregation can be distinguished in linear asci by the pattern of distribution of the ascospore genotypes with the result that centromere–locus distances can be readily obtained.

Independent segregation and linkage

When a pair of alleles is segregating the occurrence and frequency of second division segregation indicates the distance of the locus from its centromere. A single cross-over between locus and centromere involves two of the four chromatids, i.e. half of them, and so the distance from the centromere is half the percentage of the asci exhibiting second-division segregation. However, if the locus is so far from the centromere that double cross-overs also occur, they will be recorded as first-division segregations and so reduce the real incidence of recorded cross-overs. So a value of 67% is hardly ever exceeded for second-division segregations.

So, *for short distances, the distance of a locus from its centromere is half the percentage of asci which show second-division segregation.*

Moreover, when two or more pairs of alleles are segregating the relative frequencies of certain of the seven segregation classes provide additional information. The data can be utilized as follows:

Classes 1 and 2. *Equality indicates independent assortment. Inequality indicates linkage*, the largest class being the parental genotypes.

Classes 5 and 6. *Equality indicates that the loci are on opposite sides of the centromere.* Both classes arise as a consequence of double cross-overs involving either two or four chromatids but these are equally likely. *Inequality, in particular an excess of class 5, indicates that the loci are on the same side of the centromere.* Class 5 arises from a single cross-over but class 6 from triple cross-overs which are likely to be less frequent than single cross-overs; hence the excess of class 5 over class 6.

If linkage is detected the frequencies of the different classes can be used to estimate the frequency of recombination. This is particularly valuable in cases where three linked loci are segregating since gene order and distance can then be ascertained with considerable efficiency, provided that a sufficiently large number of tetrads have been analysed. In practice dissection and analysis of a sufficient number of tetrads can be exceedingly laborious. The data may also be rendered unsatisfactory through the reduced viability of ascospores with multiple nutritional requirements if protrophic markers are used. An example of a three-point cross is given later.

Example 3.9 Use of second division segregation to determine gene–centromere data when a single pair of genes is segregating

When a single pair of genes is segregating in an ordered ascus 1st division segregation is represented by Classes I + II and second division segregation by Classes III–VI in Table 3.1 (p. 55).

The earliest such segregation was observed in *N. sitophila* for mating type and since then many such observations have been made. The earliest and totalled results are given:

Cross	Total	1st-division segregation	2nd-division segregation	Frequency 2nd division, %	Gene–centromere distance
Earliest	5	0	5	100	100
Total	910	477	433	47·6	23·8

(Data of Fincham, 1951.)

Example 3.10 The use of data based on the 7 recognizable segregation classes when two pairs of genes are segregating

The joint segregation of two pairs of genes gives rise to the seven classes shown in Table 3.2 (p. 58).

Some of the earliest crosses made with *N. crassa* were of this type. They involved the mating-type factors *A/a* and two morphological mutants, *gap* with scattered conidia and *pale* with lighter coloured less abundant conidia. The data obtained are set out below.

Cross		Asci in segregation classes						Total
	1	2	3	4	5	6	7	
(a) *A gap* × *a gap*⁺	135	4	20	3	14	0	2	178
(b) *A pale* × *a pale*⁺	62	1	24	10	0	1	1	109

Note

(a) (i) Classes 1 and 2 are unequal.
    (ii) Classes 5 and 6 are unequal and, in particular, class 5 > 6.
(b) (i) Classes 1 and 2 are unequal.
    (ii) Classes 5 and 6 are small but not significantly different from equality.

Deductions

(i) That the *A/a* locus is linked with both *gap* and *pale* (Classes 1 and 2 not equal).
(ii) That *gap* is probably on the same side of the centromere as the *A/a* locus (Class 5 > 6) but *pale* could be on the opposite side of the centromere to *A/a* locus (Class 5 ⇌ 6).

Further analysis can be carried out by determining the frequency of second division segregations of the three loci.

(a) Since *A/a* and *gap/gap*⁺ are on the same arm of the centromere from Table 3.2,
    for *A/a*, 2nd-division segregation is shown by Classes 3, 5, (6) and 7;

for gap/gap^+, by Classes 4, 5, (6) and 7.
Hence for A/a,

$$20 + 14 + 2 = 36/178 \ (20 \cdot 2\%)$$

and for gap/gap^+

$$3 + 14 + 2 = 19/178 \ (10 \cdot 7\%)$$

So centromere to A/a is $10 \cdot 1$ units and to gap $5 \cdot 3$ units.

(b) Here A/a and $pale/pale^+$ are on opposite sides of the centromere and hence the segregation classes are derived by different cross-over patterns (Figure 3.1, p. 55).

for A/a, 2nd-division segregation is shown by classes 4, 5, (6) and 7;
for $pale/pale^+$ by classes 3, 5, (6) and 7.

Hence for A/a,

$$10 + 1 + 1 = 12/109 \ (11 \cdot 0\%)$$

and for $pale/pale^+$

$$34 + 1 + 1 = 36/109 \ (33 \cdot 0\%)$$

So centromere to A/a is $5 \cdot 5$ and to $pale$ is $16 \cdot 5$.

There is evidently a discrepancy between the different values obtained for the centromere–A/a locus, so that these need to be averaged and investigated further; the mean is $7 \cdot 8$.

The order on the chromosome is evidently

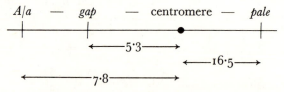

A further check on distance can be made by determining the recombination values between the A/a locus and the other two loci.

(a) The recombinants between A/a and gap/gap^+ are given by classes 2, 3, 4, (6) 7, (Figure 3.2, pp. 56, 57), i.e. $8 + 20 + 32 = 33/178 \ (18 \cdot 5\%)$ or $9 \cdot 3$ units.

(b) Similarly for A/a and $pale/pale^+$, the same classes are involved 2, 3, 4, 6, 7, i.e. $2 + 24 + 10 + 2 + 1 = 49/109 \ (45 \cdot 0\%)$ or $22 \cdot 5$ units.

Hence the best fit can be achieved from

Cross (a)	$\left\{\begin{array}{l} \\ \\ \end{array}\right.$
Mean of (a) and (b)	
Cross (b)	$\left\{\begin{array}{l} \\ \\ \end{array}\right.$

(Data from Lindegren, 1933, 1936.)

Example 3.11   Genetic analysis when three pairs of genes are segregating

Genetic analysis of three-point crosses can be made from ordered tetrads but it is excessively laborious. However, from such crosses two kinds of information can be obtained:

(i)   Whether the markers segregate at the 1st division of meiosis or the 2nd. This gives an estimate of centromere–gene distances.
(ii) The ascus type, PD, NPD or T, for genes taken two at a time.

These two sets of data can then be used to develop a linkage map.

A second procedure is to establish by inspection the likely gene order, based on the commonest parental types and those with a single cross-over in one or the other of the two intervals between the three loci—the commonest recombinant types. Coroboratory evidence is then obtained by working out the possible combinations of single, double, triple, quadruple etc. cross-overs on the basis of accounting for each ascus arrangement on the simplest hypothesis concerning crossing-over.

Both these methods are tedious and it is often sufficient to ascertain order and approximate linkage values from random ascospores. Thereafter, further linkage data can be obtained with two-point crosses etc.

Some examples are given:

Cross	Parental types	Recombinant types			Total	Recombination frequency
		Singles I	Singles II	Doubles I and II		
$\dfrac{+\ b\ c}{a\ +\ +}$	+bc a++	+++ abc	+b+ a+c	++c ab+		a–b/b–c
(a) $\dfrac{A\ +\ arg\text{-}1}{a\ ade\text{-}5\ +}$	19 26	2 1	0 0	0 0	48	6·2/0
(b) $\dfrac{+\ arg\text{-}1\ +}{leu\text{-}3\ +\ cr}$	36 25	6 6	3 0	0 0	76	15·8/3·9
(c) $\dfrac{+\ +\ al}{rg\ cr\ +}$	41 43	4 8	33 41	2 3	175	9·7/45·1

Key to symbols: A/a, mating type; *ade*, adenine-requiring; *al*, albino; *arg*, arginine-requiring; *cr*, crisp (morphological mutant); *leu*, leucine requiring; *rg*, ragged (morphological mutant).

Note

(a) No singles or doubles in II and only small number in I. Therefore genes closely linked. Recombinant frequency I 3/48.
(b) Singles in both I and II but no doubles. Recombination in I, 12/76 and in II 3/76.
(c) Singles in both I and II and also doubles. Several recombinants in II, therefore, long distance apart. Recombination in I, 12 + 5 = 17/175 and in II, 74 + 5 = 79/175.

Hence probable order is

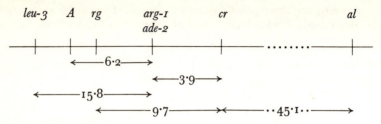

It is clear that the relationship of *A* to *rg* is ambiguous and more data are required. (Data of Houlahan, Beadle and Calhoun, 1949.)

Translocations and inversions

These can often be analysed by the use of ordered tetrads, and Heslot (1958) has given a detailed theoretical treatment for equal and unequal reciprocal translocations with 0, 1 or 2 cross-overs, and both para- and pericentric inversions with 0, 1 or 2 cross-overs, the last of these variously disposed in relation to the centromere and the inversion. Emerson (1963) gave an independent theoretical treatment for translocations but also gave the expectations for unordered tetrads. There is not much published data on the effects of chromosome aberrations on segregation. Heslot gave a few examples from *Sordaria macrospora* and Cox (Cox and Gill, 1967; Cox and Parry, 1967) has published a very full treatment of a case in *S. fimicola*. Translocations occur extensively in *A. nidulans* and they have been examined in detail by Käfer (1962, 1965). In general, if there is a high proportion of aborted asci or of ascospores in individual asci, a gross translocation may be suspected although, of course, such an appearance may be due to the segregation of a lethal gene.

Many translocations are now known for *Neurospora* but most of them are quite small.

Interference

It has been well known for many decades that the occurrence of one cross-over reduces the likelihood of another near to it (Muller, 1916). This has been termed *chiasma interference*. However, when two adjacent cross-overs form, it was pointed out that the double cross-over may involve but two of the chromatids (2-strand), all the chromatids (4-strand), or three of the chromatids (3-strand) in two different ways (Figure 3.3). Their frequency, if they form at random, is 1 : 2 : 1 for 2 : 3 : 4-strand exchanges. Any departure from randomness has been termed *chromatid interference* (Mather, 1933).

The segregation of three, or more, linked genes in ordered tetrads provides particularly valuable material for observing the direct consequences of both these phenomena if they occur. *Chiasma interference can be detected by seeing*

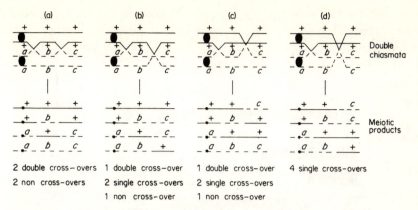

FIGURE 3.3 Types of double chiasmata and their genetic consequences. The chiasmata involve (a) only two chromatids, or (b), (c) only three chromatids, or (d) all four chromatids.

whether the observed number of double cross-overs (which can be detected genetically, see Figure 3.2) *is significantly less than that expected*, i.e. for a chromosome with loci *a–b–c*, the expected frequency of double cross-over is:

Frequency of single cross-over in region *a–b* × frequency of single cross-over in region *b–c*

Similarly, since the genotypes of 2-, 3- and 4-strand cross-overs can be distinguished in an ordered tetrad, *their departure from a ratio of* 1:2:1 *can be ascertained and thus chromatid interference detected.*

In general, there is usually evidence from fungi of chiasma interference although it appears only to operate within chromosome arms and not across the centromere. The evidence for chromatid interference is less well substantiated but there is a suggestion of a small but consistent excess of 2-strand exchanges over expectation in certain regions of a few fungi, e.g. near the centromere of *Coprinus lagopus* the ratio is 12:1:0 (Day and Swiezynski, in Fincham and Day, 1971). The significance of such findings will be discussed briefly in Chapter 14 together with problems of non-reciprocal recombination and anomalous ratios. These phenomena, while of importance for understanding the process of recombination, have little practical bearing on the practical determination of the positions of loci.

Chromosome maps

Linkage data can eventually be synthesized into chromosome maps. Because of interference, estimates of the frequency of crossing-over may be biased. Ideally, therefore, it is best to investigate recombination between reasonably closely linked markers. In practice, this is hardly possible in any fungus investigated to date save perhaps *Neurospora crassa* and *Saccharomyces cerevisiae* and even here certain chromosome regions are only marked by loci many map units apart, e.g. the ends of chromosome VII in *N. crassa* or the mating-type chromosome (III) of *S. cerevisiae*. In contrast, certain very short regions in some fungi, notably *N. crassa* and *Aspergillus nidulans*, have been mapped extremely intensively (Figure 3.4).

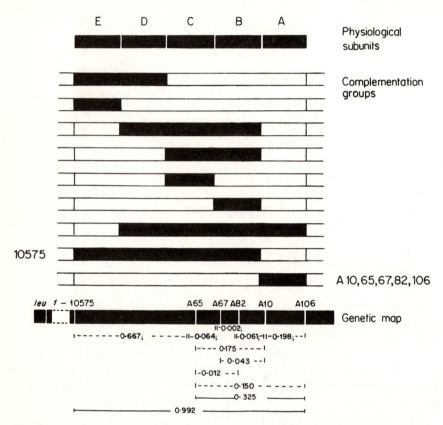

FIGURE 3.4 An example of fine structure mapping of heteroalleles at the *trp-1* locus in *Neurospora crassa*. (Based on Ahmad and others, 1964.)

In intensively studied fungi, such as *Neurospora*, yeast and *Aspergillus*, two kinds of technique have been developed to assist mapping. In the former, Barratt and others (1954) have developed mapping functions which seek to correct map distances through knowledge of the probable levels of interference as determined by experiments. This kind of technique can only be used with an organism where a good deal of detailed linkage data already exists.

The second technique is the preparation of multiply marked stocks with which any new mutant can be crossed: linkage is then looked for. In yeast, stocks with centromere-marker genes are available. In *Neurospora* a stock called *alcoy* has been introduced by Perkins and others (1969). This is a stock in which, as a result of radiation treatment, interchanges have taken place between:

Chromosomes I and II—marked with *al*—albino conidia on I
Chromosomes IV and V—marked with *col*—temperature sensitive and colonial growth on IV
Chromosomes III and VI—marked with *y*—yellow on VI.

Thus linkage with any one of these markers indicates on which of two chromosomes out of six the unmapped mutant may lie and further test crosses can readily be made. If no linkage occurs the mutant may lie on chromosome VII ($n = 7$ in *Neurospora*) or, of course, it may lie far from the markers used. In any event, this multiply marked interchange stock considerably reduces the labour of mapping a new mutant.

For most fungi, however, the preparation of detailed general chromosome maps is hardly a proposition as yet. Nevertheless, considerable progress may be made for particular purposes by developing mapped regions adjacent to some region under particular study. A good example of this is the mapping of the regions adjacent to the mating-type loci in *Schizophyllum commune* by Raper and his colleagues (see Figure 10.5, p. 212).

When maps are constructed particular care must be taken to ensure that either data from different workers can be related to a particular genetic stock, or better, that all data is derived from stocks whose genetic background is identical or very similar. Frost (1961) and others have shown that where different wild-type stocks of *N. crassa* are employed and data on locus–centromere distances pooled, there is marked heterogeneity in the recombination distances recorded. He was able to attribute this to certain stocks and when this was done the heterogeneity was eliminated. The causes of these discrepancies are largely genetic and will be discussed in Chapter 8. Here it is sufficient to emphasize that they are widespread in fungi, e.g. in *Venturia inaequalis* (Boone and Keitt, 1956), *C. lagopus* (Day, 1958) and *S. commune* (Papazian, 1951) save where a single wild-type strain has been employed, e.g. the 'Glasgow' strain of *A. nidulans* (Pontecorvo and others, 1953b).

For mapping, therefore, either *employ a single wild-type strain* or *repeatedly back-cross* markers to a *designated strain* before determining linkage values.

OTHER RECOMBINANT PROCESSES

Reciprocal meiotic recombination accounts for the vast majority of cases of recombination in fungi. The procedures already outlined would be pointless and inapplicable if this were not so and information on linkage and segregation of genes could neither be obtained nor be relied upon. That such information can be obtained and used for effective prediction indicates that it is reliable. Nevertheless, data exist which are not explicable in terms of meiotic crossing-over and segregation. Only a summary will be given here since the object is merely to draw attention to the possibility of occurrence of such anomalous situations. Some of them will be discussed further in Chapter 14.

Gene conversion

Lindegren (1949, 1953) noticed a number of cases in yeast where tetrads derived from a cross such as $a^+ \times a$ gave rise to aberrant tetrad ratios, such as $3a^+ : 1a$ or $1a^+ : 3a$, rather than that expected, namely $2a^+ : 2a$. Such ratios, including $4a^+ : oa$ and its reciprocal, can of course arise from well-understood genetical causes such as gene interaction, duplicate genes, polyploidy or polysomy (cf. Emerson, 1956), but a residium of aberrant ratios exists after these possible cases have been eliminated. Lindegren ascribed these ratios to a process of gene conversion so that a $3a^+ : 1a$ tetrad arose through the conversion of an a gene to the a^+ form. The phenomenon has since been found in several other Ascomycetes including *Neurospora*, *Sordaria*, *Ascobolus* and *Podospora*. In most of these cases, particularly in *Saccharomyces*, gene conversion took place between extremely close-linked allelomorphs and was associated with non-reciprocal recombination.

For example, Mitchell (1955) investigated the phenomenon in relation to two very closely linked loci, whose mutant alleles resulted in a requirement for pyridoxin. The alleles could also be distinguished by the fact that the requirement for pyridoxin could be remedied for one of them—*pdxp*—by changing the pH but not so for the other *pdx*. In a cross such as:

$$+ \ pdxp \times pdx \ +$$

rare tetrads arose, for example:

Observed	Expected
pdx +	*pdx* +
pdx +	*pdx pdxp*
+ *pdxp*	+ +
+ +	+ *pdxp*

Note that there is a $2 : 2$ segregation of $+ : pdx$, there is an anomalous $3 + : 1$ *pdxp* ratio and the expected double mutant, *pdx, pdxp*, is lacking.

A number of other examples are known and, in general, the phenomenon is associated with:

(i) Heteroallelic heterozygotes, i.e. where the 'alleles' are different mutants of a single locus,

(ii) both meiotic and mitotic crossing-over and

(iii) normal, reciprocal recombination of non-allelic, marker genes lying adjacent to the locus.

For most studies on linkage between adjacent non-allelomorphic genes, therefore, the phenomenon is of little significance and can be disregarded. Its interpretation however, is believed to have a bearing on the general problem of gene structure and the mechanism of crossing-over. This will be considered, in detail, in Chapter 14.

Chapter 4

Recombination, segregation and linkage II. Mitotic systems

Reciprocal Mitotic Recombination

Mitotic recombination was discovered in fungi by Roper (1952) who first demonstrated it in a synthetic diploid of *Aspergillus nidulans*, a sexually reproducing species. It was first shown to occur in a non-sexual species, *A. niger*, by Pontecorvo and others (1953a). It has now been demonstrated in a number of Ascomycetes, Basidiomycetes and Fungi Imperfecti but has not been shown to occur in Phycomycetes (e.g. *Verticillium* spp., Hastie, 1970: *Ustilago* spp., Holliday, 1961a, b; Day and Jones, 1969: *Coprinus* spp., Swiezynski, 1962; Prud'Homme, 1970).

The classic expositions of the methods of genetic analysis employing mitotic recombination are those of Käfer (1958) and Pontecorvo and Käfer (1958) employing *A. nidulans*. Since then, however, as a consequence of studies of a wider range of fungi, including several strictly asexual species, modifications can and have been introduced into the procedures. For example, with *A. nidulans*, haploidization is a rare event so that data from derived haploids are not readily available, but in other species, e.g. *Verticillium albo-atrum*, *S. cerevisiae*, the frequency of spontaneous haploidization is high (Hastie, 1964; Thornton and Johnston, 1971), and it can be induced in others by the use of appropriate chemical treatments, e.g. DL-*p*-fluorophenyl-alanine (Lhoas, 1961).

Mitotic segregation may be affected by the ploidy of a fungus. Very little is known about this phenomenon but a valuable paper deals with such a situation in autotetraploid and triploid *Saccharomyces cerevisiae* (Mackinnon and Johnston, 1972). The tetraploid arose spontaneously and was fully analysed from recombinants derived from two successive meioses. However, the spontaneous rate of mitotic segregation of the tetraploid was exceedingly low, about 1×10^{-7}, and took place in two stages: the segregating locus was for adenine deficiency, viz.

$$
\begin{array}{ccc}
\textit{ade-8} & \textit{ade-8} & \textit{ade-8} \\
\textit{ade-8} & \textit{ade-8} & \textit{ade-8} \\
\textit{ade-8}^+ \longrightarrow & \textit{ade-8} \longrightarrow & \textit{ade-8} \\
\textit{ade-8}^+ & \textit{ade-8}^+ & \textit{ade-8}
\end{array}
$$

When triploids were synthesized the frequency of segregation varied with the genotype. Triploids of constitution $(\textit{ade-8}, \textit{ade-8}^+, \textit{ade-8}^+)$ were as stable as the tetraploid but those of constitution $(\textit{ade-8}, \textit{ade-8}, \textit{ade-8}^+)$ segregated as frequently as did diploid $(\textit{ade-8}^+, \textit{ade-8})$ clones, i.e. 4×10^{-4}. Nothing appears to be known concerning the haploidization frequency of tetraploid yeasts under the influence of appropriate haploidizing agents.

The induction of diploidy, mitotic segregation and haploidization have been treated in Chapter 2. It will be convenient, therefore, to consider first the genetical consequences of mitotic recombination and then the procedures followed in the utilization of data for ascertaining linkage and gene order.

Consequences of mitotic recombination and haploidization

Mitotic recombination and haploidization are used together, whenever possible, to determine linkage groups, gene order and the position of the centromere.

Haploidization

This results in the segregation and reassortment of whole chromosomes. This is because haploidization probably occurs via aneuploidy following non-disjunction in a diploid nucleus. The hyperhaploid nuclei, $(n + (n - 1))$ or $(2n - 1)$, so formed undergo successive chromosome losses until a stable haploid (n) is reached. In the successive divisions each pair of homologous chromosomes segregate independently of each other and of all the other pairs of chromosomes. So, *marker genes in the same linkage group never show recombination at haploidization.*

Thus, the complete genotype of a diploid can be determined by isolating haploids from it, or the process can be utilized for allocating a new mutant to its linkage group, provided that in each case marker genes are known for each chromosome pair. The procedure, therefore, is to incorporate a new mutant gene '*mut*' into a synthesized diploid already carrying marker genes on each of its chromosomes. Haploids are derived from it and are examined. Those marker genes located on the chromosome carrying *mut* will remain constantly associated with it, whereas, in contrast, all other markers will reassort freely with *mut*.

Mitotic recombination

The consequence of mitotic crossing-over provides the means whereby the sequence of genes on a chromosome can be determined. All the available evidence suggests that mitotic crossing-over occurs at, or after, the time of chromosome duplication, that exchanges are reciprocal and symmetrical, and that centromeres show normal mitotic segregation. There is no cytological evidence concerning the pairing of homologous chromosomes and the mechanics of the process are obscure (Day, 1972). The apparent pairing of chromosomes in Figure 4.1 is, therefore, merely a convenient form of representation in order to illustrate the consequences of a mitotic cross-over.

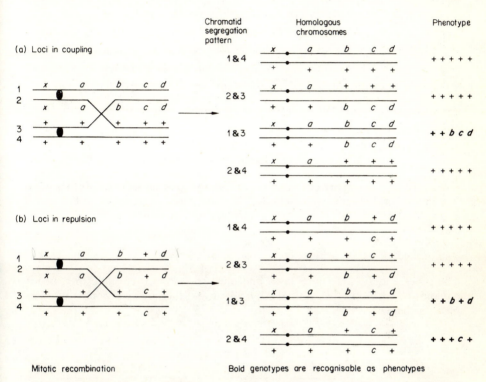

FIGURE 4.1 Diagram to illustrate the consequences of mitotic recombination between the four chromatids of a pair of homologous chromosomes. The locus $x/+$ is on one side of the centromere, $a/+$, $b/+$, $c/+$ and $d/+$ on the other. The consequences of recombination are shown when loci are in both coupling and repulsion.

It can be seen that all the heterozygous loci distal to both the centromere and the point of exchange are rendered homozygous in half of the possible mitotic products. Moreover, although three of the four possible products are genotypically different from that of the parental diploid, the number of phenotypically recognizable classes is fewer. If the loci distal to the exchange are in coupling there is but a single class: $+ + b c d$ (Figure 4.1a) or, if in repulsion, two, namely: $+ + b + d$ and $+ + + c +$ (Figure 4.1b). It will be clear, therefore, that if the loci under study are present in a heterozygous

76

diploid in coupling, the position of a cross-over can be detected and hence the sequence of the loci determined. Using the scheme in Figure 4.1a, the consequences of a single cross-over in each region separately are:

Cross-over between	Homozygosis of
c and *d*	*d* only
b and *c*	*c* and *d*
a and *b*	*b*, *c* and *d*
centromere and *a*	*a*, *b*, *c* and *d*
x and centromere	*a*, *b*, *c*, *d* and *x*

Fortunately, since multiple mitotic crossing-over is apparently rare, (Hastie, 1964; Lhoas, 1967) its possible consequence can be neglected. Gene order in a chromosome arm can, therefore, be determined in this way.

Mitotic Recombination: Examples

Example 4.1   Assignment of genes to linkage groups after haploidization

A diploid of *Aspergillus niger* including six mutant loci was induced and was then haploidized by *p*-FPA. The data obtained are set out below

Initial haploid components	*arg*	*a*	+	+	*leu*	+
of diploid	+	+	*o*	*his*	+	*put*

The genes *a* and *o* determine fawn and olive conidia, *a* is epistatic to *o*, hence haploids could be classified initially by colour and then by auxotrophic requirements as below

		a				*o*			
		arg		+		*arg*		+	
		leu	+	*leu*	+	*leu*	+	*leu*	+
his	*put*	1	6	0	1	0	0	0	8
	+	0	8	1	1	0	0	0	5
+	*put*	1	6	0	1	0	0	0	13
	+	0	8	0	0	0	0	2	13

Note

(i) In respect of *a*, *o* and *arg* that there are 30 *a arg*:4 *a* + and 0 *o arg*:41 *o* +. These are parental combinations.

(ii) There are no significant departures from random segregation between other markers.

It is, therefore, probable that *o*, *a* and *arg* are linked.
(Data of Lhoas, 1967.)

Example 4.2   Ascertaining gene order and distance

The largest body of data available is from *Aspergillus nidulans*, from which the following data have been taken.

The diploid utilized was made from two haploids, one auxotrophic for adenine, the other auxotrophic for proline and *p*-aminobenzoic acid and with yellow conidia. Hence,

$$\text{Diploid} \quad \frac{ade : + : + : +}{+ : pab : pro : y}$$

Mitotic recombinants	Numbers
$+ : + : + : y$	96
$+ : pab : + : y$	245
$+ : pab : pro : y$	30
$ade : pab : pro : y$	0
Total	371

ad pro pab y

Note that any gene distal to a mitotic cross-over between it and the centromere will be rendered homozygous in a diploid heterozygote.

Hence *y* is the most distal gene, since only $+ + + y$ diploids were recovered. Similarly, since only *pab y* not *pro y* or *ade y* recombinants were recovered, *pab* must be next to *y*. Hence *pro* is next to *pab* and *ade* beyond *pro*. The order, therefore, is

centromere—*ade*—*pro*—*pab*—*y*

Since mitotic cross-overs are rare events and doubles unusual, the recombination frequencies can be calculated directly viz.

Centromere—*ade*: No recombinants. Either this locus is closely linked to the centromere or it is beyond it. This latter possibility arises because mitotic cross-overs in both arms of a chromosome are very rare in *A. nidulans*. Further data are required.

Centromere—*pro*   30/371 or 8%.
pro—*pab*:   245/371 or 66%.
pab—*y*:   96/371 or 26%.

(Data of Pontecorvo and Käfer, 1958.)

It might be supposed that the frequency of mitotic cross-overs could be used to determine gene distance. Its absolute incidence cannot so be determined for two reasons. Firstly, if a cross-over occurs in a nucleus in a hypha, its daughters may divide several times before the recombinant phenotype is detectable. So the consequent multiplication and 'cloning' of recombinant nuclei causes a bias in the estimate of its absolute frequency. Secondly, selection procedures are usually employed to detect recombinants and these will vary for different loci and hence in different regions of the chromosomes. The efficacy of selection will, therefore, bias estimates of the frequency of recombination. The best that can normally be achieved is to measure the relative incidence of mitotic crossing-over. This is done by utilizing differently marked diploids to study the same chromosome arm in both coupling and repulsion. Characteristic and repeatable distributions of cross-overs may then be detected.

Phialide analysis

A special technique termed 'phialide analysis' has been developed for *Verticillium albo-atrum* by Hastie (1967, 1968). Because this technique determines the location of the divisions where mitotic recombination occurs with considerable precision, potentially it enables frequencies of recombination to be determined more accurately and also permits the detection of simultaneous multiple exchanges. This type of analysis is applicable to any situation where the products of successive nuclear divisions of a single parental nucleus, as in a phialide, are separable, e.g. as uninucleate conidia, provided that the products of successive divisions can be isolated successively, or as a group. The principle is illustrated in Figure 4.2 where a phialide giving rise to 4 conidia by successive divisions ($D1$–$D4$) is taken as an example and the 'phialide family' so formed is isolated as a group.

A recessive mutant gene, *sooty* (*so-2*) results in the autonomous darkening of the mycelium when homozygous. Its use has permitted rapid scoring of its segregation both in phialide families and, indeed, in the subsequent growth of a mycelium from a conidium, heterozygous for the gene. This type of analysis could be applied equally well to other fungi possessing phialides and recessive genes which show autonomous, easily detectable phenotypes such as colour, when homozygous, e.g. *pale* conidia in *A. tamarii* (Pontecorvo, 1946).

Finally, centromeres can be located by employing data derived both from mitotic crossing-over and haploidization. Mitotic recombination data do not indicate whether groups of loci linked to a centromere represent two arms of one chromosome or one arm from each of two non-homologous chromosomes. However, suppose recombination data suggest the two sequences:

centromere *a b c*

and centromere *x y z*

and haploidization that *a b c x y z* segregate as a single group,

then the centromere loci are identical and it must lie between *a* and *x* viz.:

z y x—centromere—*a b c*

If the *two groups do not segregate together* then they represent arms of non-homologous chromosomes and *the centromeres are non-identical.*

Procedures in mitotic analysis

To carry out this type of analysis it is necessary to obtain diploid nuclei and then to select diploid and, preferably, haploid segregants. Various procedures have been employed and will be briefly reviewed here.

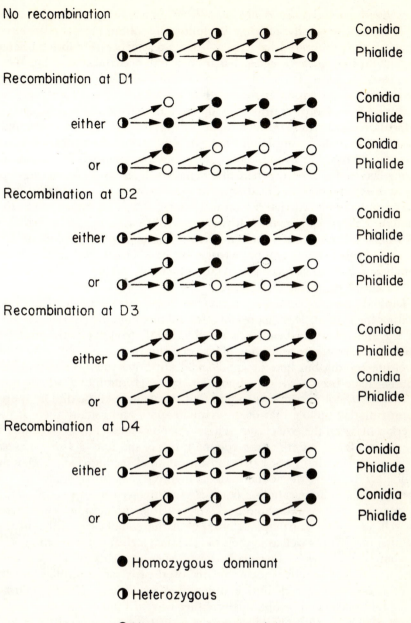

No recombination — Conidia / Phialide

Recombination at D1 — either / or — Conidia / Phialide

Recombination at D2 — either / or — Conidia / Phialide

Recombination at D3 — either / or — Conidia / Phialide

Recombination at D4 — either / or — Conidia / Phialide

● Homozygous dominant

◐ Heterozygous

○ Homozygous recessive

FIGURE 4.2 Phialide segregation. Diagram to illustrate the consequences of mitotic recombination at different divisions (D_1–D_4, first–fourth) of the phialide nucleus on the genotypes of the conidia produced. It is supposed that a pair of alleles are segregating, one being dominant to the other. The earlier in the successive division of the phialide nucleus mitotic recombination occurs, the higher the proportion of conidia with homozygous nuclei. (From Hastie, 1967.)

Establishment and isolation of diploid nuclei

The most generally effective technique of obtaining somatic diploid nuclei is that pioneered by Roper (1952) in which rarely formed heterozygous diploid nuclei are selected from a balanced heterokaryon. A prerequisite, therefore, is the establishment of a heterokaryon. The control of heterokaryon formation is discussed in detail in the next chapter; here attention need only be drawn to the fact that its establishment may be difficult, either because fusion fails, or because the fusion product is lethal or unstable. In either case, failure may be due to environmental or genetic factors, or both. For example, in *Ustilago violacea*, sporidial fusion can only occur between different, genetically determined, compatible mating-types, *A1* and *A2*, and only in the absence of certain cations which inhibit fusion, so that it has to be carried out on water agar (Day and Jones, 1968). Even though a heterokaryon may appear relatively unstable, as in *Verticillium albo-atrum* (Hastie, 1962), it often persists long enough for diploid nuclei to arise. Indeed Hastie (1970) has written 'the easiest method of demonstrating that heterokaryosis has occurred in laboratory cultures of *Verticillium* is simply to search for heterozygous diploids'.

Diploids are most frequently recovered from protrophic heterokaryons by plating their uninucleate spores on minimal medium, the component nuclei of the heterokaryon being auxotrophic. Only conidia with protrophic nuclei can germinate and grow. This procedure is not possible with multinucleate conidia but here morphological attributes such as size and DNA content have been used, e.g. *Aspergillus oryzeae* (Ishitani and others, 1956). It is, indeed, not uncommon to find that the volume of the diploid spore is approximately double that of the haploid spore and usually there is little overlap between the size classes. These differences may be altered either by the environment, e.g. the $2n:n$ volume ratios of sporidia of *Ustilago violacea* on complete medium are $2 \cdot 0 : 1$ and on minimal medium $3 \cdot 6 : 1$ (Day and Jones, 1968); or by the actual markers involved, e.g. in *Aspergillus niger* the $2n:n$ conidial volume ratio is $1 \cdot 6 : 1$ if *black* is involved but $1 \cdot 35 : 1$ if *olive* is a component (Pontecorvo and others, 1953a). If aneuploid nuclei are present as well as diploid and haploid ones, or if the conidia can differ in their number of nuclei as well as in their ploidy, then a clear cut separation of size or diameter classes is rarely obtained. Measurements of DNA content whether direct, or by measurements of density of controlled nuclear staining are, however, definitive although laborious (e.g. Roper, 1966).

Diploid nuclei can be obtained in other ways. In some cases they arise as errors of meiosis or through its failure, e.g. 1 in 100 ascospores of *A. nidulans* are diploid (Pritchard, 1954). Diploid nuclei can also be recovered in *Ustilago maydis* and the mycelia derived from them are usually solopathogenic, i.e. they can, like the dikaryon, but unlike haploid monokaryons, infect the maize plant. This was made use of by Holliday (1961b) who inoculated two compatible but auxotrophically different strains and recovered diploid sporidia from the developing gall before brandspore formation (and, therefore, possibly meiosis) had begun. In other Basidio-mycetes it has proved possible to isolate protrophic uninucleate spores from a heterokaryotic mycelium. These are normally only found on the haploid monokaryotic phase but in some groups, such as the Thelephoraceae, both monokaryons and dikaryons produce uninucleate conidia. It has proved possible to isolate protrophic uninucleate conidia from a protrophic dikaryon of *Varraria granulata* (Burnett, unpublished). Casselton (1965) used in-compatible, common-*A* heterokaryons of *Coprinus lagopus*, e.g. *A6 met-5 B6 ade-8* × *A6 adhi-1 chol-1* (requiring methionine, adenine, adenine + histi-dine and choline respectively). These, unlike true dikaryons, have the property of producing uninucleate oidia which are expected to be auxo-trophic; hence protrophic oidia are diploid. It should also be mentioned that in some fungi, notably *Neurospora crassa* itself, repeated efforts to produce stable diploid nuclei have failed. Whether this is because they are never formed or are exceedingly unstable is unknown.

No method is yet known of inducing somatic diploidy but the frequency of recovery of diploids can sometimes be increased. The oldest method is to expose the heterokaryon to the vapour of *d*-camphor. In species of *Asper-gillus* and *Penicillium* this has resulted in a several hundred-fold increase in recovery (e.g. Barron, 1962 with *P. expansum*). Using multinucleated conidial species of *Aspergillus*, Ishitani and others (1956) claim to have increased the rate of recovery of diploids with u.v. light even more than with camphor. At survival values of 10^{-2} the frequency of recovery of diploids was greater than 1 in 100, as compared with less than 1 in 10^7 in camphor-treated colonies. However, ultraviolet light has been used more often to increase the frequency of crossing-over (see p. 82) than to enhance diploid recovery.

Detection and enhancement of segregation

Because of the relatively low frequency of mitotic crossing-over, detection of segregants is difficult and this is even less probable if haploidization is rare and cannot easily be induced. A variety of ingenious techniques have, therefore, been devised to improve the detectability of the rare segregants. An especially useful procedure is to incorporate some type of colour marker so that a well-developed contrast between parental and segregant hyphae or spores is exhibited. It is obviously desirable that such marker genes should be autonomous in their action. For example, *w*, white and *y*, yellow are good visual selector genes in *A. nidulans* where the normal diploid conidia are green, and *sooty* in *Verticillium albo-atrum* causes spores and hyphae carrying the homozygous gene to be black, contrasting with the white parental form. In many ascomycetous fungi certain adenine-requiring (*ade*) strains develop a pink, red or purple pigment as a consequence of the accumulation of a precursor blocked by the genetic lesion in the biosynthetic sequence. Thus homozygous red *ade* colonies can be easily seen against the background of white heterozygous *ade+/ade* parental cells in yeast.

The detection of homozygous autotrophic segregants has, of course, much in common with general methods for detection of mutants (cf. Chapter 2). Particular examples of such techniques as have been employed are the use of recessive suppressors of particular loci, drug-resistant mutants, and the starvation and replica-plating techniques. For example in *A. nidulans*, diploids of constitution

$$\frac{ade\text{-}20 \quad su\text{-}1\,(ade\text{-}20)}{ade\text{-}20 \quad su\text{-}1\,(ade\text{-}20)^+}$$

grow without producing conidia at a slow rate in the absence of adenine. Recombinants (recessive suppressor *su-1 (ade-20)*) such as

$$\frac{ade\text{-}20 \quad su\text{-}1\,(ade\text{-}20)}{ade\text{-}20 \quad su\text{-}1\,(ade\text{-}20)}$$

show normal conidial development and growth. They can, therefore, be isolated rapidly, giving sectors from plates of the parental diploid (Pontecorvo and Käfer, 1958).

In contrast to the paucity of agents which initiate diploidy, several are known which either increase the frequency of recombination or promote segregation. Ultra-violet light increases the frequency of crossing-over in both *U. maydis* and baker's yeast and the latter is also susceptible to X-irradiation (Nakai and Mortimer, 1969) and chemical mutagens such as nitrous acid, diethyl sulphate and certain carcinogenic alkylating nitrosamines (Zimmerman and others, 1966). It will be recalled that the frequency of mitotic crossing-over of a gene is expected to increase the further it is from its centromere. It is of interest, therefore, that the frequency of X-ray induced mitotic segregation is also related in an approximately linear manner to the distance of the segregant genes from their centromere in yeast. This technique has been employed for chromosome mapping in *S. cerevisiae*, (Manney, 1964; Manney and Mortimer, 1964). Agents such as Mitomycin C or 5-fluorodeoxyuridine, which are known to inhibit DNA synthesis or repair, are effective in inducing mitotic segregation in *U. maydis* which, otherwise, is exceedingly rare as a spontaneous event. Indeed, the frequency of mitotic recombination often seems to be increased by those agents which restrict DNA synthesis but not that of RNA and protein (Putrament, 1967), although the interpretation of this finding is not clear.

Haploid segregation is, of course, a different process from diploid segregation and here DL-*p*-fluorophenylalanine shows exceptional effectiveness in *Aspergillus*, *Penicillium* and *Verticillium* (Lhoas, 1961, 1968; Garber and Beraha, 1965; Ingram, 1968). Nevertheless, this agent is remarkably species specific for while it is an effective haploidization agent of diploid *Ustilago violacea* (Day and Jones, 1971), it is quite ineffective on *U. maydis* (Holliday, 1961b). It is also ineffective on *C. lagopus* although here the component diploid nuclei of haploid–diploid heterokaryons are partially unstable, and completely unstable in diploid–diploid heterokaryons (Casselton and

Lewis, 1966). Although haploidization can thus be induced, mitotic segregation has not yet been observed.

It will be apparent that although several possibilities exist in several fungi for inducing various stages involved in mitotic recombination and haploidization, no general rules can yet be formulated. Until this is possible the use of mitotic recombination as a regular tool in genetic analysis is always worth attempting but is not necessarily possible.

NOVEL SOMATIC RECOMBINATION PROCESSES

These phenomena have been recorded in *Coprinus* and *Schizophyllum,* and have been described and discussed in detail in Ellingboe (1965) and Raper (1966). They only involve particular types of heterokaryons arising from matings described either as compatible, or incompatible di–mon matings: somatic recombination in common AB heterokaryons, i.e. ($A1B2 + A1B2$), resembles that already described (Middleton, 1964; Mills and Ellingboe, 1971). These terms refer to the mating-type factors, A and B, which occur as multiple alleles at two unlinked loci, and 'di' and 'mon' to dikaryon and monokaryon respectively. So for example, a compatible di–mon mating is:

<table>
<tr><td>Dikaryon</td><td>Monokaryon</td></tr>
<tr><td>($A1\ B2 + A2\ B1$)   ×</td><td>$A3\ B3$</td></tr>
</table>

and a non-compatible di–mon mating is:

$$(A1\ B2 + A2\ B1) \times A1\ B1$$

In such crosses, somatic recombinants of two types were recovered, i.e. no normal meiotic stage was passed through (Table 4.1).

TABLE 4.1 The types of somatic recombinants from 33 incompatible di–mon matings of *Schizophyllum commune.* (After Raper, 1966 based on Ellingboe and Raper, 1962.)

		Parental strains crossed								
Dikaryon	$A41$ +: $B42$: ura-1 + nic-2: ade-2 ade-2 + ade-1: nic-3: +									
	$A42$ +: $B41$:	+ade-3	+:	+	+	+	+	:	+	:+
Monokaryon	$A41$ × 15: $B41$: +		+:	+	+ xii	+	:	+	:+	

	Recombinant types									
Class										No.
Ia	$A42$ +: $B42$: ura-1 + nic-2: ade-2 arg-2 +	+:	+:	arg-6						52
Ib	$A42$ +: $B42$: + ade-3 +: +	+	+	+:	+:	arg-6				15
Ic	$A42$ +: $B42$: + + +: +	+	+	+:	+:	+				12
II	$A42$ x15: $B42$: + + +: +	+	xii	+:	+:	+				99

Notes:
1. Recombinants classes are defined as follows:
   Ia. Assortment and recombination of factors indistinguishable from meiotic recombination.
   Ib. Differing from one or the other of the dikaryotic components by a single mating-type factor.
   Ic. Protrophic for the linked loci ura-1 $^+$ ade-3 $^+$ nic-2 $^+$, the reciprocal, triple auxotroph not being recovered.
   II. Those possessing markers only from the monokaryon, apart from mating-type factors.
2. Key to symbols: $A41$, $A42$, $B41$, $B42$, mating-type factors of the A and B loci; ade-1, ade-2, ade-3, unlinked loci, adenine-requiring; arg-2, arginine-requiring; nic-2, nic-3, unlinked nicotinic acid requiring; ura-1, uracil-requiring; x11, x15, unlinked auxotrophs whose requirements were unknown.

Class I showed recombination and reassociation of both linked and unlinked genes with frequencies not inconsistent with those obtained from meiotic recombination data. However, they appeared to represent recombination not merely between the two constituent nuclei of the dikaryon but also involving the genotype of the monokaryon. There is thus a suggestion that a process of recombination with a frequency similar to that of meiosis can occur in the hyphae and that it can be associated with a trinuclear fusion. Class II segregants were even more puzzling and the phenomenon was termed 'specific factor transfer' since it only related to the mating-type factors. In one case there was an exchange of an A factor and its two component allelomorphs without the transfer of a marker locus, pab, which lay between them, viz.

$$
\begin{array}{cc}
\textit{Dikaryon} & \textit{Monokaryon} \\
\begin{array}{ll} A41\alpha1 + \beta1 & B42 \\ A42\alpha3\,pab\beta5 & B41 \end{array} \times \quad A41\alpha1 + \beta1 \quad B41\,xII \\[2ex]
A42\alpha3 + \beta5 \qquad B42\,xII
\end{array}
$$

i.e. $A42\alpha3$ and $A42\beta5$ transferred without pab. Once again, no rational explanation can be given at this time.

Attention is drawn here to these novel somatic exchange phenomena in order that investigators may be alert to their possible occurrence elsewhere. It is evident that somatic recombination in fungi is not yet wholly understood although, equally clearly, the great majority of such recombination falls into the well-established pattern of reciprocal mitotic recombination.

CHAPTER 5

Recombination and segregation of nuclei

Heterokaryons are mycelia containing more than one genetically distinct type of nucleus in an effectively common cytoplasm. They occur spontaneously or can be induced. Sometimes they are stable or they may break down either into their component homokaryons or into new heterokaryotic recombinations. They can originate through mutation in an initially homokaryotic mycelium or, as a consequence of hyphal fusion followed by intermingling of nuclei, between genetically distinct homokaryons. They are perpetuated through the growth of heterokaryotic apical cells of hyphae or by the abstriction of multinucleate heterokaryotic propagules of various types, or by repeated hyphal fusions. Conversely, they can be dissociated either by the abstriction of uninucleate (homokaryotic) propagules, by the concentration of identical nuclei in an apical cell or in multinucleate propagules.

Heterokaryosis, therefore, confers great potential genetic plasticity on a mycelium and enables the recombination and segregation of its component whole nuclei to occur and be subjected to selection (cf. Chapter 9).

In principle, those same processes which result in the association or dissociation of genetically diverse nuclei, can bring about the same situations in respect of cytoplasmic components, whether cell organelles, cytoplasmic genetic determinants or infective agents, such as viruses. By analogy with the nuclear condition the terms heteroplasmon and homoplasmon can be used for cytoplasmic entities.

Since the possible biological significance of heterokaryosis as an explanation for variation in phytopathogenic fungi was first suggested (Brierley, 1929; Hansen and Smith, 1932) it has often been studied but, until recently, the possibility of parallel cytoplasmic heterogeneity has been neglected or ignored.

There is now much evidence to support the view that somatic segregation

85

of extranuclear elements is not uncommon in some fungi at least (e.g. Jinks, 1963), so that this possibility should never be ignored when investigating heterokaryons. The 'heterokaryon test' described by Jinks (1958) enables a distinction between nuclear and cytoplasmic segregation to be made in certain circumstances (see p. 103). In this chapter the possibility of genetically heterogeneous cytoplasm will be ignored.

Heterokaryons, dikaryons and their nomenclature

A special case of a heterokaryon is one with nuclei which differ in their mating-type factors. In Phycomycetes and many Ascomycetes such heterokaryons either can not be established, or do not show stable vegetative growth. In the Basidiomycetes, however, such heterokaryons are regularly formed and, because their two complementary nuclei are usually closely associated in each hyphal compartment, frequently dividing synchronously, the term *dikaryon* has been applied to them.

Clearly such heterokaryons are of a rather different nature from the more usual heterokaryons. It is desirable that care should be taken when describing them (Jinks and Simchen, 1966). For example, a heterokaryon can be developed between two strains carrying genetically different nuclei of the same mating type or between strains carrying different nuclei which also differ in mating type. The former is best described as a heterokaryon (or heterokaryotic monokaryon) the latter as a heterokaryotic dikaryon. Even so, ambiguity may arise in certain fungi where the regularly binucleate condition of the hyphal compartments has either never arisen or has been lost in the course of evolution. For example, in Ascomycetes such as *Neurospora sitophila* the compartments are multinucleate and may contain nuclei of both mating types, *A* and *a*, or a single mating type, as well as nuclei which differ at other loci. In Basidiomycetes a similar condition may occur, e.g. *Agaricus bisporus*, the cultivated mushroom, whereas in others the binucleated compartments may include only genetically identical nuclei, e.g. *C. sterquilinus*, which is, therefore, phenotypically dikaryotic but genetically homokaryotic. Figure 5.1 illustrates some of the possibilities diagrammatically.

The establishment of a heterokaryon

Heterokaryosis as a result of mutation is a rare but virtual certainty in any multinucleate mycelium. Whether or not it will be detected depends upon:

(a) the nature of the phenotype expression of the mutant,
(b) the relative rate of nuclear division of mutant nuclei compared with that of original nuclei,
(c) the probability of a localized, predominantly mutant nuclear cluster occurring, so enabling the mutant phenotype to be manifest.

If the phenotype is not easily detectable, if the rate of division of mutant nuclei is significantly less than that of normal nuclei, or if mutant nuclei

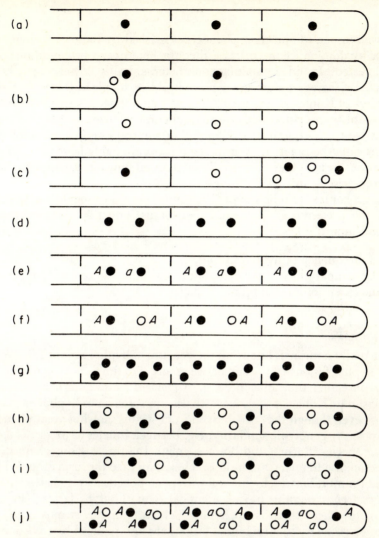

FIGURE 5.1 Diagrams to illustrate the different kinds of nuclear condition which can occur in fungal hyphae. (a) Monokaryotic homokaryon. (b) Transient heterokaryon, e.g. *Verticillium dahliae*. (c) Apical heterokaryon; only the apical cell is heterokaryotic, septa lack central pores, e.g. *Fusarium oxysporum*. (d) Dikaryotic homokaryon. (e) Mating-type dikaryon, each cell carries two mating-type factors, one in each nucleus, e.g. *Coprinus*. (f) Heterokaryotic dikaryon, homokaryotic for mating type. (g) Multinucleate homokaryon, e.g. *Neurospora crassa*. (h) Multinucleate heterokaryon, e.g. *N. crassa*. (i) Coenocytic multinucleate heterokaryon, e.g. *Mucor*. (j) Multinucleate heterokaryon, heterokaryotic also for mating type, e.g. *N. tetrasperma*. Many of these states are given simpler designations when no confusion can arise. Thus (a) is termed a monokaryon; (d), (e) and (f), dikaryons and (b), (c), (g)–(i), heterokaryons.

are well dispersed and their phenotypes masked by those of normal nuclei, either by dominance or epistasis, then the heterokaryotic condition will not be detected. In effect, spontaneous heterokaryosis is most usually only detected if segregation occurs. An admirable example of this is available in the work of Köhler (1935) with *Mucor mucedo*, a fungus where hyphal fusion is probably non-existent. By repeated subculturing from spores of an original strain '*40+*', he obtained evidence that some presumptive homokaryotic strains could be isolated; their nuclei must presumably have arisen through mutation. In fact, there is virtually no data on how often detectable heterokaryons arise in this way in fungi.

Heterokaryosis through hyphal fusion is better documented since this has been the preferred experimental tool for its induction. Its establishment depends upon:

(a) the frequency of hyphal fusions,
(b) the compatibility of the cytoplasms and nuclei so brought into contact,
(c) the extent of intermingling of nuclei.

Hyphal fusion

Virtually nothing is known of the genetics of hyphal fusion. It is probable that hyphal fusions can occur between any two fungi, however, taxonomically unrelated they may be, e.g. *Neurospora crassa* and *Botrytis allii* fuse successfully (Köhler, 1930). Nevertheless, in some cases hyphae never seem to fuse even though they are derived from strains showing great phenotypic and possibly genotypic similarity, e.g. between isolates of *Mycocalia denudata* sibling species (Burnett and Boulter, 1963). In other cases mechanisms exist to reduce the frequency of fusions. They may take the form of some kind of zone of inhibition which only a few hyphae penetrate, e.g. between *Schizophyllum commune* strains having identical *B* mating-type factors (Papazian, 1950), or where there is no hyphal penetration, e.g. strains of *Diaporthe perniciosa* (Cayley, 1923). Hyphal fusions are also reduced by environmental factors, e.g. the composition and concentration of the medium (Bourchier, 1957). There is good evidence in *A. nidulans* that the frequency of hyphal fusions is under genetical control (Jinks and others, 1966). Their technique was to make heterokaryons between strains differing by a single, mutant spore-colour marker gene whose expression was autonomous. Thus any uninucleate conidium carrying the mutant gene had a mutant phenotype in respect of its colour. Heterokaryons could be assessed by detecting the number of conidial heads with mixed spore colours. Strains tested showed varying degrees of phenotypic expression and fell into four groups which appeared to reflect their degree of genetic divergence. For example, a wild strain and its spore-colour mutant formed heterokaryons readily whereas wild isolates from different localities differing greatly in their phenotypes formed heterokaryons only rarely, i.e. they showed 'heterokaryon incompatibility'. It was

TABLE 5.1 The heterokaryon compatibility of the sexual progenies of crosses between isolates belonging to the same or different heterokaryon compatibility (hc) groups of *Aspergillus nidulans* when tested against their respective parental strains. (Based on data from Jinks, Caten, Simchen and Croft, 1966.)

Cross	Sample size	Compatibility of progeny		
		with both parents	with one parent	with neither parent
Derived from same isolate	40	40	0	0
Derived from same hc group and locality	40	40*	0	0
Derived from same hc group but different localities	40	32*	8	0
Derived from different hc groups and different localities	153	0	8*	145

Note: The * indicates reduced compatibility (recorded as frequency of mixed heterokaryotic conidial heads) as compared with the parental combinations.

possible, however, to obtain sexual progeny and these were tested for their ability to form heterokaryons with their parents (Table 5.1).

It can be seen that the reactions differed markedly and in the way expected from the original tests. Jinks and others (1966) assumed that for an offspring to form a heterokaryon with one of its parents it must carry alleles identical to that of the parent at a number of loci. For n unlinked loci the proportion capable of forming heterokaryons in the progeny will be $2(\frac{1}{2})^n$.

Application of this formula to their extensive data suggested that some 8 loci were necessary in each partner for effective heterokaryon formation and a minimum of 5 loci for any heterokaryosis to occur at all.

In *Venturia inaequalis* a gene has been detected which promotes the fusion of hyphae by determining the production of a heat-stable, dialysable and non-volatile substance. Production occurred over a range of temperatures but response was only detected at 24°C or below. The significance of this substance in *V. inaequalis* is obscure since heterokaryosis has not been detected in this species despite experiments to induce it (Leu, 1967, cited in Boone, 1971).

Compatibility of cytoplasm and nuclei

Neurospora crassa provides evidence which supports this genetic situation although here the success of a heterokaryon depends upon the compatibility of the cytoplasm and nuclei after hyphal fusion has brought them into contact. Garnjobst and Wilson (1956) have shown that two allelomorphic pairs, *Cc* and *Dd* are responsible and that a necessary condition for heterokaryon formation is that fusing hyphae must carry nuclei with identical genotypes in respect of these genes, i.e. nuclei of constitution *CD* determine heterokaryon formation with *CD* but not with *Cd-*, *cD-* or *cd*-carrying hyphae, etc. Inappropriate combinations do not prevent hyphal fusion but the fusion cell is rapidly isolated by complete septa. In a short time bubble-like vesicles appear within and the cell dies and autolyses. Extracted cytoplasm micropipetted into hyphae of unlike genotypes has a similar effect and a protein moiety seems to be responsible (Garnjobst, 1953, 1955; Garnjobst and Wilson, 1956; Wilson and others, 1961). Wilson (1963) has since shown that transplantation by microinjection of nuclei alone can bring about this reaction. Evidence exists for a similar genetic system in strains of different origins of *N. crassa*. Perkins (1969) has demonstrated a third allele, c^3, in the Panama *a* strain. At least 4 pairs of genes determined the ability to form a heterokaryon in another strain but whether any of these were identical with the *C* and *D* loci is not known, nor are physiological details of heterokaryon breakdown available (Holloway, 1955). It should also be recalled that, save under most abnormal circumstances, mating-type heterokaryons, *A/a*, cannot be formed in *N. crassa* (Gross, 1952) so that it will be seen that some 3–7 allelomorphic pairs of genes may well be involved in regulating successful heterokaryon formation in that species. The related species, *N. sitophila* and *N. tetrasperma* are not so restricted and form *A/a* heterokaryons readily but, in the former species an allelic pair het^+/het^- determine successful heterokaryosis which only occurs when opposed hyphae both carry het^+ (Mishra, 1971); in the latter species heterokaryosis is unrestricted.

An apparently quite different system operates in the basidiomycete, *Thanetophorus cucumeris* of the 'Praticola type' (= *T. praticola* of some authors). Here heterokaryon formation occurs at the junction of opposed homokaryotic mycelia as fast growing tufts of hyphae. Their development is determined by the occurrence of different *H* factors in the opposed mycelia and some 15 such factors have been identified. The *H* factor appears to be a complex locus since in four cases it has proved divisible by recombination over the range 0·6–2·2%. *H*-Factor behaviour, therefore, is determined by two, closely linked loci, the $H\alpha$ and $H\beta$ loci, acting in concert (Anderson and others, 1972). Although structurally it resembles the mating-type factor locus of unifactorial diaphoromictic fungi, i.e. one (complex) locus with several alleles, it is not such a system since *T. cucumeris* is basically homomictic (Flentje and Stretton, 1964; see also Chapter 8). In many, but not all cases, the opposition of mycelia carrying identical *H* factors results in a 'barrage zone' developing. Here a clear area between the mycelia arises from the death and autolysis of a region of fused cells.

Nuclear intermingling

Nucleo-cytoplasmic interactions of a less drastic type may affect the subsequent intermingling of nuclei after a successful fusion but intermingling may also be affected by the regulation of nuclear migration. Very little is known of the genetic control exercised on these processes.

When there is no nuclear migration after hyphal fusion, the mycelium as a whole may behave as if it were heterokaryotic. In *Verticillium dahliae* fusion cells develop 1–2 mm back from the edge of a mycelium but the nuclei do not appear to migrate from them. Nevertheless, if two auxotrophic mycelia are used to establish such a 'transient heterokaryon', complementation can occur and the whole colony behaves as if protrophic. If hyphal fusion is impaired, as occurs when mycelia are grown at 30°C, then the heterokaryons fail to grow. Heterokaryosis here depends, therefore, on the continued establishment through hyphal fusions of scattered individually heterokaryotic cells which sustain the whole colony (Puhalla and Mayfield, 1974).

A common phenomenon is unilateral nuclear migration in which one nucleus enters the other opposed mycelium whose nuclei do not behave reciprocally. No generalization concerning this phenomenon can be made but there is some evidence that hyphae containing nuclei which cause

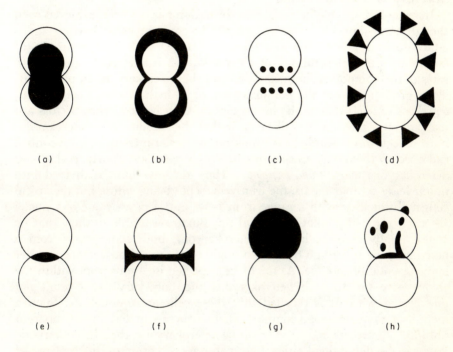

FIGURE 5.2 Patterns of heterokaryotization. In each diagram two homokaryotic mycelia carrying different nuclei are confronted, the blackened area represents the extent of the heterokaryotic mycelium developed. In (a), (e) and (f) limited heterokaryotic regions develop in the zone of confrontation; in the last, heterokaryotic mycelium grows away from the sides of the confrontation zone. In (b), (c) and (d) nuclei migrate but only the apical cells maintain the heterokaryotic condition in (b) and (d). Heterokaryosis in (c) is transient. In (g) and (h) only unilateral nuclear migration occurs from one mycelium, in (g) heterokaryotization is uniform but in (h) it is sporadic and irregular. Similar patterns are shown at dikaryon formation when monokaryons are confronted.

severe morphological aberrations in the homokaryon can donate nuclei but cannot receive them (Raper and San Antonio, 1954). Even if heterokaryotization is reciprocal its extent may be very limited. Figure 5.2 illustrates some examples. Although based on patters of dikaryotization i.e., mating-type heterokaryons in Basidiomycetes, similar patterns can also be found in heterokaryon formation in all groups.

It is important, if heterokaryons are to be studied that their pattern of development is known. Frequently even if the heterokaryotic zone is localized when mycelia are opposed, subculture from such an area will give a balanced stable mycelium, heterokaryotic throughout. The stability of a heterokaryon is, therefore, for genetic studies a matter of considerable importance.

Stability of heterokaryons

In some fungi stable heterokaryon formation has not been achieved even though there appear to be no genetic or environmental barriers to its establishment. Pontecorvo and Gemmell (1944) and Pontecorvo (1946) have shown that the stability of a heterokaryon is affected by the growth rates of the component homokaryons and the heterokaryon. In particular, if the growth rates of the homokaryons, or at least one such component, equals or exceeds that of the heterokaryon, then homokaryotic sectors are likely to segregate out. It will be clear that the stability of a heterokaryon must reflect, in very large part, the kinds of nuclei present and the control exercised by them over apical growth. This is seen most clearly, perhaps, in heterokaryons of *Fusarium oxysporum*. Here the mycelium is divided into uninucleate compartments, the transverse septa being entire, but the apical cells multinucleate with a range from 1–15, and a mean value of 7 nuclei (Buxton, 1954). By isolating hyphal tips Buxton was able to show that he could usually recover, in dual heterokaryons, both components from a heterokaryotic mycelium but only a single component from sectors of either homokaryotic phenotype. As might be expected isolations from within the colony give rise to both heterokaryons and homokaryons; a result not unexpected when it is recalled that these cells are uninucleate. Such a mycelium, therefore, has a complex genetic structure underlying its apparent stability. The apices may or may not be heterokaryotic, their heterokaryotic state can be maintained either by synchronous division of different nuclei in the apical cell, or through the restoration of the heterokaryotic condition as a consequence of lateral fusions between adjacent homokaryotic hyphae. Failure of either of these mechanisms can result in homokaryosis. The heterokaryotic apices not only determine the general growth rate of the heterokaryon but also the general features of the hyphae even though in the subapical regions, the cells become uninucleate and hence automatically homokaryotic.

Buxton's evidence suggested that a stable heterokaryotic phenotype could be maintained across a fairly wide range of ratios of different nuclear components, e.g. 1·35:1 to 7:1.

This particular example is a model for stabilizing processes in all hetero-karyons. In those fungi (the majority) where the transverse septa are per-forated and compartments are multinucleate, the heterokaryotic condition may extend throughout the mycelium and apical cell heterokaryosis could be maintained by nuclear migration from subapical cells. In fact, there is extremely little direct evidence of just how heterokaryosis is maintained and a stable phenotype expressed.

Instability can be induced either internally or by external selective agen-cies; the latter aspect will be considered in detail in Chapter 9. Perhaps the best examples of inherent internal stability are provided by the I/i and $pan/pan-m$ loci in *Neurospora crassa*. In the former case, Pittenger and Brawner (1961) showed that if the I nuclei exceeded 30% of the total in the hetero-karyon it became unstable and homokaryotic. Since the effects of these genes are to enhance the action of the *am* locus (amination deficient), the homokaryons are nutritionally deficient and so fail to grow. Thus, as homo-karyosis increases, growth declines and finally ceases. Nuclear ratios could not be altered by varying the concentration of the exogenous medium. The mechanism is not understood but it could reflect competition for intracellular metabolites, a situation which occurs in $pan/pan-m$ heterokaryons. Here, both homokaryotic strains require exogenous pantothenate but $pan-m$ homo-karyons can grow at far lower concentrations than pan homokaryons. At low pantothenate concentrations, however, the mycelium shows a cyclical pattern of growth in which rapid growth alternates with a virtual cessation of growth. In the former phase $pan-m$ nuclei predominate, in the latter phase pan nuclei increase. Evidently $pan-m$ nuclei determine a more effective uptake of pantothenate which is then utilized more effectively by pan nuclei. This utilization is so effective that intracellular depletion of pantothenate occurs and growth is checked until a predominance of $pan-m$ nuclei leads to its resoration through enhanced uptake (Davis, 1960a and b).

Similar changes in the ratios of nuclear components can be brought about in some fungi by changes in nutrient concentrations. The most notable example is *Penicillium cyclopium* where the heterokaryon is stable on Czapek–Dox apple-pulp medium. Progressive reduction of the apple-pulp component in 5% steps to zero is paralleled by a decline in growth rate of the heterokaryon associated with changes in the ratio of the two nuclear com-ponents, $4A$ and $4B$. When the growth rate of the heterokaryon had declined to a value equal to that achieved by homokaryotic $4A$ strains alone on medium of the same composition, then homokaryotic sectors arose (Jinks, 1952 and pp. 184–185). Thus Pontecorvo's generalization concerning the stability of a heterokaryon in relation to the growth rate of its component homokaryons, mentioned earlier, applies to this case but the conditions for instability arise from a selective cause exogenous to the mycelium, namely the composition of the medium. (Selective effects of this nature are discussed more fully in Chapter 9.)

Finally, it may be observed that, although few experiments have been made, it seems extremely difficult to render a heterokaryon wholly homo-karyotic. This was certainly Pittenger and Atwood's (1956) experience with *N. crassa* heterokaryons even though homokaryotic sectors could normally be obtained. The detection of the heterokaryotic state is related to the processes of segregation and reassortment of whole nuclei and this will now be considered briefly.

Segregation and reassortment in heterokaryons

Segregation is clearly related in part to the stability of a heterokaryon and, also, to the frequency of statistical accidents which result in the pro-duction of localized homokaryotic apices or propagules. The detection of such segregations is not necessarily straightforward. It will depend upon many factors including the relative rates of division of genetically different nuclei, the mode of growth or spore formation of the fungus and gene-dosage effects between allelomorphic and non-allelomorphic genes. In fact, it will depend largely upon internal, interacting factors which govern the popula-tion densities of nuclei and their expression. Very little is known about such matters. There is little evidence for gross differences in rates of division of genetically different nuclei although this has been suggested more than once as a possible cause of homokaryotic sectors (e.g. Beadle and Coonradt, 1944; Jinks, 1952). Gene-dosage effects have been demonstrated more often and, indeed, it is a common experience that a particular phenotype is often exhibited when far fewer than 100% of the nuclei which give rise to that phenotype are present. For example, *N. crassa* heterokaryons containing wild-type (*pan-1*$^+$) and pantothenate-requiring (*pan-1*) nuclei, still exhibit normal growth and a wild-type phenotype when their nuclear ratio is 5% *pan-1*$^+$ to 95% *pan-1* nuclei (Pittenger and Atwood, 1956). Figure 5.3 illustrates how different ratios of combinations of nuclei carrying non-allelomorphic albino mutants *al-1*$^+$ *al-2* and *al-1* *al-2*$^+$, result in different carotenoid contents of conidial masses in *N. crassa*. The optimum value, at 40% *al-1* *al-2*$^+$:60% *al-1*$^+$ *al-2* is equal to that in normal wild-type strains *al-1*$^+$ *al-2*$^+$ (Davis, 1966). In fact, it is clear that nuclear expression and behaviour will reflect the common cutoplasmic milieu, which is itself affected by both nuclear the cytoplasmic determinants.

Even when the mechanics of spore formation result in uninucleate (hence homokaryotic) spores arising, segregation may not be detected. For example, in the green-conidial species *Aspergillus heterothallicus*, a number of conidial colour mutants are known, e.g. white (*W*), brown (*B*), yellow (*Y*) and pinkish-tan (*PT*). Heterokaryons can be made between any of these, provided they are of the same mating type, and this was done. Combinations of the first three gave rise to striated conidial heads. These are parallel chains of uninucleate conidia on the same head expressing different mutant colour genes, i.e. gene action is autonomous. Here, therefore, segregation can be detected directly. In contrast, any combination which included *PT*-carrying nuclei gave rise either to green conidial heads resembling the

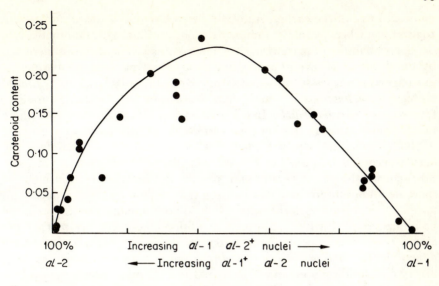

FIGURE 5.3 The relationship of carotenoid content to nuclear ratio in two-member heterokaryons of *Neurospora crassa*. One nucleus carries the albino mutant *al-1* the other, the non-allelomorphic *al-2*; the heterokaryon, therefore, has the genotype *al-1 al-2⁺/al-1⁺ al-2*. Carotenoid content measured as absorbance at 470 nm in arbitrary units corrected for comparable extracts. (From Davis, 1966.)

wild type, or green/colour mutant striated heads, i.e. green/*W* or *B* or *Y* heads, but never green/*PT* heads. On plating the conidia, however, all the green conidia gave rise to colonies with exclusively pinkish-tan conidia. The expression of *PT* in the heterokaryon was, therefore, non-autonomous, being masked by components in the heterokaryotic cytoplasm, in which it resided (Kwon and Raper, 1967b). In general, non-autonomous mutants cannot be detected directly but must be given this further plating test before their true genotype is exhibited phenotypically (see Figure 2.1, p. 28).

The reassortment of whole nuclei through heterokaryons is effected by chance as much as is their segregation. Somatic reassortment depends partly upon the frequency of hyphal fusion, to which some consideration has already been given (pp. 88–89) and thereafter, processes as little understood as nuclear migration or the advantage in growth rate of a heterokaryotic apex compared with a homokaryotic one. The structure and biology of a particular fungus will also influence reassortment. For example, multinucleate spores and conidia which do not pass through a uninucleate phase in their development may give rise to propagules with different nuclear ratios.

Circumstantial evidence for this phenomenon was provided from *Botrytis cinereria* in one of the earliest studies of heterokaryons (Hansen and Smith, 1932). Even when spores are uninucleate, or multinucleate and homokaryotic, the probability of reassortment of new heterokaryotic strains is increased by the dispersal of several spores in a drop of viscous fluid or in adherent mucilagenous masses as in the Myxosporae of Mason (1937), e.g. *Graphium, Cephalosporium, Fusarium,* or when groups of ascospores are discharged in slime threads, e.g. *Sordaria fimicola.* In dikaryons and in those Ascomycetes where nuclei which differ in mating type can coexist satisfactorily in vegetative hyphae, special mechanisms may exist to reconstitute or maintain the heterokaryotic condition. For example, in *Neurospora tetrasperma* and *Podospora anserina* the orientation of 2nd meiotic division spindles and mode of spore delimitation are such as to ensure that *A-* and *a*-carrying nuclei are normally incorporated in the same spore. Meiotic spindle orientation in the basidium coupled with a 2-spored condition may also ensure that the majority of basidiospores from 2-spored species of Basidiomycetes are dikaryotic for mating type, e.g. *Galera tenera* f. *bispora* gives rise to 94% dikaryotic basidiospores and 6% monokaryotic ones (Sass, 1929). Less effective, but nevertheless still very efficient, is the situation where genetically different basidiospores germinate rapidly in close proximity after their release and so may fuse, e.g. in many Ustilaginales.

There seems little doubt that many of the morphological and biological features which fungi exhibit can be related to the selective advantage inherent in maintaining a heterokaryotic mycelium (see also Chapter 9).

Complementation and allelism

When the phenotype of a heterokaryon is closer to the non-mutant (wild-type) phenotype than any component homokaryon is when grown alone, the homokaryons are said to complement each other. Complementation then is the ability of a defective mutant to compensate for a different defective mutant when brought into the same cytoplasm, or nucleus. If *two homokaryons differ by each having a single mutant locus, complementation provides evidence that the mutant loci are non-allelic, its absence suggests, prima facie, that the mutant loci are identical, i.e. homoallelic.*

In most cases the results of the complementation test are reasonably clear-cut. If the heterokaryon is unstable, its development, even as small sectors, is highly suggestive evidence for complementation. It is even more probable if the components can be recovered from the phenotypically non-mutant region, e.g. the white mutants *w2* and *w3* of *Penicillium notatum* give green sectors from which both components can be isolated. Cases are also known where the phenotype is only partially restored in the heterokaryon. An interesting example of this has been given by Prasad (1969) for the two mutant strains of *Aspergillus niger Y1 his* and *Y1 hypox.* The *Y1* strain is a yellow conidial protrophic mutant derived by u.v.-irradiation, while *his* and *hypox* are two non-allelic auxotrophic mutants requiring histidine and hypoxanthine, respectively. When the two component strains were incubated together on minimal medium, normally growing heterokaryotic mycelium was isolated but it failed to conidiate even though grown for three weeks and subcultured. Both nuclear components could be recovered from it. A diploid strain was synthesized from the heterokaryon by exposure to *d*-camphor. This was comparable in every way to $Y1$ and conidiated equally profusely. Here, therefore, there was partial complementation in the heterokaryon but complete complementation occurred only when the mutant genotypes were within the same diploid nucleus. The reasons for partial complementation are not clear. It could be that there was some great disproportion in nuclear ratios within the heterokaryon which restricted conidiation but not growth as a whole. Alternatively, there could have been a cytoplasmic dilution effect on a secondary gene product arising from the interaction of the genes; within a diploid nucleus, this interaction could occur without dilution, because of the initially closer proximity of the products. This latter explanation is supported by the observations of Casselton and Lewis (1967) who compared complementation in heterokaryons, dikaryons and diploids of *Coprinus lagopus*. Heterokaryon complementation was poor but that in dikaryons and diploids good. It will be recalled that the two nuclei in each compartment of a dikaryon usually remain close together; this relationship does not obtain in heterokaryons.

A particularly valuable use of the complementation test is to distinguish between mutants of similar phenotypes. Indeed, this has become a regular and rapid method with *Neurospora* workers, especially for auxotrophic mutants. Cross-streaking, or superimposed spore suspensions of the two mutants on minimal medium followed by incubation are the usual procedures. The plates are then examined after 24–36 hours for wild-type growth. The test can be made more precise by including, in addition to the two mutants under test two non-allelic auxotrophic mutants, one in each homokaryon, which are known to give a wild-type heterokaryon. The heterokaryon is established on a medium lacking the necessary nutrients for the latter two mutants but containing that for the mutants under test. In this way the formation of a heterokaryon is ensured. It can then be subcultured on truly minimal medium and if it grows no better than either of the mutant homokaryons, it is clear that the mutants are non-complementing and so are probably determined by identical allelomorphs.

This technique of 'forcing' a heterokaryon is obviously of great value when the mutants under test have by themselves failed to develop a heterokaryon.

The complementation test has proved of value in assigning large numbers of apparently identical auxotrophic mutants to their respective loci. Such data are often presented as a complementation matrix and the allocation of 224 arginine-requiring mutants (*arg*) in *N. crassa* to some six loci is presented as an example in Figure 5.4a.

The complementation test, therefore, provides a rapid method of sorting phenotypically similar mutants, especially auxotrophs into their non-allelic

98

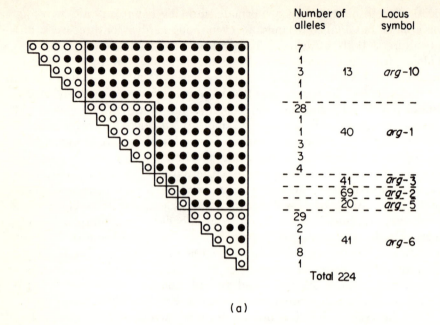

Number of alleles		Locus symbol
7		
1		
3	13	*arg*-10
1		
1		
28		
1		
1	40	*arg*-1
3		
3		
4		
	41	*arg*-3
	69	*arg*-2
	20	*arg*-5
29		
2		
1	41	*arg*-6
8		
1		

Total 224

(a)

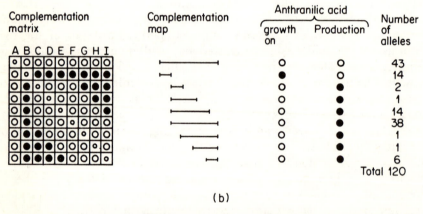

Complementation matrix	Complementation map	Anthranilic acid		Number of alleles
A B C D E F G H I		growth on	Production	
		○	○	43
		●	○	14
		○	●	2
		○	●	1
		○	●	14
		○	●	38
		○	●	1
		○	●	1
		○	●	6

Total 120

(b)

FIGURE 5.4 Two complementation matrices derived from data from auxotrophs of *Neurospora crassa*. (a) Complementation matrix for 224 independently obtained auxotrophs requiring arginine. Six main groups occur, three (*arg-3*, *arg-2*, *arg-5*) in which no complementation occurs at all and three (*arg-10*, *arg-1*, *arg-6*) in which there is only limited complementation. Complementation occurs between auxotrophs from different groups. (b) Complementation matrix and complementation map derived from it of 120 *trp-1* mutants (tryptophan-requiring). Their ability to accumulate or to grow upon anthranilic acid is also shown. Note the large groups of 43 mutants which show no complementation with each other nor with any member of the other 8 groups. Key: ● complementation; ○ non-complementation. (From Catcheside, 1966.)

groups. It can be seen that mutants within a locus fall into two patterns. Some show no complementation between all possible test pairs, e.g. *arg-2*, *arg-3* and *arg-5*; others, a minority, show a greater or lesser degree of complementation, e.g. *arg-1*, *arg-6*, and *arg-10* where 6/21, 3/15 and 3/15 confrontations respectively show complementation. The detailed study of complementation between alleles from the same locus has sometimes been expressed in the form of a complementation map. This is a diagrammatic representation of the complementation matrix. The normal, fully functional locus can be represented by a line and the various classes of mutants by shorter lines representing regions for which they are defective in function. Thus two non-complementing mutants will be defective in the same region and so their representations will overlap, but two complementing mutants are defective in different regions and so their representations are non-overlapping. A simple example of this is shown for 120 tryptophan-requiring mutants (*trp-1*) in Figure 5.4b.

In many cases the mutant order along the linear complementation map is found to be identical with the order of alleles mapped by meiotic recombination but this collinearity is by no means universal. Moreover, in some cases the complementation map is non-linear; circular, circular + linear, and even S-shaped examples are known. It will be recalled that the two kinds of mapping represent quite different things. The chromosome map locates the position on the chromosome of an allele with a particular function; the complementation map indicates which alleles, defective in function, can in concert with other alleles defective in a different function, restore the full function of the locus. *Complementation between mutants, therefore, at the intralocus level, only provides information concerning their functional activities within the locus, not on their allelism.* Such functional activities will be considered more fully in Chapter 15.

CHAPTER 6

Recombination and segregation of extrachromosomal elements

In fungi there are three situations where the operation of extrachromosomal elements may be suspected, namely, non-mendelian segregation at sexual reproduction, somatic segregation, and invasive spread in a mycelium subsequent to hyphal anastomoses. Confirmation of the presence of such elements may result either from the recognition of their physical basis, e.g. a cell organelle, or by the induction of a specific mutation at an extrachromosomal site.

In many cases all, or several, of these criteria may be applicable, whereas in other cases the recognition and identification of extrachromosomal elements and their effects is difficult because the phenotype is determined by an unpredictable interaction of chromosomal genes and such elements. Nevertheless, because of the range of situations in which the operation of extrachromosomal elements can be detected in fungi, they are of particular value for such studies. The operation of these determinants is equally important for the biology of many fungi where it bears on problems of strain maintenance and pathogenic expression.

Extrachromosomal elements in sexual reproduction

The offspring of reciprocal crosses may show constant and persistent differences when the parental, conjugant structures have contributed unequal amounts of cytoplasm, but identical amounts of chromosomal material, to the cross. An early, striking example was the pathogenic behaviour of rust fungi on their hosts. When physiological Races 34 and 14 of *Puccinia graminis tritici* were crossed reciprocally and leaves of the wheat cultivar Marquis infected, the reaction developed resembled that of the aecidial parent only. Moreover, in subsequent generations, to the F_3 at least, such differences persisted (Newton and others, 1930b; Johnson and others, 1934). In this fungus fertilization occurs between a pycnidiospore—a small spore almost wholly occupied by the nucleus—and the flexuous hyphae of a pycnidial mycelium of opposite mating type (see Figure 1.8, p. 18). It has been claimed that at fusion only the nucleus penetrates the receptor mycelium but this is clearly a technically difficult matter to demonstrate (Lamb, 1935). Nevertheless, the donor:receptor cytoplasmic ratio is clearly highly disparate so that the cytoplasm of aecidiospores is almost wholly of receptor origin.

A similar situation is shown in respect of the *mi* (maternal inheritance) mutants of *N. crassa*. The *mi* mycelia grow more slowly than wild type, reaching their maximum dry weight of only about $\frac{1}{3}$ the normal maximum weight in 10 days instead of 5. Reciprocal crosses between *mi-1* (= *poky*) and wild-type mycelia were made by Mitchell and Mitchell (1952). When the protoperithecial parent, contributing a large volume of cytoplasm, was of *mi-1* phenotype, all 8 ascospores in each ascus of every perithecium exhibited this phenotype. When the protoperithecial parent was normal so were all the ascospore progeny. In each cross, allelic chromosomal genes segregated 1:1, quite normally in each ascus.

Several cases of this type are now well authenticated and this situation is easily recognized. In the case of the *mi* mutants, it has proved possible to locate the site of action of the extrachromosomal elements (see p. 115).

In many cases in fungi, however, there is no great disparity in the amount of cytoplasm contributed by each conjugant to a cross. Nevertheless, in such cases extrachromosomal elements can be detected. The most thoroughly studied and understood examples are the *petite* mutants of baker's yeast (*S. cerevisiae*) and a number of other yeasts which show repression of respiratory enzymes by high, exogenous glucose concentrations (Ephrussi, 1953 for summary of early work; De Deken, 1966). *Petite* mutants may be spontaneous arising at characteristic low frequencies, or induced at very high frequencies by specific agents such as acriflavines. Phenotypically, colonies derived from mutant cells can be recognized because their rate of growth is never more than half that of normal colonies and they fail to give a blue colour reaction to the Nadi reagent. The cells lack respiratory enzymes, notably cytochrome *c* oxidase and cytochromes *a* and *b*. It has been claimed that the mitochondria of mutants show ultrastructural abnormalities but similar defects have been found in normal yeast cells with both glucose and melibiose as carbohydrate substrates (Yotsuyanagi, 1962; Smith and others, 1969).

The mutant condition can be determined by a number of different non-allelic chromosomal genes. These mutants are termed *segregational petites* and crosses between one of them and a normal cell give normal diploids which segregate 2:2 ratios of normal:*petite* haploids at ascus formation. However, crosses of other *petite* mutants with normal, give one of two possible results (Figure 6.1a, b).

In one (Figure 6.1a) the *petite* condition is neither exhibited in the diploid zygote nor by the haploids derived from ascospores after sporulation.

102

Subsequently, in cultures from budding diploids or haploids, *petite* mutants may be recovered with the characteristic, low spontaneous mutant frequency of the strain. Here then the *petite* mutant phenotype is masked by the normal and such mutants are termed *neutral petites*. In the other case (Figure 6.1b), however, it is effectively the mutant phenotype which suppresses the normal phenotype, hence they are termed *suppressive petites*. In these crosses the diploid zygote exhibits a *petite* phenotype, as do anything up to 95% of the diploid cells budded from it. If zygotes are placed in sporulation conditions immediately after their formation they will form ascospores. The majority of derived haploid colonies are *petite* in phenotype but rarely, in

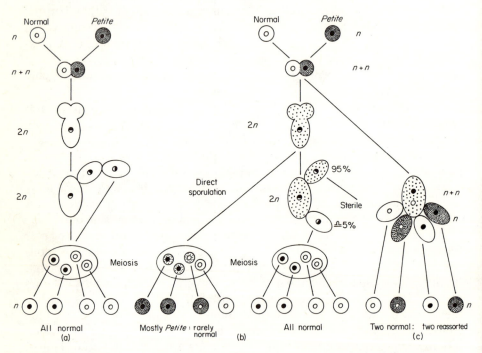

FIGURE 6.1 Diagram to illustrate the segregation patterns derived from crosses between haploid *petite* and normal cells of *Saccharomyces cerevisiae*. (a) Cross of neutral *petite* and normal cell with nuclear marker. Note the nuclear markers segregate but the *petite* phenotype is lost after conjugation. (b) Cross of *suppressive petite* and normal cell with nuclear marker. If the diploid zygote is sporulated immediately some asci may be formed in which some ascospores progeny will exhibit the *petite* phenotype; it may or may not reassort with the nuclear marker which segregates normally. Most diploids if budded are sterile (95%) but some can sporulate when the *petite* character is lost. (c) Cross of *suppressive petite* and normal cell with nuclear marker without fusion of nuclei. If the derived heterokaryotic cell is allowed to bud the extrachromosomal elements determining the *petite* character may show reassortment with the nuclear markers. This is a special form of the heterokaryon test.

some asci, normal colonies result either from all or from 3, 2 or 1 of the ascospores. *Petite* diploid cells budded from zygotes, or zygotes not permitted to sporulate immediately are incapable of sporulation and only the rare, phenotypically normal, budded cells can sporulate; all their progeny are normal in phenotype. In both types of cross several chromosomal marker genes were included and all segregated completely normally (Ephrussi and others, 1949, 1955).

These irregular types of segregation are indicative of the intervention of one or more extrachromosomal elements which, as will be discussed later (see pp. 114–116), are believed to reside in the mitochondrial DNA. Further evidence for the existence of such elements can be obtained from somatic segregation, which will now be considered.

Somatic segregation of extrachromosomal elements

The most direct test for the occurrence of somatic segregation of extra-chromosomal elements is that termed by Jinks (1958) the 'heterokaryon test'. At its simplest it involves uniting two homokaryons, each marked by a single contrasting chromosomal gene, to form a heterokaryon which is then resolved into its original components as assessed by the chromosomal markers. If extrachromosomal elements are present they will not necessarily reassort with the parental, homokaryotic marker genes and so the pheno-types may differ from those of the original parents. For example, suppose chromosomal genes A and a are associated in homokaryons with $[\alpha]$ and $[\beta]$, representing extrachromosomal elements, then:

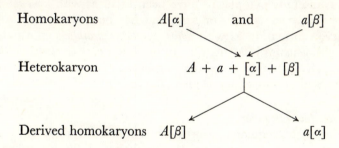

i.e. independent segregation of A/a and $[\alpha/\beta]$.

This expectation assumes that

(a) There are no interactions between the elements $[\alpha]$ and $[\beta]$.
(b) There are no nucleo-cytoplasmic interactions.
(c) Extrachromosomal elements are capable of clear-cut segregation, i.e. there is no $[\alpha + \beta]$ cytoplasmic component etc.

Pittenger (1956) was able to show just such reassortment of *mi-4* and normal phenotypes with the chromosomal genes for lysine dependence and independence in *N. crassa*. The homokaryons *lys mi-4* and *lys⁺* normal were combined into a heterokaryon, spores from which gave rise through reassortment to four types of stable homokaryon, namely:

lys mi- 4	*lys* normal
lys+ normal	*lys*+ *mi-4*
Parental	Reassorted

Other linked chromosomal markers in this cross showed no recombination and segregated with *lys* and *lys*+.

A similar test has been done, although it is technically more difficult, with a *petite* mutant of *S. cerevisiae* var. *ellipsoideus*. Fowell (1951) had shown that in some strains when two haploids were permitted to fuse, nuclear fusion was not immediate and the dikaryotic cell could bud off haploid cells. Wright and Lederberg (1957) made use of this situation to combine a *suppressive petite* and a normal cell, each containing different, recognizable, chromosomal marker genes. The haploid cells budded off included both parental types but also cells with a *suppressive petite* phenotype associated with the chromosomal markers of a normal cell (Figure 6.1c).

Whenever a heterokaryon test is made which involves an extrachromosomal element that is suppressive in action there will be, as with *suppressive petites* in yeast, deficiencies in the expected, reassorted phenotypes. For example, if either *mi-1* or *mi-3* are combined with a normal strain of *N. crassa* the phenotype is normal at first but eventually a mutant phenotype tends to appear and then reassorted phenotypes can be detected. Suppression is the simplest kind of interaction between extrachromosomal elements which may obscure expected reassortments. To avoid such situations different ratios of combining cytoplasms have been used, as with *mi-1/mi-3* heteroplasmons (Pittenger, 1956), or sampling has been done very soon after heterokaryon formation as with *white, vegetative death/buff*, normal, heterokaryons of *A. glaucus* (Jinks, 1959b). In this latter case *vegetative death* showed up phenotypically within 5 days of heterokaryon formation and, of the many spores sampled at 7 days, less than 1% germinated.

Unstable somatic segregation

Apart from the regular reassortment of stable chromosomal and extrachromosomal determinants under the conditions of the heterokaryon test, unstable somatic segregation may occur spontaneously or be induced by various agents. The most thoroughly documented case of such a condition is undoubtedly the *red* mutant of *Aspergillus nidulans* (Arlett, 1957; Arlett and others, 1962; Grindle, 1964; Jinks, 1966). This condition arose in a colony derived from a single uninucleate conidium of the Birmingham Isolate 1 of *A. nidulans* whose conidia had been exposed to u.v.-irradiation. The green colony sectored and the mutant *red* sectors showed the following features:

(a) a growth rate, 27% on average, faster than normal,
(b) mostly small and abnormal conidiophores with a few large ones near the perithecia,

(c) a reduced density of perithecia, averaging 51% of normal and lacking viable ascospores,

(d) a red pigment both in the mycelium and suffusing into the medium.

Subcultures from these sectors maintained their parental features for a time but, sooner or later, themselves sectored. Normal subcultures frequently gave rise to *red* sectors whereas *red* subcultures gave rise to normal sectors rather infrequently. Even more remarkably, propagation from uninucleate conidia taken from each sector gave similar results.

Conidia from *red* sectors never gave less than 5% *red* colonies and usually 10–90%; moreover, the progeny of the abnormal conidiophores were predominantly of *red* phenotype, while those from large conidiophores gave more phenotypically normal colonies. In contrast, conidia from normal sectors never gave rise to more than 20% *red* colonies and usually 90% normal colonies. There is rarely any difficulty in distinguishing the two types but neither is entirely uniform. *Red* sectors vary in their intensity of colouration, their growth rates and the proportion of *red*-sectoring colonies derived from their conidia. The normal sectors are more uniform but they too differ in the frequency with which their conidial offspring sector. In fact, it seems that there is a continuous distribution of *red*-sectoring colonies amongst the offspring of both normal and *red* sectors.

The condition has persisted since it originated in 1955 so that, for example, after almost 500 fortnightly subcultures from *red* sectors employing uninucleate conidial transfer, sectoring colonies still develop. Stable *red* colonies have never developed from such stocks whatever means of transfer or isolation have been used. By contrast, a proportion of normal sectors do give rise, through conidial transfer, to stable, persistent normal colonies.

In heterokaryon tests there is clear evidence that *red* is determined by an extrachromosomal element even though the heterokaryotic state is of very short duration and heterokaryotic conidiophores are rare. For example, where w_3^+ represents the original wild-type gene determining green-coloured conidia and w_3 an allelomorph determining white-coloured conidia, the following results were obtained:

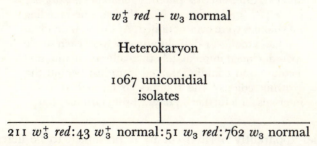

$$w_3^+ \ red + w_3 \ normal$$

$$|$$

$$\text{Heterokaryon}$$

$$|$$

$$1067 \ \text{uniconidial isolates}$$

$$|$$

211 w_3^+ *red* : 43 w_3^+ normal : 51 w_3 *red* : 762 w_3 normal

In a number of such tests there was a statistically significant shortage of white-spored *red* (w_3 *red*) phenotypes. There is, indeed, other evidence that the *red* condition results in reduced viability despite its faster growth rate.

It has been suggested that the basis of the phenomenon originated in the mutation of an extrachromosomal element, [R]—a normal component of the cytoplasm of the wild type, to a mutant form, [r]. It was supposed that [r] elements increased either by multiplication or recurrent mutation. There is evidence that the former supposition is more probable. If *red* sectors arising from apparently normal colonies are cloned and their conidia sampled daily, the proportion of *red*-sectoring progeny increases daily (Figure 6.2).

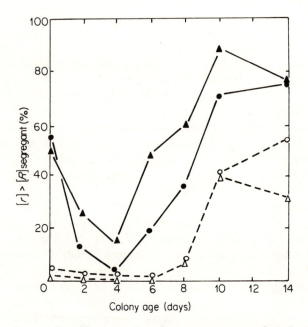

FIGURE 6.2 Changes in the percentage of *red* segregants of *Aspergillus nidulans* during the growth of single colonies of two different green ('normal') colonies, genotype $w_3^+ m_3^+ m_4^+$ [R] > [r] (solid lines) and two different '*red*' parents, genotypes $w_3^+ m_3^+ m_4^+$ [r] > [R] (broken lines). Note that no true-breeding *red* colonies were ever obtained. At time 0 individual colonies among 14-day old progeny were each single spored, cloned immediately thereafter on to minimal medium and conidia sampled from just behind the growing edge of the colonies so formed at 2-day intervals for a further 14 days. (From Grindle, 1964.)

The initial apparently normal colony is presumed to be heteroplasmic for [R] and [r] and the numerical properties of the elements presumed to be [R] > [r]. In a *red* sector this balance has shifted to [r] > [R] and the subsequent increase in *red*-sectoring progeny results in a further increase of

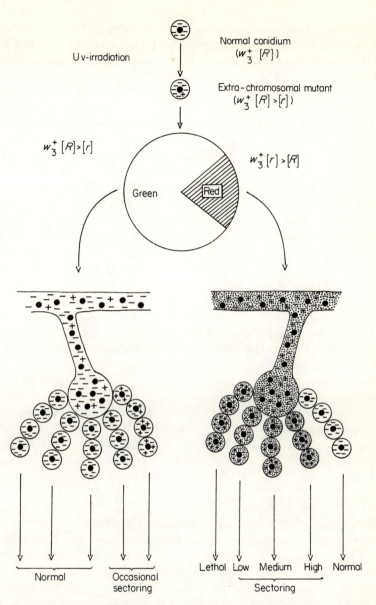

FIGURE 6.3 Diagram to illustrate the behaviour of the *red* variant of *Aspergillus nidulans*. The origin of a mutant extrachromosomal element (+) is shown in a conidium after u.v.-irradiation. The colony derived from it sectors and typical conidiophores from normal and sectored areas are shown. Hyphae from the sector carry more mutant extrachromosomal elements and their frequency in the conidia determines the ability of colonies derived from them to sector. Note that in w_3^+ strains pure-breeding *red* colonies are not obtained; conidia potentially giving rise to them are inviable.

$[r]$ relative to $[R]$, i.e. a more rapid multiplication of $[r]$ elements than of $[R]$ elements. Since, however, *red* colonies are known which can give rise to 90% phenotypically normal colonies via conidial transfer, it must be further supposed that quite a small number of $[r]$ elements can suppress the expression of $[R]$ elements. These notions are expressed diagrammatically in Figure 6.3.

So the range of phenotypes encompassed by normal and its *red* variant are explicable, on this hypothesis, by the balance between the numbers of $[R]$ and $[r]$ elements within the cytoplasm. A stable normal is homoplasmic for $[R]$; unstable normal or *red* colonies are heteroplasmic for $[R]$ and $[r]$, the exact phenotype being determined by the $[R/r]$ ratio. There remains one unresolved issue; why have stable *red* cultures, homoplasmic for $[r]$ not been recovered from these stocks? The solution to this problem lies in the observation that in the heterokaryon test the w_3 *red* class was significantly deficient. The persistence and expression of the $[R]$ and $[r]$ elements is conditioned by the nuclear genes. Detailed consideration will be given to this situation later (see p. 113). Here it need only be said that, if combined with w_3co (white spores; compact colony) or either of two gene mutants m_3 and m_5, whose action was not precisely defined but resulted in extremely small colonies, stable *red* colonies, homoplasmic for $[r]$ elements were obtained. In these situations the $[R]$ elements had apparently been lost or totally suppressed.

Somatic segregation is an especially valuable tool for the study of extra-chromosomal elements. In the heterokaryon test it provides reliable evidence for the action of such an element in situations where sexual reproduction may not be possible. Indeed, even in limiting conditions where heterokaryosis is transient as in the case of *red*, it can provide an insight into the action and nature of such elements. In particular, it suggests in some cases at least, that the basis of the phenomenon may be a multiplying, particulate element capable of spreading via hyphal fusions, i.e. by invasive spread.

Invasive spread of extrachromosomal elements

Invasive spread between mycelia that have fused has been known for several decades. In 1936 Dickson showed for *Coprinus macrorhizus* that as a consequence of hyphal contact between a *fluffy* variant and a normal mycelium, regardless of mating type, the receptor mycelium acquired the *fluffy* habit. This transfer was unlikely to have been mediated by nuclear genes since nuclear exchange is determined in agarics by mating compatibility. In Ascomycetes a number of moribund conditions, *senescent* in *Podospora anserina* (Rizet, 1957) and *vegetative death* in *A. glaucus* (Jinks, 1959b) are known. Marcou and Schecroun (1959) demonstrated most elegantly by a combination of micrurgery and the heterokaryon test that *senescent* was due to an infective extrachromosomal element (Figure 6.4).

Other experiments showed that minute fragments taken close to, and immediately after a hyphal fusion could show the mutant phenotype while

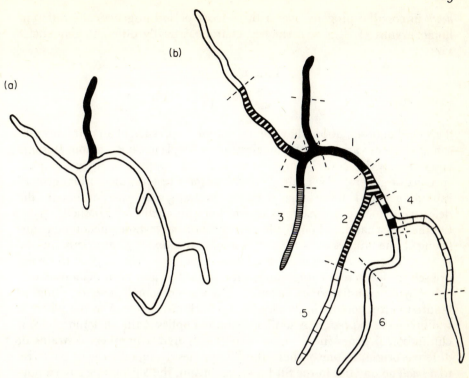

FIGURE 6.4 Invasive spread of an extrachromosomal factor in *Podospora anserina*. (a) Fusion between a *senescent*, recessive (*i*) coloured strain and a colourless (*I*) strain. (b) The fusion mycelium was cut up into 6 fragments and subcultured to test their growth. The closer to the point of fusion the lower the growth viz.

Fragment No.	1	2	3	4	5	6
Growth (cm)	0	1	4	8	12	12

i.e. the extrachromosomal element determining *senescent* had shown invasive spread. (After Marcou and Schecroun, 1959.)

later and farther from the point of anastomosis, fragments might or might not show the character. This suggests that the infective agent is particulate.

Invasive extrachromosomal elements are not necessarily lethal, indeed the 'A' type variant of *A. glaucus*, often now referred to as *conidial*, exhibits five modified mycelial characters including conidiation, which was greatly enhanced (Subak-Sharpe, 1958).

Interaction of chromosomal and extrachromosomal elements

The existence of such interactions has already been referred to, e.g. *red* in *A. nidulans* (p. 108), and it seems probable that they occur commonly and show a wide range of expression. For example, the incidence of *vegetative death* in conidial progeny over a three-week period in genetically different homokaryons of *A. glaucus* showed characteristically different frequencies viz:

Line 1	0·9
Line 3	1·6
Line 6	15·2

This kind of observation, that the phenotypic expression of an extrachromosomal element varies with the nuclear genotype, is not uncommon. In some cases it may be entirely suppressed and this suppression may be quite specific, or general. For example, in *N. crassa*, *f* restores the normal growth rate to group I respiratory deficient mutants, e.g. *mi-1*, although the defective cytochrome system is unaffected (Mitchell and Mitchell, 1956). On the other hand, although *f* has no effect on *mi-3* whose phenotype is unchanged, the suppressor *su-1* (*mi-3*) is similar in action to *f* but strictly specific to *mi-3*. The reverse situation, the failure of a chromosomal gene to be expressed in a different cytoplasm from normal has also been demonstrated. Srb (1958) transferred *ac*—aconidial, a chromosomal gene resulting in failure to form conidia in *N. sitophila*, by back-crossing to *N. crassa* where it was no longer expressed. A similar situation applies to the allelomorphs *S*, *s*, chromosomal genes in *N. crassa*. Srb (1958) had a number of strains of different origin. In one of these, the Philippines strain, either gene gave rise to a *small* colony but in the St. Lawrence strain, the *S* genotype gave a normal colony, the *s* genotype continued to determine a *small* colony. The differences between the strains were exclusively cytoplasmic.

One or two cases of interaction are sufficiently well understood to present the possibility of analysing them at the molecular level. The most remarkable of these is perhaps [*psi*] in yeast (Cox, 1965, 1971; Young and Cox, 1971, 1972).

It is known that there are several chromosomal genes in *S. cerevisiae*, the super-suppressors, whose action is to nullify the gene expression of specific mutant alleles at a number of different loci (Hawthorne and Mortimer, 1963, 1969). For example the super-suppressor Sup_{Q5} renders cells carrying the mutant alleles *ade 2–1*, *met* and *tyr* (normally auxotrophic for adenine, methionine and tyrosine respectively) protrophic. However, the action of the super-suppressors themselves are in some cases, namely Sup_{Q5} and Sup_{Q2}, determined by the presence in the cells of either [*psi*$^+$] or [*psi*$^-$]—two alternative extrachromosomal elements. For example, Sup_{Q5} has its usual action in the presence of [*psi*$^+$] but not in the presence of [*psi*$^-$]. So the genotypes can be recognized by the phenotypes in the following way:

ade 2-1$^+$ Sup_{Q5} [*psi*$^+$]—a normal white protroph
ade 2-1 sup_{Q5} [*psi*$^+$]—a red, adenine-requiring auxotroph
ade 2-1 Sup_{Q5} [*psi*$^+$]—a normal white protroph
ade 2-1 Sup_{Q5} [*psi*$^-$]—a red, adenine-requiring auxotroph

The issue is complicated by the presence of a further pair of chromosomal allelomorphs, r and R, the latter being a dominant mutant allele. In cells of genotype R, R/R, or R/r the $[psi^+]$ elements are not expressed and they appear to be progressively eliminated. This was shown by making diploid cells heteroplasmic for $[psi^+]$ and $[psi^-]$ and heterozygous for R/r. All cells carried $ade\ 2$-$1\ Sup_{Q5}$. These cells were transferred to sporulation media at successive periods for 4 days during which time the percentage of diploid clones which gave rise to increasing proportions of tetrads with non-suppressed cultures, i.e. mutant, increased. This suggests that in the presence of R, $[psi^+]$ is progressively eliminated so that the red, auxotrophic ade-2-1, $Sup_{Q5}\ [psi^-]$ phenotype increased.

The specificity involved in these interactions is remarkable. Chromosomal super-suppressors are quite specific for the alleles they affect, e.g. Sup_{Q5} acts on ade-2-1 but not on its allelomorph ade-2-c and R is specific for $[psi^+]$. It is true that $[psi]$ does affect another super-suppressor Sup_{Q2} but it acts on it differently from its action on Sup_{Q5}. The differential relationships are set out below:

	Haploid	Diploid
$[psi^+]$ ade-2-1 Sup_{Q2}	lethal	red, auxotrophic
$[psi^+]$ ade-2-1 Sup_{Q5}	white, protrophic	white, protrophic
$[psi^-]$ ade-2-1 Sup_{Q2} or Sup_{Q5}	always red and auxotrophic	

It seems possible that this specificity can be analysed further since there is some understanding of the molecular basis of the action of super-suppressors. By analogy with a similar situation in *E. coli* it has been suggested that the so-called 'stop' codons UAA, UAG or UGA, which occur in the messenger RNA transcribed from the mutant alleles, prevent the production of a complete, normal protein. Super-suppressor genes are supposed to be able to alter the appropriate transfer RNA so that the codon is not translated as a 'stop' signal and thus the normal protein is synthesized. This implies that $[psi]$ elements are involved in some way in the mechanism of protein synthesis.

That the degree of expression of extrachromosomal elements can be controlled by the chromosomal genes either by affecting the stability or the rate of multiplication of such elements, has already been mentioned in connexion with the red variant of *A. nidulans* (p. 106), and $[psi]$ and the R gene in yeast. The situation has been more fully studied in *A. nidulans* than in yeast and it will now be outlined.

The *red* phenotype is supposed to be determined by a preponderance of $[r]$ elements over the normal extrachromosomal elements $[R]$. Colonies that sector are supposedly heteroplasmic for $[R/r]$ elements. Grindle (1964) introduced some 11 mutant chromosomal genes in various combinations into the wild-type genotype and was able to show that the balance of $[R/r]$ elements was affected by them. The results are given in Table 6.1 and Figure 6.5.

TABLE 6.1 Effects of interaction of various nuclear genes with cytoplasm heteroplasmic for [R] and [r] factors in *Aspergillus nidulans*. Samples were subcultured from sectors in plate culture. (Based on Grindle, 1964.)

Nuclear constitution	Phenotype of sector samples in cytoplasm $[R] > [r]$ $[r] > [R]$	% *red* segregants from conidial sample	Pure breeding [r] strains	Phenotype after 8 days mass subculture
w_3^+	Green	1–20	—	Mosaic
	Red	16–95	—	Red
$w_3^+ m_2$	Green	14–44	—	Mosaic
	Red	23–90	—	Red
$w_3^+ m_4$	Orange	2–22	—	Mosaic
	Red-brown	78–92	—	Red
w_3	Pale-pink	0–1	—	Pale-pink
	Red	0–55	—	Pale-pink
$w_3 m_1$	Green	0	—	Green
	Pale-red	0–4	—	Green
w_3 ade	Green	0	—	Green
	Pink	0–15	—	Green
w_3 rib	Orange	0	—	Orange
	Brown	0	—	Orange
$w_3 m_3$	Pale-brown	15–89	+	Red-brown
	Red-brown	100	+	Red-brown
$w_3^+ m_5$	Green	0–90	+	Pale-red
	Pale-red	100	+	Pale-red
$w_3 co$	White	0	—	White
	Red	100	+	Red
$w_3 df$	White	0	—	White
	Red	100	+	Red

Note: Apart from w_3^+ the nuclear condition is not shown in respect of wild-type genes.
Key to symbols: w_3^+—wild type; w_3—white; m_1–m_5—minute colony; df—diffuse colony; co—compact colony; ade—adenine-requiring; rib—riboflavin-requiring.

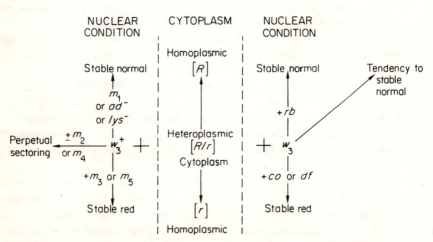

FIGURE 6.5 Diagram to illustrate the interaction of the normal [R] and mutant [r] extrachromosomal elements with different nuclear genes in *Aspergillus nidulans*.

The most plausible explanation of these interactions is that nuclear genes affect the balance of $[R]:[r]$ elements by modifying their multiplication. Suggestive evidence comes from the observation that the change from the sectoring to the stable state is both a rapid yet a gradual process. For example, when m_3 was introduced into a sectoring colony of normal phenotype and successive conidial samples were taken from the sectors of normal appearance, the numbers of *red* variant progeny increased up to 4 days when they were exclusively stable and of *red* phenotype. Moreover, during this period conidial progeny with a normal phenotype sectored in a way different from their parental colony. Derived colonies grew more slowly than parental ones but the *red* sectors which arose grew far more rapidly than usual and soon took over the whole growing front of the colony. Thereafter hyphal-tip isolations gave rise only to stable *red* colonies.

How changes in rates of multiplication are affected is not known. The rate of multiplication of one element could increase over that of the other, it could decrease, or there could be a simultaneous increase of one and a reduction of the rate for the other. In any of these cases the ratio of extra-chromosomal elements in the apical cytoplasm would change and there would be an increased probability that a conidial head or apical cell would become homoplasmic. It seems that in the *red* variant the presumed mutant $[r]$ elements are not stable in chromosomally determined auxotrophic strains. This suggests that $[r]$ elements are defective when compared with $[R]$ elements.

The rapidity of such changes suggests that the number of $[R]$ elements in any one cell is quite small and other evidence bears upon this. Both the homoplasmic and the heteroplasmic conditions can be transmitted by conidia. Provided that a constant number of elements enters each conidium and that they are distributed at random, the proportions of homoplasmic and differently constituted heteroplasmic conidia should approximate to the expansion of the binomial $(p + q)^n$ where p = proportion of $[R]$ elements, q = proportion of $[r]$ elements and n = the number of elements per conidium. An initial minimal estimate of n was 5·8 but since then Jinks (1954) has observed that from w_3^+ *red* variants 7 classes of conidia could be distinguished. These classes are based on the mean relative frequency with which they give rise to daughter *red* colonies, namely 0, 13, 40, 57, 66, 79 and 92% respectively, but they never include stable *red* colonies. So the minimal, theoretical class number is 8, i.e. to include a stable *red* colony class, and the conidia could range in constitution from $7[R]:0[r]$—homoplasmic normal, to $0[R]:7[r]$—homoplasmic *red*. He estimated the expected frequencies of each class using the binomial and compared it with that found in a random sample of 25 conidia; the agreement was reasonably good. Despite the limitations of the data and the somewhat arbitrary division of the 7 conidial classes recognized this is an interesting and, indeed, unique estimate of the numbers of extrachromosomal elements in a cell. Nevertheless, it is not incompatible with the views of those who suppose that some extrachromosomal elements are associated with, if not located in, cell organelles. It is convenient, therefore, at this point to consider the nature and properties of such elements.

Nature and properties of extrachromosomal elements

The properties common to the extrachromosomal elements described are that they are persistent, particulate entities which arise by spontaneous or induced mutation, are capable of multiplication and, in several cases, of invasive spread via hyphal anastomoses.

A number of cellular components would fit all, or some of these characteristics, those most probable are:

(a) cell organelles or some component of them,
(b) undefined but particulate elements, best described by the genetic term plasmagene,
(c) viruses or other extraneous agencies,
(d) persistent molecules probably involved in protein synthesis, of which RNA is the most plausible.

Cell organelles

In yeast there is now good circumstantial evidence that some extra-chromosomal elements are located in the DNA of the mitochondria (M-DNA). The evidence is of three kinds, mutational, chemical and genetic. Firstly, the u.v.-induction of *petite* variants is dependent on wavelength and corresponds to the absorption spectrum of nucleic acid with a maximum mutagenic effectiveness at 265 nm (Raut and Simpson, 1955; Wilkie, 1963). Secondly, in *neutral* and *suppressive petites* the M-DNA differs from that in normal cells. There is little or no such DNA in *neutral petites* while the buoyant density of mitochondria from *suppressive petites* in CsCl solution differs from that of normal and also between different variants. This is due to its increased content of adenine and thymine compared with about 77% AT in normal mitochondria, indeed, in one strain the M-DNA is largely pure AT.

Finally, there is evidence that other extrachromosomal elements are associated with the hypothetical [*rho*] elements which determine the normal/*petite* phenotypes. These include induced resistance to various antibiotics, e.g. erythromycin *ery-r*, chloramphenicol *chl-r*, compared with the normal sensitive condition due to *ery-s*, *chl-s* elements, etc. Deutsch and others (1970) employed the same mutagen ethidium bromide, known to produce *petite* phenotypes with a high frequency, to induce antibiotic resistance, which it did most effectively. Moreover, they were able to show that when either *ery-s* or *chl-s* were lost then [*rho*+] was also lost. This suggests that these three types of element were associated with the same entity, M-DNA. Formal evidence of linkage between the [*rho*] element and other elements has been provided by several workers but, recently, Young and Cox (1971) have not only demonstrated linkage between *ery* and [*rho*] but also that [*psi*] is inherited independently of either. Two hypotheses could account for this, either [*psi*] is so distant from [*rho*] in the M-DNA linkage group that it appears to segregate independently, or else [*psi*] is located elsewhere. No decision can yet be made.

No evidence exists to support the location of other extrachromosomal elements in M-DNA either in yeast or in other fungi. Suggestive evidence for this condition does, however, exist for the respiratory-deficient variants of *Neurospora*. In *N. crassa* it has proved possible to inject either nuclei, mitochondria or mitochondrial fractions from the *abn-1* variant into hyphae of normal fungi. This results with the last two components only in a reduction in the rate of hyphal growth and a modification of the cytochromes, although this was not as extreme as in *abn-1* itself. In contrast, microinjection of *mi-1* mitochondria and subsequent culture of the infected segments resulted in a phenotype remarkably similar to that of *mi-1* (Diacumakos and others, 1965; Wilson, 1969). In addition, Woodward and Munkres (1966) showed that the structural protein of the mitochondria (M-SP) of *mi-1* and *mi-3* strains differed from normal by the addition of a cysteine and lack of a tryptophan in the former, and one less tryptophan in the latter. In contrast, M-SP of two chromosomal mutant strains showing respiratory deficiency did not differ from normal M-SP. An alternative view is that the many defects associated with the *mi* phenotype are due to errors in transcription during M-SP synthesis (Zollinger and Woodward, 1972). These data suggest that the typical basis of respiratory-deficient variants in *Neurospora* is likely to be M-DNA as it is in yeast. These elements do, however, show complementation, a phenomenon not shown by [*rho*] elements. Complementation only occurs between elements of different groups, not between elements of the same group (Table 6.2, p. 116). This could be interpreted to mean that each group represents either a distinct locus or a different extrachromosomal element.

The [*R*] element in *A. nidulans* has not been located. The fact that it may only be represented by some 7 sites per cell is not very helpful in locating it since no data are available as to the number of organelles, other than nuclei, in apical cells. In any case there does not seem to be a close correlation between the number of mitochondria in yeast, for example, and the number of extrachromosomal sites capable of mutation or inactivation by radiation or chemical mutagens. Estimates of mutable [*rho*] sites in anaerobic yeast cells vary from 3 to 6 and in aerobic yeast cells up to 20 although the number of mitochondria in a diploid yeast cell is probably 40–50 (Avers and others, 1964; Slonimski and others, 1968; Allen and McQuillan, 1969). This difference in numbers of sites is not unexpected since mitochondria degenerate in cells grown anaerobically, although it is not clear whether any are actually lost (Matile and others, 1969). Nevertheless, there is not a 1 : 1 relationship between mutable sites and the number of mitochondria per cell. One possible explanation of the discrepancy is that only a limited number of

TABLE 6.2 The three complementation groups between extrachromosomal mutants of *Neurospora crassa* and some of their properties. (Data from Bertrand and Pittenger, 1972.)

Group	Symbol	Growth habit	Female fertility	Mitochondrial cytochromes $a_1 + a_2$	b	c	Suppression by f	$su\text{-}1$	Origin
	mi-1	Poky	+	Deficient	Deficient	Excess	+	−	Spontaneous
	SG-1	Poky	+	Deficient	Deficient	Excess	+	−	Acriflavine
	SG-3	Poky	+	Deficient	Deficient	Excess	+	−	Spontaneous
I	*stpR1*	Poky	+	Deficient	Deficient	Excess	+	−	Spontaneous
	exn-1	Poky	+	Deficient	Deficient	Excess	+	−	NG
	exn-2	Poky	+	Deficient	Deficient	Excess	+	−	NG
	exn-3	Poky	+	Deficient	Deficient	Excess	+	−	Spontaneous
	exn-4	Poky	+	Deficient	Deficient	Excess	+	−	NG
II	*mi-3*	Inter-mediate	+	None	Normal	Excess	−	+	Spontaneous
	stp	Stop–start	−	Deficient	Deficient	Excess	−	−	U.v.-irradiation
	stpA	Stop–start	−	Deficient	Deficient	Excess	−	−	Spontaneous
	stpA18	Stop–start	−	Deficient	Deficient	Excess	−	−	Spontaneous
III	*stpB2*	Stop–start	−	Deficient	Deficient	Excess	−	−	Spontaneous
	stpC	Stop–start	−	Deficient	Deficient	Excess	−	−	NG
	abn-1	Stop–start	−	Deficient	Deficient	Excess	?	?	Spontaneous
	abn-2	Stop–start	−	Deficient	Deficient	Excess	?	?	Spontaneous
	mi-4	Stop–start	−	Deficient	Deficient	Excess	?	? ?	Spontaneous

Note: 1. ? indicates not known. 2. The mutant *mi-4* is now lost.

mitochondria pass into each daughter cell, so, provided just a few M-DNA sites carry the mutant element, some buds would be homoplasmic for the mutation. Indeed, earlier studies by Mundkur (1953), although probably underestimating the total number of mitochondria per cell, showed clearly that buds could become detached with low or even single mitochondrial complements. There was no regular correlation between mitochondrial numbers in the bud and the mother yeast cell, regardless of its size. In this context it is also significant that Coen and others (1971) have observed that if a zygote includes a diversity of M-DNA elements and is allowed to develop into a clone, the cells will be homoplasmic for their M-DNA elements after some 20 cell divisions. This, in part at least, reflects the mechanical sorting of different mitochondrial types at successive bud formations.

Plasmagenes

No other elements have been unequivocally located on sites within the fungal cell. Elements which can apparently be expressed when present in relatively small numbers could well be associated with organelles which exist in relatively small numbers in cells, e.g. mitochondria. However, several phenotypes show more or less continuous variation, e.g. growth rate, density of sporulation, perithecial frequency in Aspergilli (see, for example, Jinks, 1966). Here the numbers of hypothetical elements are presumably much larger and their location becomes even more problematical.

A possible approach to this problem is the direct one of examining and comparing the ultrastructural morphology of different hyphal regions. For example, if *A. nidulans* is subcultured exclusively by hyphal-tip transfer or conidia it rapidly loses the ability to produce perithecia but, if at any stage an ascospore subculture is employed, perithecial production is fully restored (Mather and Jinks, 1958; Figure 6.6).

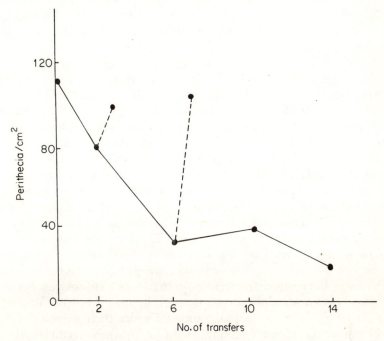

FIGURE 6.6 Perithecial production by successive transfers of conidia only in *Aspergillus nidulans*. The broken lines indicate perithecial production when an ascospore was transferred instead. (From Mather and Jinks, 1958.)

One or two possibilities could account for this observation, either, the hyphae which give rise to perithecia are those with the smallest population of mutant extrachromosomal elements or else there is a reorganization and selection of such elements in the ascus. Comparisons of the ultrastructure of these different regions could perhaps suggest a likely candidate for the physical basis of these unlocated elements. As yet, however, the physical basis of most plasmagenes is obscure.

Viruses

In some cases, the extrachromosomal element could be a virus. A class particularly susceptible to this explanation is that of degenerative phenotypes whose hypothesized elements show invasive spread, e.g. *senescent* in

P. anserina; *vegetative death* in *A. glaucus*. Although a search by electron microscopy failed to reveal any virus-like particles in *A. amstelodami* exhibiting the *vegetative death* phenotype (Caten, 1972), cellular degeneration was associated in *S. commune* with the presence of virus-like particles (Koltin and others, 1973). Mycoviruses are becoming increasingly well known in all groups of fungi and several occur in Penicillia and Aspergilli as well as at least three in the cultivated mushroom (*Agaricus bisporus*). (Review: Buck and Kempson-Jones, 1970; Hollings and Stone, 1971; Dielemon-von Zaayen, 1972.) Most of those studied to date have been double-stranded RNA. Although the mycoviruses of the mushroom were first discovered as a consequence of the degenerative changes induced in the sporophores and mycelium many, especially in the Penicillia, appear to be symptomless. However, it is clear that in *P. chrysogenum* the development of lytic plaques depends upon high titres of virus (Lemke and others, 1973). The identification of mycoviruses is either by electron microscopic investigation or by biochemical isolation. Nevertheless, it is significant that Lhoas (1971) has been able to demonstrate the transmission of a mycovirus in *P. stoloniferum* via hyphal anastomoses. Hyphal anastomoses, therefore, not only promote heterokaryon and heteroplasmon formation but also act as routes for invasive infection. Day (1970) and Caten (1971, 1972) have suggested that heterokaryon incompatibility is in fact a device to reduce the spread of infective agencies both genetic and viral. Despite these observations two facts militate against the view that all invasive agents are necessarily viruses in nature. Firstly, no extrachromosomal element has, in fact been isolated and shown to have the characteristic properties of a mycovirus. Secondly, cell organelles can certainly spread through hyphae. Most of the observations refer, of course, to nuclear migration for which there is good direct and indirect genetical evidence (e.g. Snider, 1963a, 1965). Nevertheless, data on *mi* heteroplasmons in *N. crassa* suggests that mitochondria too can migrate and this has been observed directly in *Rhizoctonia solani* (Sanford and Skoropad, 1955) while cytoplasmic streaming can, presumably, transport smaller organelles. In contrast it has been shown in *Coprinus lagopus* that the *acu-10* phenotype (exhibiting cytochrome deficiency and inability to use acetate as sole carbon source) is probably due to an element in the mitochondria and that variant mitochondria are incapable of migration during dikaryotization when nuclear migration occurs actively (Casselton and Condit, 1972). This bears out Niederpruem's (1969) claim that he could observe no evidence of cytoplasmic streaming or organelle movement during nuclear migration in *Schizophyllum commune*. It is possible, of course, that invasive elements of various kinds behave differently in different fungi: the dolipore septum of the Homobasidiomycetes may perhaps fill the same prophylactic role as heterokaryotic incompatibility in Asco- and Deuteromycetes.

The demonstration of a double-stranded RNA species in a fungal cell, even when it can form virus-like particles, is not evidence that it is, in fact, a virus, unless an infective cycle can be demonstrated. Such a situation occurs in the strains designated 'killers' and 'sensitives' in yeast by Bevan and his coworkers. Briefly, two species of ds-RNA have been detected, the larger P1 (mol. wt. $2 \cdot 5 \times 10^6$) occurs in most cells. It can be isolated from some strains as isometric particles 39 nm in diameter. A smaller RNA species P2 (mol. wt. $1 \cdot 4 \times 10^6$) occurs, in addition, in 'killer' cells only. The 'killer' phenotype is exhibited only in cells carrying the nuclear gene M plus P1 and P2 i.e. MP1P2, but not its allele m. A toxin is then produced which kills sensitives. These latter are genotypically either m or M, may carry P1 or not but never carry P2, i.e. mP1 $-$, m $-$ $-$, MP1 $-$, M $-$ $-$. Three sensitive strains were found to carry P1 and a second ds-RNA species of even lighter molecular weight than P2; two of them had been derived from killer strains by treatment with the mutagen 5-fluorouracil. It may well be that many fungi will be found to carry virus-like particles which are not necessarily infectious and which, like the P2-RNA, show integrated action with nuclear genes (Bevan and Somers, 1969; Somers and Bevan, 1969; Mitchell and others, 1973; Herring and Bevan, 1974).

The existence of apparently non-infectious, cytoplasmically located, persistent RNA molecules in yeast suggests that the occurrence of other persistent molecules in the cytoplasm should not be discounted as potential extrachromosomal elements.

Persistent molecules

Just as there is no unequivocal evidence for the role of viruses as extrachromosomal elements so there is none for any molecular species other than M-DNA. Only two examples suggest that an extrachromosomal element may conceivably be involved with a different molecule; one is the [*rho*] element, the other is the [*psi*] element. Zollinger and Woodward (1972) have suggested that transcription errors could account for the abnormal M-SP which occurs in *mi-1* strains. Such errors could arise in DNA and be copied by messenger RNA or they could arise in the messenger RNA itself. However, since m-RNA can clearly migrate from the nucleus and probably from mitochondria to the ribosomes (Küntzel, 1969), there seems to be no absolute objection to its migration from cell to cell. The evidence is, however, that messenger RNA, unlike double-stranded viral RNA, does not persist longer than a few hours at the most. The evidence relating to the [*psi*] element is more circumstantial since it depends upon an understanding of the action of the super-suppressors which is not yet defined completely. If, however, super-suppressors do act through altering transfer RNA, then presumably the [*psi*] element could act directly in the same way. Since, of course, both messenger and transfer RNAs are specified by DNA, both these extrachromosomal elements may ultimately be found to reside in extrachromosomal DNA but, in fungi, this would, to date, implicate the mitochondria alone.

It is clear that the study of fungal RNA is a desirable objective in relation to the physical basis of extrachromosomal elements. In this connexion it might be worth focusing attention on the most stable and widely distributed form of RNA in the fungal cell—ribosomal RNA. Not only does this occur in *N. crassa* in free ribosomes in the cytoplasm but also in the mitochondria (Küntzel and Noll, 1967; Küntzel, 1969; Noll, 1970). The size and composition of the latter differs strikingly from that of the former. It is disappointing, therefore, that mutant elements have not yet been obtained which can be located in ribosomes, cytoplasmic or mitochondria. A possibility, the occurrence of short pieces of persistent, transmissible, single-stranded RNA, the so-called viroids of higher plants, has not been investigated in fungi.

Extrachromosomal elements in fungal biology

It is clear that, provided appropriate techniques are employed, it is not difficult to detect and demonstrate extrachromosomal heredity in fungi.

Rather simple selective procedures, discussed in detail in Chapter 9, can bring about striking extrachromosomally determined changes in cultured fungi. The simple demonstration that strain characteristics can be altered in respect of characteristic features such as growth rate, habit, sporulation and sexual reproductive ability by selecting hyphal tips, asexual spores or sexual spores in both Ascomycetes and Deuteromycetes is of considerable significance for the maintenance of culture collections (Jinks, 1956). In a similar way, the fact that degenerative changes which have an extra-chromosomal basis, such as *vegetative death*, arise in stocks cultured for 2 or more years, is perhaps a pointer to explain the well-known degenerative changes which often arise in stock cultures, particularly if subculturing intervals are long. The justification for the common empirical practice in many culture collections of changing the medium from time to time, as well as utilizing as wide-ranging a subcultural sample as possible for inoculating, evidently lies in the fact that such practices compensate for different selective pressures whether on nuclear or cytoplasmic determinants.

A particular manifestation of this kind of change which may arise in culture is the 'dual phenomenon'. This was first described for several Deuteromycetes by Hansen (Hansen and Smith, 1935; Hansen, 1938). When propagated successively by single spores, mainly mycelial, *M*, abundant conidial, *C*, and intermediate types, *MC*, are produced. In the case of some pathogens the different types differed in their pathogenicity, e.g. that of *Phoma terrestris* on onion roots decreased in the order *M*, *MC*, *C*. Sometimes the differences were so great that cultures could be assigned to different form genera, e.g. *M* and *MC* types produce typical *P. terrestris* pycnidia, but pycnidia of the *C* type have beaks several times as long as the pycnidial diameter quite unlike true *Phoma* pycnidia. It has often been held that this is due to heterokaryotic segregation (e.g. Snyder, 1961) but in most examples rigorous proof of this has not been provided: Indeed, in fungi with uninucleate conidia it is difficult to see how the phenomenon can arise unless binucleate spores are not infrequent. Heteroplasmic assortment has, however, been demonstrated as the underlying cause of the dual phenomenon in two out of four freshly made, wild isolates of Penicillia of the Assymetrica group (Jinks, 1959a). If any case of this phenomenon arises, a heterokaryon test is the only sure arbiter of its basis.

It seems not improbable that many cases of 'saltation', especially 'ever-saltating strains', as they were described in the early literature, will turn out to be examples of the assortment of extrachromosomal elements; *red* in *A. nidulans* providing an admirable model for an 'ever-saltating' condition. Examples of such situations will be found in Brown (1926)—saltation in *Fusarium*; Horne and Das Gupta (1929)—an 'ever-saltating' strain of *Diaporthe*; and a useful review has been provided by Das Gupta (1935).

The role of extrachromosomal elements in pathogenicity should never be disregarded. This is obviously of particular importance in the rust fungi where plasmogamy is associated with very unequal parental contributions of cytoplasm and where, indeed, most examples of 'maternal' effects on pathogenicity have been detected. There is a quite different aspect, however, which needs examination—the 'training' of fungi either to resistant or toxic agents, or to hitherto resistant host varieties. So far as the former is concerned, it will suffice to observe that both nuclear and non-chromosomal resistance towards toxic agents have been demonstrated, e.g. copper-resistant yeasts—dominant mutants *CUP-1* etc. (Hawthorne and Mortimer, 1960), and mercury-resistant *A. glaucus*—an extrachromosomally inherited character (Jinks, 1959b). The latter type of adaptation is less well understood and little investigated. An example is the acquired ability of a strain of *Fusarium oxysporum pisi* Race 1, originating from a single uninucleate conidium, to infect the wilt-resistant pea *cv.* Alaska. This followed incubation in root exudate of the cultivar for 14 days and, it was claimed, involved no nuclear change. The change was apparently permanent and persisted through seven reisolations and reinoculations of *cv.* Alaska (Buxton, 1959). The critical genetical study of such situations is clearly of importance to pathology, especially in relation to the breeding of disease-resistant plants. It will be considered further in Chapters 11 and 12.

CHAPTER 7

Quantitative inheritance

Genetic control of metric characters

Most of the marker genes employed in genetical analysis described so far in this book have been ones which showed clear-cut, non-overlapping phenotypes, differing in kind rather than degree, and exhibiting unambiguous segregations in crosses, e.g. conidial versus aconidial, normal versus colonial growth, ability to ferment maltose or inability to ferment it. On the other hand, if numbers of strains of the same fungus are examined they will usually be found to differ in respect of several of their characters in degree rather than in kind, e.g. densely to sparsely conidial, fast, medium and slow growers, rapid or slow fermenters. These characters are quantitative or metric in nature, i.e. they are quantifiable in some way or other, and a population sampled for such a character exhibits a more-or-less continuous range between extreme values, the frequencies of the different steps in the range approximating to the Normal Curve. Moreover, strains differing in the mean value of expression of a metric character do not show clear-cut segregations when crossed but rather repeat the Normal Curve although its limits may now exceed those of the parental generation (Figure 7.1).

Analogous behaviour was detected in higher plants in the early years of this century where the F_1 in respect of a metric character was often found to be midway between the parental values but the F_2 showed a Normal Distribution with a wider range than either the F_1 or parental types. Bateson and Saunders (1902) suggested that the segregation of a large number of genes could account for traits that showed continuous variation rather than clear-cut segregations but it was the experimental work of Nilsson-Ehle (1910) and East (1910), coupled with the clear distinction by Johannsen (1909) between genotype and phenotype, that provided an adequate working hypothesis consistent with 'Mendelian' heredity. This was the multiple factor hypothesis. Stated briefly the hypothesis was that:

(i)   Metric characters were determined by numerous allelomorphic pairs of genes.

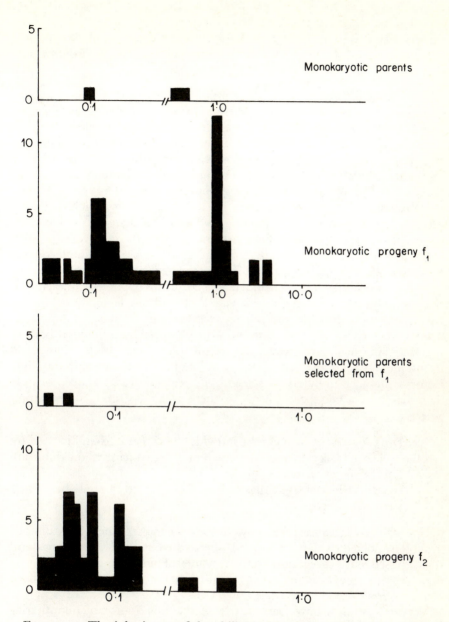

FIGURE 7.1 The inheritance of the ability to hydrolyse cellulose by monokaryons of *Polyporus betulinus*. Cellulase activity was estimated from the
ability to reduce carboxymethylcellulose (CMC) to glucose; scale in
arbitrary units. (Bell and Burnett, unpublished.)

(ii) These genes duplicated each other's effects, which were additive; dominance might be complete or incomplete.

(iii) The contribution of any individual allele to the net phenotype was likely to be so small that it could be obscured by environmentally caused variation.

The evidence for this hypothesis was indirect at the time. Nilsson-Ehle had described *polymeric genes* in wheat. These were four pairs of allelomorphs each of which duplicated the effects of the others precisely and whose effects were additive. The dominant allelomorph in each case determined red pigmentation in wheat kernels and the more dominant genes there were present in an individual, the more intense was the pigmentation. Johannsen had already shown that *Phaseolus* beans of similar genotypes exhibited a range of phenotypes because of slight differences in the environments in which they were growing but the beans, nevertheless, bred true. So there was evidence in favour of (ii) and (iii) above. East then showed that the range of expression of various metric characters in different crosses was predictable on the hypothesis that numerous genes of duplicate effect were segregating.

There is now direct experimental evidence for this hypothesis. If quantitative variation is determined in this way then multiple factors should show the usual genetic attributes of linkage, segregation, dominance and interaction. Mather and his associates have been especially influential both in carrying out such experiments and in devising appropriate biometrical techniques for their analysis. These techniques are necessary since multiple factors cannot be individually recognized, because of (iii) above, for only their net effects can be studied. So there is an operational distinction between

(a) *major* or *oligogenes* which can be individually recognized and located by their clear-cut segregation and which only infrequently show duplicate and additive action, and

(b) *polygenes*, as Mather (1941) has termed them, which can only be studied by their net effects.

Nevertheless, because they show linkage to major genes, for example, groups of polygenes can be located approximately on chromosomes and their subsequent segregation studied. Units of linked polygenes are the operationally recognizable units of quantitative inheritance; they have been termed *effective factors* by Mather (1949).

Very few studies have been made of the inheritance of quantitative characters in fungi. Many of the earlier ones, e.g. Hanna's (1926) study of the inheritance of spore size in *Coprinus sterquilinus* (and earlier work referred thereto), got little further than had Johannsen almost 20 years earlier. This was principally because methods had not been developed at that time to enable a more informative analysis to be made. Since then progress has been made in two ways. Firstly, the detailed studies initiated by Winge on the fermentation of sugars by yeasts have revealed a situation similar to that revealed by Nilsson-Ehle in wheat, namely the occurrence of major genes with duplicate and cumulative effects. Detailed evidence for such polymeric genes is considered in the following section. Secondly, methods developed for the analysis of quantitative variation in diploid organisms have been applied to haploids, initially by Chovnick and Fox (1953) who investigated the problem of estimating the number of loci (in fact, the number of effective factors) determining quantitative variation, and later by Jinks and his colleagues in an important series of papers. These methods were based largely on Mather's (1949) *Biometrical Genetics,* which developed the initial biometrical approach and which in their present form, including applications to haploid organisms, have recently been summarized in a new edition (Mather and Jinks, 1971). These methods will be outlined and illustrated after consideration has been given to data bearing on polymeric genes.

Before proceeding to these considerations, however, it should be recalled that fungi differ from higher plants in two features. Firstly, most of them are haploid most of the time; secondly, the effects of extrachromosomal characters are more apparent, and are more readily detected, in fungi. How far might such considerations affect the study of quantitative variation in fungi? As to the first point, it seems that haploidy is likely to cause differences of degree rather than those of kind, for example, dominance is not likely to play a part but linkage and gene interaction would still be expected to do so. In addition, in those fungi with regular diploid or dikaryotic phases, i.e. Oomycetes, yeasts, Basidiomycetes, the condition is either identical or analogous to the diploid condition of higher organisms and this could even be true of heterokaryons. Extrachromosomal elements are certainly likely to affect the expression of metric characters, indeed, in connexion with the *red* variant of *A. nidulans,* data have already been given to illustrate a range of phenotypes showing continuous variation (see pp. 105–108). The fungi, in fact, have advantages over most diploid organisms for the study of metric characters. Since the duration of their haploid and diploid (or dikaryotic) phases can often be indefinitely extended and alternated, the effects of polygenes in the two phases can be compared and, in appropriate circumstances, related to the effects of genetic elements in the cytoplasm. Moreover, the environments of fungi can be controlled more easily and precisely than those of large animals and plants and this provides an ideal opportunity to investigate the subtler interactions of environment and genotype which modify phenotypic expression. Thus fungi also provide material for such experiments.

Polymeric genes

The best examples of polymeric genes in fungi are those in yeasts which determine the ability to ferment sugars. One of the most thoroughly studied cases is that of the 4 allelomorphic pairs in *S. cerevisiae* which regulate maltose fermentation by the production or non-production of α-glucosidase. In diploid yeasts it can be shown that there are 4 dominant genes *MAL-1–*

MAL-4, any one of which enables the enzyme to be produced but in which the homozygous recessive, i.e. *mal-1 mal-1*:*mal-2 mal-2*:*mal-3 mal-3*:*mal- 4 mal- 4* is incapable of enzyme production. By appropriate crosses it proved possible to demonstrate that the 4 genes segregated independently and that each was sufficiently far from the centromere to allow 50% crossing-over to occur. It did not prove possible to establish any biochemical differences between the 4 genes although yeasts carrying *MAL-1* adapted to maltose fermentation more rapidly than those possessing *MAL-3* (Winge and Roberts, 1948, 1950). Subsequently, the same genes have been studied biochemically by Halvorson (Halvorson and others, 1963; Rudert and Halvorson, 1963). The α-glucosidases were extracted from haploid cells of differing genotypes each carrying one *MAL* allele, purified 16–35 fold and their properties studied. In each case only a single enzyme was produced and they were found to be indistinguishable by electrophoresis, on CM and DEAE-cellulose columns and in their pH optima, substrate specifications and reactions to a specific antiserum made to one of them. It was concluded that 'These findings are strongly suggestive, but not conclusive, that proteins having identical amino acid sequences and tertiary structures are involved.' It was shown, however, that different genes induced different levels of enzyme although in the presence of more than one dominant *MAL* allele the enzyme levels were very close to the expected cumulative values (Table 7.1).

TABLE 7.1 The effects of gene dosage on α-glucosidase synthesis in diploid *Saccharomyces cerevisiae*. (From Rudert and Halvorson, 1963.)

Genotypes	Differential rate of enzyme synthesis (units enzyme/mg protein)	
	Observed	Expected on a cumulative basis
MAL-1 MAL-1	2090	
MAL-2 MAL-2	300	
MAL-3 MAL-3	1000	
MAL-4 MAL-4	1100	
MAL-1/mal-1 MAL-2/mal-2	1280	1195
MAL-1/mal-1 MAL-2/mal-2 MAL-3/mal-3	1625	1695
MAL-1/mal-1 MAL-3/mal-3	1450	1545
MAL-2/mal-2 MAL-4/mal-4	1790	700
MAL-3/mal-3 MAL-4/mal-4	950	1050

Apart from these examples of polymeric genes, it has been suggested that super-suppressors in yeast also form a polymeric gene series. It must be admitted, however, that few other examples of polymeric genes in fungi have been clearly recognized although they have not been searched for systematically.

Nevertheless, the occurrence of polymeric genes analogous in their behaviour and actions to those in higher plants suggests that it is not unreasonable to suppose that the same spectrum of genic behaviour—oligo-genes:polymeric genes:polygenes—which occurs in higher organisms should also occur in fungi.

POLYGENES AND THE ANALYSIS OF QUANTITATIVE VARIATION

General considerations

The methods developed for studying polygenically determined variation in diploids can be extended to haploids. Hypothetical 'effective factors' have been known to exhibit properties of linkage with major genes, inter-action, dominance and segregation. Moreover, this has been studied both in haploid monokaryons and haploid dikaryons; existing methods are, pre-sumably, applicable to diploid fungi. Such studies raise a point of ter-minology which it is important to clarify since so much of the analyses are based on analogy with diploid higher plants. In these the nuclear cycle and terminology can be represented by:

$$2n \longrightarrow n \longrightarrow 2n \longrightarrow n \longrightarrow 2n \longrightarrow \text{etc.}$$
$$\text{P} \quad \text{gamete} \quad \text{F}_1 \quad \text{gamete} \quad \text{F}_2 \quad \quad \text{etc.}$$

but this is hardly applicable to haploids, haploid-dikaryotic or haploid–diploid forms. Accordingly, the convention of Dodge (1931) will be applied in an extended form viz.:

In haploids
$$n \longrightarrow 2n \longrightarrow n \longrightarrow 2n \longrightarrow n \longrightarrow$$
$$\text{p} \quad \text{zygote} \quad \text{f}_1 \quad \text{zygote} \quad \text{f}_2 \quad \text{etc.}$$

In haploid–dikaryotic forms

$$n \longrightarrow n+n \longrightarrow 2n \longrightarrow n \longrightarrow n+n \longrightarrow 2n \longrightarrow n \longrightarrow n+n \longrightarrow$$
$$\text{p} \quad \text{P} \quad \text{zygote} \quad \text{f}_1 \quad \quad \text{F}_1 \quad \text{zygote} \quad \text{f}_2 \quad \quad \text{F}_2 \quad \quad \text{etc.}$$

In haploid–diploid forms

$$n \longrightarrow 2n \longrightarrow n \longrightarrow 2n \longrightarrow n \longrightarrow 2n \longrightarrow$$
$$\text{p} \quad \text{P} \quad \text{f}_1 \quad \text{F}_1 \quad \text{f}_2 \quad \text{F}_2 \quad \text{etc.}$$

i.e. haploid phases will be distinguished from dikaryotic/diploid phases by lower v. upper case symbols.

Before considering specific problems a few general considerations are desirable. The techniques developed for biometrical analysis are based essentially upon the ability to ascertain the means, and to partition the variances amongst different components. For example, consider the data in Figure 7.1. Here the f_1 basidiospore progeny show a wide range of variation in their ability to break down carboxymethylcellulose. This variation may arise from several causes. Some may be due to environmental (error) variation, resulting from each assay being done individually and, therefore, being exposed to microenvironmental differences. Some variation may be due to genetic causes. Amongst the latter it is useful to know how many effective factors are segregating, how much is due to cumulative action (additive effects) and how much to gene interaction, i.e. epistasis. This information is valuable for predicting the outcome of future crosses as is a knowledge of the location of the effective factors. It is also extremely important to know whether similar phenotypes represent similar genotypes. This is a problem which can be resolved for oligogenes by appropriate crosses and a study of the resultant segregations. In the polygenic situation this is not so easily resolved, for *differently constituted genotypes can give rise to the same phenotype, in a fixed environment, because the phenotype is an expression of the net effects of the polygenes.* For example, if the polygenes which represent an increase in the metric character above the mean value are indicated by $+$ and those having the opposite effect by $-$, any of the genotypes set out below, amongst others could give rise to the same phenotype, assuming $n = 3$:

$$+ + + + \quad + + + + \quad - - - - \qquad + + - - \quad - - + + \quad + + + +$$
$$8 + : 4 - \qquad\qquad\qquad\qquad\qquad 8 + : 4 -$$

$$+ + + + \quad - - - - \quad + + + + \qquad - - + + \quad + + + + \quad - - + +$$
$$8 + : 4 - \qquad\qquad\qquad\qquad\qquad 8 + : 4 -$$

In each case there is an $8 + : 4 -$ balance of polygenes but they are achieved in different ways between the three chromosomes. If there are tight linkages between polygenes and marker oligogenes, for example, it is possible in principle to follow the segregation of the latter and compare it with that of the metric character, but this can rarely be achieved. A quite different technique can be employed, based upon the response of the genotype to the environment. Strains of similar phenotype can be compared with each other in a range of environments and their changed responses studied. It can be argued plausibly that those strains which exhibit the most comparable phenotypes in each environment over the range as a whole are likely to be those strains with the most similar genotypes. Such a test, by implication, also enables an investigation to be made of the effects of genotype–environment interactions on the phenotype.

The methods described subsequently, therefore, seek to discriminate and quantify the following components of the total variation exhibited:

(a) environmentally caused variation,
(b) additive genetic variation,
(c) variation due to gene interaction, allelic and non-allelic,
(d) variation due to interaction between the genotype and the environment,

while providing, in addition, information concerning:

(a) the numbers of effective factors,
(b) the average contribution of each effective factor to the character,
(c) the approximate location and behaviour of effective factors.

The methods applicable to haploids will be considered first.

Analysis of haploids

Components of means

(a) Suppose there are two haploid strains p_1 and p_2 which differ in respect of a metric trait and, specifically, in respect of a pair of alleles A and a of equal and additive effect.

The metric values of the phenotypes of the strains can be represented diagrammatically as

$$\underset{+d_a \quad -d_a}{\bar{p}_1 \quad\quad m \quad\quad \bar{p}_2}$$

where $\bar{p}_1$ and $\bar{p}_2$ represent the mean values of each strain, m is the midpoint between them, $+d_a$ and $-d_a$ represent the respective values of the contributions of the alleles A and a to total value.

So, by definition

$$m = \frac{\bar{p}_1 + \bar{p}_2}{2}$$

hence $\qquad \bar{p}_1 = m + d_a \quad \text{and} \quad \bar{p}_2 = m - d_a \qquad\qquad (7.1)$

so, rearranging $\qquad d_a = \dfrac{\bar{p}_1 - \bar{p}_2}{2}.$

(b) Consider a cross between these strains and their f_1 progeny. The progeny segregate equal numbers of $A:a$. So the mean value of f_1 is

$$\bar{f}_1 = m \qquad\qquad (7.2)$$

since the allelic contributions $+d_a$ and $-d_a$ cancel each other out.

(c) If the f_1 are crossed at random then for the reasons enunciated in (b) above $\bar{f}_2 = m$ also.

(d) If the f_1 are back-crossed to either parent the mean values of each back-cross $\bar{b}_1(p_1 \times f_1)$ and $\bar{b}_2(p_2 \times f_1)$ are:

$$\bar{b}_1 = \frac{\bar{p}_1 + \bar{f}_1}{2} = \frac{m + d_a + m}{2} = m + \tfrac{1}{2}d_a$$

and $\qquad \bar{b}_2 = \dfrac{\bar{p}_2 + \bar{f}_1}{2} = \dfrac{m - d_a + m}{2} = m - \tfrac{1}{2}d_a$

Note that $\bar{b}_1 - \bar{b}_2 = d_a$.

Estimating the numbers of effective factors

(a) Suppose that two haploid strains differ in respect of k alleles which are:

(i) unlinked,
(ii) are so distributed that all the $+$ alleles are in p_1 and all the $-$ alleles in p_2,

(iii) are of equal and additive effects,
(iv) show no interactions.

By analogy with the arguments already adduced

$$\bar{p}_1 = m + S(d) \qquad \bar{p}_2 = m - S(d)$$

where $d_a = d_b = d_c = \ldots = d_k$ and $S(d) = d_a + d_b + d_c + \ldots + d_k$ because of restriction (iii). The differences between the strains, therefore is

$$\begin{aligned}\bar{p}_1 - \bar{p}_2 &= m + S(d) - (m - S(d)) \\ &= 2S(d)\end{aligned}$$

because of restriction (ii). Since there are k alleles, however,

$$\bar{p}_1 - \bar{p}_2 = 2kd \qquad (7.3)$$

(b) In practice, the initial restrictions are unlikely to obtain, so that it is necessary to use estimates for k. Suppose there are k' alleles in one strain and $(k - k')$ alleles in the other. Then

$$\bar{p}_1 - \bar{p}_2 = 2(k - k')d$$

or

$$= 2rkd \quad \text{where} \quad r = \frac{k - k'}{k}$$

Here r represents the extent of the association of genes of like effects in each parent. However, the number of polygenes cannot be determined, nor normally, can their precise locations and associations be ascertained. Therefore their individual contributions to each d value cannot be estimated. So, rewriting

$$\bar{p}_1 - \bar{p}_2 = 2k_1\bar{d} \qquad (7.4)$$

where $k_1 =$ estimate of k effective factors, $\bar{d} =$ mean increment of any effective factor.

(c) To estimate k_1 and $\bar{d}$ the variance of the f_1 generation has to be introduced to provide another relationship of k_1. Recall that, by definition, Heritable component $= Sd^2$, for k_1 factors. The genetic variance V_G, i.e. total variance less environmental variance, is

$$V_G = k_1\bar{d}^2$$

and $\bar{p}_1 - \bar{p}_2 = 2k_1\bar{d}$ or, squaring, $(\bar{p}_1 - \bar{p}_2)^2 = 4k_1^2\bar{d}^2$, hence

$$k_1\bar{d}^2 = \frac{k_1(\bar{p}_1 - \bar{p}_2)^2}{4k_1{}^2} = \frac{(\bar{p}_1 - \bar{p}_2)^2}{4k_1}$$

so

$$k_1 = \frac{(\bar{p}_1 - \bar{p}_2)^2}{4k_1\bar{d}^2} = \frac{(\bar{p}_1 - \bar{p}_2)^2}{4V_G} \qquad (7.5)$$

and

$$\bar{d} = \frac{\bar{p}_1 - \bar{p}_2}{2k_1} \qquad (7.6)$$

(d) Because of the segregation and assortment of polygenes at meiosis, f_1 progeny may include individuals with more or fewer $+$ and $-$ alleles than their parents, i.e. they may approach the situation originally envisaged in restriction (a) (ii). Hence, by using the differences between the extreme values of f_1 individuals in place of $(\bar{p}_1 - \bar{p}_2)$ in equation (7.5), the estimate k_1 will be more likely to approach k.

(e) *To estimate the number of effective factors,* k_1, *involved in an* f_1 *cross, employ* *either*

$$k_1 = \frac{(\bar{p}_1 - \bar{p}_2)^2}{4V_G}$$

or

$$k_1 = \frac{(\text{Highest} - \text{lowest progeny value})^2}{4V_G}$$

Similarly, *to estimate the mean contribution,* $\bar{d}$, *of each effective factor, employ either*

$$\bar{d} = \frac{\bar{p}_1 - \bar{p}_2}{2k_1}$$

or

$$\bar{d} = \frac{(\text{highest} - \text{lowest progeny value})}{2k_1}$$

(Chovnick and Fox, 1953; Croft and Simchen, 1965; Mather and Jinks, 1971.)

Detecting and estimating non-allelic interactions

Both of these phenomena are likely to be encountered in practice. Consider two pairs of alleles A, a and B, b and let the interactions between them, i_{ab}, be defined as the product $(d_a \times d_b)$. Then, by analogy with equation (7.1) (p. 129) the mean value of each phenotype is

$$\bar{p}_{AB} = m + d_a + d_b + i_{ab}$$
$$\bar{p}_{Ab} = m + d_a - d_b - i_{ab}$$
$$\bar{p}_{aB} = m - d_a + d_b - i_{ab}$$
$$\bar{p}_{ab} = m - d_a - d_b + i_{ab}$$

Now there may be association of these genes because they are in the same genotype, i.e. AB, ab, or they may be dispersed in separate genotypes, i.e. Ab and aB. So crosses may be made of either $AB \times ab$ in the association phase, or in the dispersion phase $Ab \times aB$. In either case the f_1 will consist of the same four genotypes already adumbrated but, by analogy with equation (7.2) (p. 129) $\bar{f}_1 = m$.

However, from the difference between the parental means and the f_1 mean an estimate can be made of i_{ab}.

Consider parents p_{AB} and p_{ab}. Then

$$\frac{\bar{p}_{AB} + \bar{p}_{ab}}{2} = \frac{m + d_a + d_b + i_{ab} + m - d_a - d_b + i_{ab}}{2}$$

$$= m + i_{ab}$$

and similarly for parents p_{Ab} and p_{aB}.

In general, however, there will be more than two alleles, say k.

Then if $[i]$ = sum of all i's, $[d]$ = sum of all d's and

$$\frac{\bar{p}_1 + \bar{p}_2}{2} - \bar{f}_1 = m + [i] - m$$

or

$$[i] = \frac{\bar{p}_1 + \bar{p}_2}{2} - f_1 \qquad (7.7)$$

and, by analogy with equation (7.4)

$$[d] = \frac{\bar{p}_1 - \bar{p}_2}{2}. \qquad (7.8)$$

Therefore, *significant non-allelomorphic interactions (epistasis) are present when the difference between the parental means and the f_1 mean is, within the limits of error, positive or negative. If zero, all the genetic variation is additive.* (Jinks and others, 1966; Butcher, 1969; Mather and Jinks, 1971.)

QUANTITATIVE INHERITANCE: EXAMPLES

Example 7.1 Estimating the numbers of effective factors, k_1, segregating in a cross involving polygenes

The growth rate of monokaryons derived from various dikaryons of *Collybia velutipes* was studied and the following data obtained.

Parental dikaryon	Genetic variance	Difference between parental means	k_1	Difference between progeny extremes	k_1
11	44.29	21.55	2.6	35.00	6.9
12	95.49	40.25	4.2	48.50	6.2
13	164.73	38.94	2.3	66.50	6.7

Note that estimates of k_1 by the two methods differ. Presumably the assumptions underlying the estimate based on the parental means, i.e. equal effects of each factor, all + factors in one parent and all − in the other and no linkage, do not hold. The second estimate is, therefore, probably more reliable but could be biased upwards by the occurrence of rare mutants or exceptional environmental conditions affecting growth.

(Data of Croft and Simchen, 1965.)

Example 7.2   The estimation of non-allelic interactions

A number of crosses were made between different isolates of *Aspergillus nidulans* and the mean growth rate of the parents and their progeny measured. From these $[i]$ was computed as $\frac{1}{2}(p_1 + p_2) - \bar{f}_1$

Cross	$\frac{1}{2}(p_1 + p_2)$	$\bar{f}_1$	$[i]$
(a) Derivatives of same isolate	6·00	5·93	+0·07
(b) Derivatives of same isolate same locality	6·60	6·76	−0·16
(c) Derivatives of same isolate different locality	6·00	6·57	−0·57
(d) Different isolates showing heterokaryon incompatibility	5·30	4·30	+1·00

Note

(i)   Absence of non-allelic interaction indicates all genetic variation is additive.

(ii) Reduced vigour in progeny indicates positive non-allelic interactions but increased vigour indicates negative interactions.

Deduction: There is little non-allelic interaction in crosses between close relatives but this increases as the isolates crossed represent genetically more diverse isolates. Thus the most diverse cross results in reduced vigour of progeny due to high, positive non-allelic interactions.

(Data of Butcher, 1969.)

Detecting linkage between polygenes and a major gene

Consider a cross between two strains which differ both in respect of a metric character and a pair of allelomorphs A and a. Then the f_1 can be classified in respect of the two major genes A and a and, if there are n progeny, the expected values of the mean squares in an analysis of variance are:

Linkage component to A/a	$\sigma_E^2 + 2\sigma_R^2 + 2n\sigma_L^2,$
Remainder of genetic component	$\sigma_E^2 + 2\sigma_R^2,$
Non-heritable (error) component	$\sigma_E^2,$

where σ_L^2 estimates the variance of the linkage component, σ_R^2 that of the genetic remainder and σ_E^2 the error variance.

If linkage is present then it can be seen that the linkage component of the mean squares will exceed that due to other genetic effects by $2n\sigma_L^2$.

Hence *to test for linkage:*

(i)   *Compare mean squares expectation for linkage component with that for the remainder of the heritable variation by the standard Variance Ratio (F) test. A statistically significant difference indicates linkage.*

(ii) *Its magnitude is*

$$\frac{(\sigma_E^2 + 2\sigma_R^2 + 2n\sigma_L^2) - (\sigma_E^2 + 2\sigma_R^2)}{2n}.$$

(Simchen, 1966a; Connolly and Simchen, 1968.)

Estimating additive and interactive components

Equations (7.7) and (7.8) set out the quantities required to specify, the generation means in respect of $[d]$ and $[i]$. The derived components of variation in the total variance of an f_1 in a cross involving all pairs of polygenes can be formulated as

$$V_{[d]} = \tfrac{1}{4}V_{\bar{p}_1} + \tfrac{1}{4}V_{\bar{p}_2}$$

and

$$V_{[i]} = \tfrac{1}{4}V_{\bar{p}_1} + \tfrac{1}{4}V_{\bar{p}_2} + V_{f_1}.$$

If D and I represent $S(d)^2$ and $S(i)^2$ respectively then the total variation of an f_1 generation, in the absence of linkage, is

$$V_T = (D + I) + E$$

where $(D + I)$ represents the total heritable variation which cannot be partitioned save by obtaining further estimates from other crosses: the most convenient are back-crosses. Consider, for example, the two back-crosses

$$b_1: AB \times AB \quad \text{and} \quad b_2: ab \times AB$$
$$Ab \qquad\qquad\qquad Ab$$
$$aB \qquad\qquad\qquad aB$$
$$ab \qquad\qquad\qquad ab$$

derived from the f_1 considered on p. 132.

When AB represents the alleles which increase the manifestation of the character and are in association then their contributions to the variation of the back-cross b_1, allowing for interaction, are

$$\tfrac{1}{4}d_a^2 + \tfrac{1}{4}d_b^2 + \tfrac{3}{16}i_{ab}^2 + \tfrac{1}{4}(d_a + d_b)i_{ab}.$$

Similarly for b_2, the contribution to the variance is

$$\tfrac{1}{4}d_a^2 + \tfrac{1}{4}d_b^2 + \tfrac{3}{16}i_{ab}^2 - \tfrac{1}{4}(d_a + d_b)i_{ab}.$$

Summing over the two back-crosses, the contribution to the variance is

$$\tfrac{1}{2}d_a^2 + \tfrac{1}{2}d_b^2 + \tfrac{3}{8}i_{ab}^2.$$

A similar result is obtained for the equivalent back-crosses to parental strains with the two genes in the dispersed phase, i.e. Ab and aB.

If, now the general case is considered of all pairs of interacting genes,

$$\text{Heritable component of variance} = \tfrac{1}{2}D + \tfrac{3}{8}I.$$

Hence, *there are two estimates of* D *and* I *from which their magnitude can be obtained since:*

Heritable component of $f_1 = D + I$.
Summed heritable components of $b_1 + b_2 = \frac{1}{2}D + \frac{3}{8}I$.

(Mather and Jinks, 1971.)

Analysis of dikaryons and diploids

The methods outlined so far are applicable to fungi whose predominant phase is haploid and monokaryotic but quantitative traits are also exhibited by those fungi with a persistent dikaryotic or diploid phase. These will now be considered.

The methods developed by Mather and Jinks (1971) for the analysis of quantitative inheritance in diploid plants can be applied with little modification to fungi which possess a long-lived diploid or dikaryotic phase. In fact, all published work to date has dealt with dikaryotic Basidiomycetes. Here it has proved possible to partition the variation between genetic and environmental components, and further, to separate genetic variation into additive and interactive effects. However, unlike the situation already described for monokaryotic fungi, gene interaction may take two forms, arising from allelic dominance effects between the two haploid nuclei in each cell, from non-allelomorphic gene interaction, or indeed from both causes. In addition, both monokaryons and dikaryons can be grown simultaneously

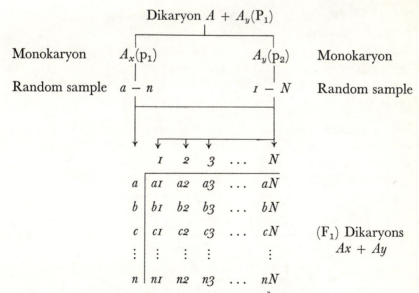

FIGURE 7.2 Design of an experiment for investigating the components of variation of a metric character in a unifactorial diaphoromictic fungus. (Based on Simchen and Jinks, 1964.)

and, since the latter are derived from the former, it is possible to compare the genetic control of a metric character in a monokaryon with that exercised on the same character in a dikaryon in the same environment. The methods are all based on a fundamental study by Simchen and Jinks (1964) of the inheritance of growth rate in dikaryotic *Schizophyllum commune*.

Components of variation in a dikaryon's progeny

The simplest situation is to consider the analysis of the dikaryotic progeny of a single parental dikaryon. A dikaryotic basidiomycete is induced to developed a basidiocarp and a sample of basidiospores from this is isolated as monokaryons. These can then be combined to form further dikaryons, allowing for restrictions imposed by the mating system of the species. A typical procedure was set out in Figure 7.2 (p. 135).

The scheme in Figure 7.2 is suitable for a unifactorial diaphoromictic fungus or a dimictic haploid–diploid species but can be readily adapted to a bifactorial form where there will be four kinds of parental monokaryon giving rise to two duplicate sets of F_1 dikaryons, i.e.

$$A_x A_y B_x B_y \longrightarrow \begin{cases} \left. \begin{matrix} A_x B_x \\ A_y B_y \end{matrix} \right\} \longrightarrow A_x A_y B_x B_y \\ \\ \left. \begin{matrix} A_x B_y \\ A_y B_x \end{matrix} \right\} \longrightarrow A_x A_y B_x B_y \end{cases}$$

| Parental dikaryon (P) | Mono-karyons (p) | Dikaryons (F_1) |

Note that in this type of experiment the monokaryons, although haploid, are regarded as the parents which give rise to dikaryotic F_1 progeny. The F_1 data can now be analysed into the components of variation, where

$$V_T = V_G + V_E = V_A + V_I + V_E \qquad (7.9)$$

and V_T = total variance, V_E = Environmental variance, V_G = genetic variance comprising: V_A = additive component and V_I = interactive component.

These can be computed from estimates of the mean squares

$$V_A = \sigma_{p_1}^2 + \sigma_{p_2}^2 \qquad (7.10)$$

$$\sigma_{p_1}^2 = \frac{(\sigma_E^2 + \sigma_I^2 + (a - n)\sigma_{p_1}^2) - (\sigma_E^2 + \sigma_I^2)}{(a - n)} \qquad (7.10a)$$

and

$$\sigma_{p_2}^2 = \frac{(\sigma_E^2 + \sigma_I^2 + (1 - N)\sigma_{p_2}^2) - (\sigma_E^2 + \sigma_I^2)}{(1 - N)} \qquad (7.10b)$$

then

$$V_I = \sigma_I^2 = (\sigma_E^2 - \sigma_I^2) - (\sigma_E^2) \qquad (7.11)$$

and

$$V_E = \sigma_E^2 \qquad (7.12)$$

This analysis does not distinguish between allelic and non-allelic inter-action. It is a difficult difference to compute save for certain special cases of non-allelic interaction such as complementary or duplicate gene action (Mather, 1967). However, a modification of the method developed by Jinks (1954) enables an assessment to be made of whether the genetic variance can be wholly accounted for by additive effects and those due to dominance. If it cannot, then non-allelic interaction is also present. In this method two further statistics are required, namely:

(a) the Variance V_r within each array, defined as all the crosses involving a monokaryotic 'parent', e.g. in Figure 7.2, an array would be $a1$, $a2$, $a3$, ..., aN, etc.,
(b) the covariances W_r' of members of each array with the means of those arrays whose common parents are the non-common parents of the members of arrays defined in (a).

Variances and covariances computed in this way for each array are then plotted as the regression of W_r' on V_r.

It can be shown (Simchen and Jinks, 1964) that if:

(i) the frequencies of the alleles at each locus which affect the same character are equal and
(ii) the genetic variation is wholly due to additivity and complete domi-nance then:
    (a) The regression ($W_r'V_r$) will be a straight line.
    (b) Its slope will be 0·5.
    (c) It will pass through the origin.

If, therefore, the ($W_r'V_r$) regression has these properties it is probable that the genetic variance is wholly due to dominance and additivity.

Departures from any of these conditions, notably, unequal gene-frequency, incomplete dominance, or non-allelic interactions of any kind, alter the slope or position of the regression line (Figure 7.3, p. 138). It is not, of course, possible with polygenes to determine whether the frequencies of alleles at the appropriate loci are equal although inequality is to be suspected if the slope of the regression line departs significantly from 0·5. The line may, however, have a slope of 0·5 but cut the axis above or below zero; this gives a measure of the average level of dominance over all loci involved. Thus, if the regression line crosses the W_r' axis above zero then dominance is incom-plete. Different dikaryons may differ in their relative numbers of dominant or recessive polygenes. Because of the relationship just discussed the position of the point for each array on the regression line gives an indication of the relative number of dominant to recessive polygenes in it. Those arrays whose common parents carried most dominant polygenes will plot nearest

138

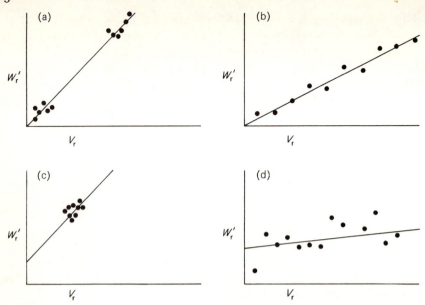

FIGURE 7.3 Graphs of the regression of W_r' on V_r to illustrate certain situations. (a) An example where dominance and additivity are the sole components involved in determining the character, i.e. straight line, slope 0·5, through the origin. The distribution of the data indicates that the character is largely determined by arrays of predominantly dominant genes (groups near the origin) and of recessive genes (grouped distally). (b) An example where non-allelic interaction may be involved (or possibly the gene frequencies are unequal i.e. slope differs significantly from 0·5 although through origin. (c) An example where dominance is incomplete although there is little interaction, i.e. slope of 0·5 but cutting W'' axis above origin. The cluster of the data suggests equality more or less between dominants and recessives. (d) An example where the character may show non-allelic interaction, unequal gene frequencies or low heritability, i.e. slope not significantly different from zero, data scattered.

to the origin, those arrays carrying least dominant polygenes will plot away from the origin towards the limiting point. An array in which dominants and recessives are more or less equal will assume an intermediate position (Figure 7.3). Note that this gives no information concerning which polygenes result in an increase in the expression of the character or *vice versa*. Dominant polygenes may increase or decrease a character's expression as may recessive polygenes.

So to analyse the components of variation involved in quantitative variation in dikaryotic fungi:

(i)　*Carry out an analysis of variance on F_1 data from a design such as that in Figure 7.2 to estimate V_A, V_I and V_E (equations 7.10–7.12).*

(ii)   *Calculate W'_r and V_r for each array and plot the regression of W'_r on V_r.*

(iii) *Dominance and additivity as sole components of V_1 and V_A are shown by the plot having a slope of $0 \cdot 5$ and passing through the origin.*

(iv)   *If the plot has a slope of $0 \cdot 5$ but does not pass through the origin, dominance is incomplete.*

(v)   *If the plot has a slope differing significantly from $0 \cdot 5$ or is non-linear, suspect non-allelic interaction.*

Components of variation for several parental dikaryons

The previous analysis has estimated the components, total genetic variation, additive and dominance effects and environmental variation and provided a test for non-allelic interaction for the dikaryotic progeny of a single parental dikaryon. If several parental dikaryons are available their monokaryotic progeny can be combined in dikaryons which represent crosses between the original dikaryotic parents. This enables the test for allelic versus non-allelic interaction to be carried out in a slightly different way; the analysis of variance to determine the initial components of variation is carried out as before.

The modification in the regression line test is this. The original dikaryons are regarded as the parental generation P_1, P_2 etc. but the monokaryons derived from them are considered as 'gametes' so that the derived dikaryons (F_1) can be regarded as 'families'. In such an arrangement each original parental dikaryon has to be regarded as a 'selfed' cross, since true selfing of a monokaryon is not possible. In this way variances V_r and covariances—now denoted by W_r—can be estimated. When plotted the regression of W_r on V_r is a straight line of slope $1 \cdot 0$, passing through the axis in the absence of non-allelic interactions. If the parents are highly heterozygous the regression line can be used to detect this, since the dikaryotic combinations of one monokaryotic parent can be compared with those of another in the W'_r/V_r graph while the points on the W_r/V_r graph should fall within the triangle formed by the W_r^2 parabola cutting the regression line (Simchen, 1965, 1966b).

These relationships can be used in another way. If W'_r/V_r graphs are obtained and found to meet the expectations of linearity and a slope of $0 \cdot 5$, then it may be assumed that the frequencies of the genes affecting the character are equal. Hence, if crosses are now made between monokaryons derived from different parents and the W'_r/V_r slope deviates from $0 \cdot 5$, non-allelic interaction between the different isolates is highly probable.

In general, unless heterozygosity is thought to be an important source of error in the interpretation of the regression line, the W'_r/V_r regression is satisfactory as a test of allelic versus non-allelic interaction.

Correlated characters in monokaryons and dikaryons

In fungi both with persistent monokaryotic and dikaryotic (or diploid) phases the same metric character may be exhibited in both phases, e.g. growth rate. It may be of value to know whether the same polygenes are affecting a trait expressed in both phases. An approach to this problem is to use the F_1 variance data from a design such as that set out in Figure 7.2 to partition the genetic variation V_G into two components, V_M, that part correlated with the mean expression of the trait in the p monokaryons, and V_{NM}, that which cannot so be allocated. It can be seen that this can be expressed in terms of a correlation coefficient r between the value of the trait in the dikaryons and the mid-values of their parental monokaryons; r can be derived from a regression analysis of the data.

Let the mid-values of the parental monokaryons be m and the values of the dikaryons d, then

$$\text{Sum of squares of regression of } d \text{ on } m = \frac{(S(d - \bar{d})\,(m - \bar{m}))^2}{S(m - \bar{m})^2}$$

and

$$r^2 = \frac{(S(d - \bar{d})(m - \bar{m}))^2}{S(m - \bar{m})^2 \times S(d - \bar{d})} \tag{7.13}$$

Hence,

$$V_M = r^2 V_G \tag{7.14}$$

and

$$V_{NM} = (1 - r^2) V_G. \tag{7.14a}$$

So, by this method of partitioning

$$V_T = V_M + V_{NM} + V_E. \tag{7.14b}$$

If characters are, in fact, found to be correlated then two interpretations are possible:

(a) the polygenes are showing pleiotropic action in affecting the expression of more than one character,

(b) the polygenes are associated either because particular frequencies have a high selective value, or because they are linked.

A decision between alternatives cannot be made save by further experimentation. So far as linkage is concerned the simplest type of test, if suitable major genes are available as markers, is to investigate the possibility of linkage between such markers and the effective factors determining the quantitative trait (cf. p. 133). The degree of expression will vary both with coupling and repulsion combinations and in different environments, where the different linked genes need not be expressed identically. However, *since pleiotropy is due to the same genes controlling two different traits, it is not improbable that correlations between the traits should be maintained over a range of environmental changes.* An examination of the expression of the traits in different environments may, therefore, enable a plausible decision to be made. In this respect the study of genotype–environment interactions is of importance.

Genotype–environment interactions in fungi

It has already been observed that the environments of fungi can be readily controlled with precision and that fungi can exist over a wide range

of environmental conditions. So the study of genotype–environment interactions is particularly applicable to work with fungi and is of value not only
because it throws light on the adaptative responses of fungi but also because
of the broader biological issues it can illuminate. As with other studies of
quantitative variation in fungi, the methods employed to study genotype–
environment interactions in them have been derived from studies with
higher diploid plants. As a consequence, the application of such methods to
diploid or dikaryotic fungi is fairly straightforward but modifications are
required to apply the methods to predominantly haploid monokaryotic
forms.

The experimental procedures are relatively simple. Appropriately
replicated strains representing different isolates, parents, or the results of
crosses, are grown simultaneously, so far as possible in the same range of
environments, and the same metric traits measured in each. The underlying
assumption in such studies is that individuals possessing similar genotypes
will respond similarly to a range of environments. Phenotypic differences
between individuals over a range of environments are, therefore, due to their
genotypes differing in either gene content or gene action.

Empirically, it had been found in other organisms that phenotypic
responses in different environments were often related in a linear manner to
some quantified assessment of those environments. This suggested that such
responses could be treated as regressions, but Freeman and Perkins (1971)
showed that this was not necessarily valid. They developed a modified
regression approach, however, which they believed to be valid. It required:

(a) that the environment should be assessed independently of the data
being analysed,
(b) that the phenotypic response be thought of as divisible into three
components:
  (i)   its mean value,
  (ii) its linear response to the range of environments,
  (iii) any other, i.e. its non-linear, response to the range of environments.

They expressed the relationship between the value of a phenotype to these
parameters by the regression equation:

$$\gamma_{ij} = \mu + d_i + \bar{\beta}z_j + \delta_j + \beta_{di}z_j + \delta_{dij} \qquad (7.15)$$

where

γ_{ij} = the mean phenotype of the ith genotype in the jth environment;
μ = grand mean over all genotypes and environments;
d_i = genetical contribution of the ith genotype;
$\bar{\beta}$ = the mean of all the regression coefficients of the genotypes for the
regression of γ_{ij} on $z_j(\beta_i$s$)$;
z_j = an independent assessment of the effect of the jth environment;
δ_j = deviation of the mean of all the genotypes in the jth environment
from the combined regression line;

β_{di} = difference between the regression coefficient of the ith genotype and the combined regression coefficient, i.e. $(\beta_i - \bar\beta)$. It is, therefore, the coefficient of regression of the genotype–environment interaction of the ith genotype in the jth environment on z_j;

δ_{dij} = the deviation of the ith genotype from its linear regression on z_j in the jth environment $- \delta_j$, i.e. $\delta_{ij} - \delta_j$.

Equation (7.15) has been simplified by Fripp and Caten (1971) to:

$$\gamma_{ij} = \mu + d_i + \beta_i z_j + \delta_{ij} \qquad (7.16)$$

here γ_{ij}, μ, d_i and z_j have the same meanings as above and

β_i = regression coefficient of the ith genotype for the regression of γ_{ij} on z_j;

δ_{ij} = deviation, in jth environment, of the ith genotype from its linear regression on z_j.

From this it can be seen that

the mean expression of the phenotype is represented by $\qquad \mu + d_i$,
the linear response to change in environment by $\qquad \beta_i$,
and the non-linear response to change in environment by $\qquad \sum_j \delta_{ij}(s - 2)$
for s environments.

Fripp (1972) and Fripp and Caten (1971, 1973) have investigated the application of this regression approach to the effects of different environments on the growth rates of dikaryons of *S. commune*. They have been able to show that, in practice, independent assessment of the environments can be achieved but that the results seem to differ only a little from an analysis based on non-independent assessments. The independent assessment can readily be obtained by the use of a single genotype, or a limited number, provided that it is as closely related to the genotypes under investigation as possible. If nothing is known about the trial genotypes then they should be effectively replicated. They have also observed that values of z_j are liable to error and have described ways in which this may be assessed. Finally, they have provided evidence that the tripartite division of the phenotypic response is of value since the three components represent the responses of independent sets of polygenes to environmental change and they are not, therefore, necessarily correlated. Their papers should be consulted for details.

Butcher and others (1972) have carried out a similar analysis on several isolates of *A. nidulans*—a haploid monokaryotic fungus. They used a slightly different form of the regression equation (equation 7.16), namely:

$$\gamma_{ij} = \mu + d_i + (\bar\beta - \beta_i)z_j + \delta_j + \delta_{ij} \qquad (7.17)$$

where all the symbols have the meanings already given in equations (7.15) and (7.16).

This analysis was of particular interest since the isolates had already been classified into groups on the basis of their heterokaryon-incompatibility reactions (cf. p. 88). It will be recalled that in considering this situation Jinks and others (1966) had supposed that strains capable of effectively forming heterokaryons must have something of the order of 8 loci in common. Thus the genotypes of strains of a heterokaryon-compatible group have at least part of their genotypes in common.

Butcher and others (1972) were able to show that most of the variation was related to differences between the heterokaryon-compatible groups and less between strains within a group, and the same difference obtained in relation to the genotype–environment interactions. Most of the variation attributed to these latter effects was associated with non-linear responses of the phenotypes to different environments. They were also able to show that in plots of β_i against d_i (that is, the linear response to the different environments against the additive genetical effects of each strain) the points for strains within a compatibility group tended to be clustered (Figure 7.4(a), p. 144). Exceptional strains showed up as points separated from their expected cluster. Although this analysis, the first of its kind, has given useful information on the nature of the genotype–environment interaction, it must be confessed that plots of strain means against environment means for each group, and of group means against environmental means, gave a rapid and reasonably clear indication of the similarities and differences in response to the different environments of isolates and groups as a whole (Figure 7.4(c)–(d)). The regression analysis, however, indicates the underlying causes.

POLYGENIC ACTION IN FUNGI

Although the number of investigations and the number of fungi investigated to date are exceedingly small there is already sufficient evidence to support the view that polygenes occur in fungi and show the basic properties common to all chromosomal genes. Moreover, a beginning has been made in studies of the ways in which quantitative characters can be assessed and utilized.

Firstly, some of the best evidence available from any organism to support the view that polygenes should show both segregation and linkage is available in *Neurospora crassa*. This is the work of Pateman and Lee on the genetic control of ascospore size and its inheritance (Pateman, 1955, 1959a, b; Pateman and Lee, 1960; Lee and Pateman, 1959, 1961; Lee, 1962). They showed that by selecting for large ascospores in an 8-spored strain of *N. crassa* for eight generations, they could gradually select for a larger ascospore which became associated with a 4-spored ascus in the sixteenth generation. Crosses of the sixteenth generation with the wild type gave 8-spored asci whose ascospores were significantly larger than those of the wild type (c. 1 μm longer). This generation back-crossed to the large ascospore selection showed segregation for ascospore size. Ordered tetrads

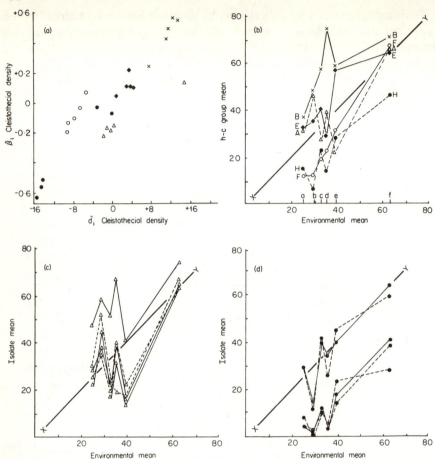

FIGURE 7.4 Comparisons of data for the cleistothecial density developed in culture by different heterokaryon-compatibility groups of *Aspergillus nidulans* in different environments. (a) Comparison of the additive genetical effects for growth rate and cleistothecial density (estimated by d_i—the mean deviation over all environments due to the ith isolate) illustrates how isolates from different heterokaryon-compatibility (h-c) groups tend to cluster, i.e. resemble each other in these attributes also. (b) Group means for each h-c group, A, B, E, F, H plotted against environmental means. (c) Means of h-c group A isolates plotted as environmental means : cleistothecial density. One isolate appears to show a somewhat different genotype–environmental response. (d) Similar data to C for h-i group H. There is more diversity of response here. Note the similarity in behaviour of the isolates of groups A in (a) and (c) and H in (a) and (d). Keys: △ group A; × group B; ◆ group E; ○ group F; ● group H. The line X—Y indicates the overall isolate means plotted against the environmental means. Environments shown in (b): a, 35°C, pH 5; b, 35°C, pH 9; c, 25°C, pH 5; d, 35°C, pH 7; e, 25°C, pH 9; f, 25°C, pH 7, all on Czapek–Dox medium. (From Butcher and others, (1972.)

were dissected and it was shown that significant differences in size occurred between ascospores representing the four initial meiotic products but *not* between the pairs of spores derived mitotically in the post-meiotic division. Moreover, since the mean value of the products of meiosis as exhibited in ascospore length differed in different asci derived from identical diploid (zygotic) nuclei, polygenic interaction had occurred. Finally, they were able to demonstrate that about one-sixth of the increased ascospore length was controlled by polygenes on chromosome I some 13 cM from an *albino* (*al-2*) locus and distal to it. Linkage with, rather than pleiotropy of, the *al-2* locus was confirmed by detecting recombination between the effective factor and the locus.

Linkage between effective factors determining growth rate and major genes has also been detected in monokaryons of *S. commune*, the *A* mating-type factor being the major gene locus (Simchen, 1966a; Connolly and Simchen, 1968). There is little evidence from *S. commune* for non-allelic interaction in respect of the control of growth rate in either monokaryons or dikaryons of common parentage, although dominance effects have been detected in the latter (Simchen and Jinks, 1964; Simchen, 1966b). In contrast, however, interaction effects are involved in the control of basidiocarp maturity and weight in the same species (Simchen, 1966b). Similar behaviour in respect of growth rate and fruiting characters, such as number, time to initiation of basidiocarps and their maturity, is shown by another agaric *Collybia velutipes* (Simchen, 1965). The ascomycete *Aspergillus nidulans* having only a monokaryotic phase showed no dominance phenomena and growth rate was largely due to additive effects in and between related strains. However, in crosses between strains non-allelic interactions occurred and were correlated with genetic differences (Jinks and others, 1966; Butcher, 1969; Examples 7.1 and 7.2, p. 132). This phenomenon was also of significance in dikaryons formed from monokaryons derived from parents of different geographical origin in *S. commune* (Simchen, 1967). It seems not improbable that this correlation between genetic differences and non-allelic interaction effects will prove to be a regular feature of such situations. This topic will be discussed further in relation to natural selection in Chapter 10.

Estimates of the number of effective factors determining monokaryotic growth rate are now available for two agaric species. The best estimates for *C. velutipes* range from 6·2 to 11·0 and 6+ is the mode; for *S. commune* the range is 5·6–9·9 and the mode about 7. These estimates are of particular interest when compared with the haploid chromosome numbers which are $n = 3$ and $n = 8$, respectively. It may be inferred that each chromosome probably carries at least one effective factor. In this connexion it is of interest also to recall that in *S. commune* it has been estimated that 50% or more of the variance in respect of growth rate in the dikaryons is due to polygenes other than those which control growth rate in the monokaryotic phase, or else to their acting in quite a different way (Simchen and Jinks, 1964).

Finally, the value of utilizing quantitative characters in an assessment of similarity and differences between genotypes and in assessing the wide range of variation in a species has been well attested in *S. commune* (Brasier, 1970; Freeman and Perkins, 1971; Fripp and Caten, 1971, 1973; Fripp, 1972). It is already clear that dikaryotic agarics are valuable material for testing techniques of analysis of genotype–environment interactions. In *S. commune*, the empirically derived but often employed procedure, of supposing that a linear regression relates the performance of genotypes to some measure of the environment, appears to be valid. However, it has been shown that the mean value, the linear response and non-linear response are all involved and can respond differently in different environments. An important consequence of this finding is that there can never be a simple rule of thumb decision by breeders to select for specific or general adaptation, since the response will depend upon the kinds of environments for which selection is being practised. Reliance upon a linear relationship between phenotypic expression and environmental assessment is an equally suspect activity as a general procedure. This is brought out by the finding that in *A. nidulans* most of the variation in genotype–environment interactions was not linearly related to the environment (Butcher and others, 1972). Despite the limited number of experiments carried out it already looks as though each species and every environment must in fact be considered as a unique situation and at least tested before any general relationship is assumed.

One other aspect of quantitative variation in fungi remains to be examined—variation due to extrachromosomal factors—this will now be considered.

Extrachromosomal Elements and Quantitative Variation

When investigating growth rate in *Collybia velutipes*, Croft and Simchen, (1965) isolated both basidiospores and oidia from dikaryons. The latter are normally uninucleate and, in effect, when taken from a single dikaryon represent samples of its cytoplasm. In respect of growth rate oidial clones of identical genotypes showed considerable variation and when this was partitioned by an analysis of variance an appreciable amount could be allocated to the within-clone variation, after the error component had been allowed for. More surprisingly, perhaps, when only two cycles of selection had been exercised for clones showing the highest and lowest growth rates, the means of the differing selections were significantly different. It is of course possible that the dikaryons from which the oidia were derived were heterokaryotic and that mitotic recombination occurred at oidium formation in a manner analogous to that at conidium formation in *Verticillium albo-atrum* (see p. 78, Chapter 4) so that the genotypes of oidia were different. But this explanation requires a considerable level of heterokaryosis and a high frequency of mitotic recombination, for neither of which is there any evidence in *C. velutipes*. An alternative explanation is that the range in growth rates is determined by extrachromosomal elements. There is considerable supporting evidence for this view from comparisons of growth rates of hyphal tips and conidia derived from the same mycelium and having identical genotypes in various Penicillia and Aspergilli (Jinks, 1956 and see p. 178, Chapter 9). Indeed, various workers have demonstrated that extra-chromosomally determined quantitative variation can account for a good deal of the variation shown by clones of Ascomycetes in growth rate, frequency of cleistothecium formation, sporulation intensity and pigmentation (Jinks, 1966). The basis of such quantitative variation is unknown but must, presumably, arise from differing combinations of extrachromosomal elements and their interaction both amongst themselves and with the nuclear genotype. The selective forces which result in different combinations of such elements may operate at an early stage in development. For example, when young monokaryotic mycelia of *S. commune* were allowed to develop from basidiospores it proved possible to divide them into duplicate portions as early as 21–22 hours after germination. In several cases when subcultured and grown on under identical conditions there were significant differences in growth rate between 'duplicates'. It was suggested that very young mycelia were particularly susceptible to induced environmental effects which then persisted (Connolly and Simchen, 1968). If true, this is a most remarkable phenomenon with far-reaching implications for physiological and other studies.

As yet there are no clear-cut rules, such as can be applied to polygenically determined quantitative characters, to predict or regulate extrachromosomally determined variation. There is little doubt that its investigation at the biochemical and physiological level might well prove rewarding in relation to an understanding of differentiation and senescence. Indeed, if there is truth in the current dogma that the genotype remains constant despite the manifold changes shown by metric characters during the life-span, they must reflect changes in the cytoplasm. It then becomes of importance, if such changes are to be understood or regulated, to discover what proportion are genically determined and which are determined by extrachromosomal factors.

It will be apparent that the extension of studies on quantitative variation in fungi is desirable for a variety of purposes. What is needed now is firstly, an extension of existing techniques to many more fungi and to different situations, e.g. extracellular enzyme production (cf. Figure 7.1), in order to see if the genetic principles are well grounded. Once that basis is assured the manipulation of quantitative inheritance to provide desirable phenotypes becomes feasible, while the analysis of natural situations appertaining to adaptation, selection and speciation, for example, should become more rewarding. Some of these possibilities will be discussed in later chapters.

Section 3

Population genetics of fungi

The genetic structure of populations of organisms is determined by many factors but four are of especial and universal importance. These are:

1. The generation of variation by mutation and recombination.
2. The genetic system which regulates recombination and defines the breeding population.
3. The consequences of selection upon gene frequencies in the breeding population.
4. The origin and extent of reproductive isolation in a population, or between populations.

These factors are intimately interrelated but for no species of fungus is there an adequate understanding of all of them. Indeed, for the fungi as a whole, in contrast to most other major groups of organisms, little is known of most of these factors. It is only possible, therefore, to present an outline of the subject matter at this time. Nevertheless, for an understanding of fungal speciation and evolution, of the adaptation of pathogenic forms, or for the manipulation of economically valuable species, extensions to existing knowledge of the population genetics of fungi are essential.

CHAPTER 8

The generation of variation

Mutation is the inception of heritable variation and the ultimate source of new variation but recombination, however effected, promotes the generation of many novel genotypes even with a relatively small number of genes. In the fungi, however, the importance of asexual reproduction and hence the role of mutation, as opposed to recombination, should not be underestimated. It will be recalled that one third to one quarter of all fungi, namely the Fungi Imperfecti, employ asexual reproduction exclusively. Moreover, in many sexually reproducing fungi, rare, sexual reproductive phases are separated by several, numerically large, successive asexual populations. It is, therefore, of interest to consider the relative speed and efficiency with which novel genotypes can be generated by mutation alone or by mutation followed by recombination.

MUTATION VERSUS RECOMBINATION

Recently, Bodmer (1970) has adapted arguments on this matter (developed originally for diploid organisms by Muller (1964) and others) to haploid prokaryotes. Many of his arguments are relevant to haploid eukaryotes such as fungi.

He investigated the number of generations, n, required to produce an individual of genotype AB from a population comprising aB and Ab individuals. This number by mutation alone is,

$$n_m = \log \frac{[1 + s/(2x_0\mu N)]}{\log (1 + s)} - 1 \qquad (8.1)$$

and by recombination is,

$$n_r = \frac{1}{2} \left[\frac{\log \left(1 + s \dfrac{(2 + s)}{Nrx_0^2}\right)}{\log (1 + s)} - 1 \right] \qquad (8.2)$$

where,

> N = size of the (breeding) population;
> x_0 = the genotype frequencies, which are equal, of aB and Ab;
> μ = the identical mutation rates of $a \to A$ and $b \to B$;
> s = the small selective advantage of aB and Ab over ab such that, either aB or $Ab:ab$ is as $1 + s:1$;
> r = the recombination frequency between a/A and b/B loci.

Adapting Bodmer's arguments for prokaryotes, the consequences of these relationships for fungi include:

(a) For a given initial genotype frequency of x_0, both n_m and n_r decrease as s and N increase.

(b) If $x_0 = 1/N$, as when but a single individual of each initial genotype occurs in the population, then n_m is independent of N but n_r increases with increasing population size.

(c) The value of n_m decreases with increase of μ while that of n_r increases with decrease of r.

So, provided that linkage is loose and $x_0 > 8 \mu$ (as will usually be the case) AB individuals will arise at least twice as fast in sexually reproducing populations as in those depending on mutation alone and the smaller the population, the more favourable will be the effect of sexual reproduction.

Note that these considerations apply only to the *initial* rate of production of new genotypes. The future fate of the AB genotype is independent of those of the single-mutant genotypes, aB and Ab, from which it arose, and not necessarily related to the occurrence of sexual or asexual reproduction in the population.

It seems probable that the behaviour of many fungi, namely those where large successive asexual populations are interspersed by a rare recombinant population, represents an effective biological compromise between the need to generate variability effectively and other requirements. Large asexual populations provide the most favourable conditions for the generation of new genotypes by mutation. They also favour dispersal and this exposes the population to a wider range of habitats and so may result in increased selection pressure, s (cf. (a) above). On the other hand, the rare recombinant populations may well be a good deal smaller than the asexual ones ((b), (c) above); hence the maximum effectiveness of recombination can be achieved. The patterns of life of many mildews and rust fungi seem to provide just such an alternation and so appear to be extremely well adapted to the continuous generation of variability. By contrast, large populations of asexually reproducing individuals can be equated in biological effectiveness with small recombinant populations in respect of the generation of new genotypes. So frequent and abundantly sporulating populations of Fungi Imperfecti may represent both a biological compensation for their lack of recombination processes and also a means of ensuring effective dispersal.

Migration is an important source of new variation in populations of fungi which possess effective dispersal mechanisms. Its scale may be considerable. For example, an annual dispersal of air-borne spores over several hundreds of kilometres per year is not uncommon and, exceptionally, after its initial invasion of Africa in 1949, *Puccinia polysora* spread eastwards across the continent at a rate of *c.* 1000 km a year (Gregory, 1973). An example of the consequences of such migration is the origin and consequent selection of the aggressive strain, Race 126–6,7 of *P. graminis tritici* in W. Australia in 1926. Waterhouse (1952) suggested that this biotype probably entered Australia as a wind-borne inoculum from Southern Africa, and then spread across the 3000 km wide central desert area to the Eastern Australian region.

McIntosh and others (1973a, b) have also summarized extremely convincing evidence for the mutational origin of new virulent genes in *P. graminis*

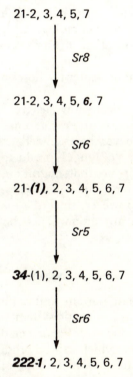

FIGURE 8.1 Successive EMS-induced mutants, in *Puccinia graminis tritici* Race 21, each acquiring a new virulence gene in opposition to the resistance genes *Sr8*, *Sr6* and *Sr5* in wheat. (From McIntosh and others, 1973a.)

tritici in Australia where the sexual phase is largely absent. They have been able to parallel changes which have occurred spontaneously in the field by mutagenic treatments. For example, mutation to virulence on hosts carrying the resistance gene $Sr5$ can be induced easily by EMS; it is also relatively frequent in the field. Successive changes in the virulence of a physiological race comparable with those which have occurred in the field have been achieved by recurrent treatments with EMS. Typically, in both cases, additional virulent mutant loci arise in successive single-step changes (Figure 8.1).

This is the most convincing data available that changes of importance for natural populations of fungi can be brought about by mutation. Many other claims have been made, and continue to be made, that new forms arise in nature by mutation, but few have been substantiated. This is because, in the past, the controls were inadequate to exclude parasexual phenomena following heterokaryosis. Nevertheless, there is evidence for the origin of new forms through mutation in *Helminthosporium sativum* (Christensen, 1940), *Phytophthora infestans* (Gallegly and Eichenmuller, 1959) and *Cladosporium fulvum* (Bailey, 1950). A clear implication of Flor's 'gene-for-gene' hypothesis is that new genes for virulence will arise by mutation from their avirulent alleles if new genes for resistance enter the host population (see Chapter 10, pp. 204).

The data discussed here indicate how meagre is the information concerning the relative roles of mutation and recombination in fungal populations. Clearly, mutation should never be excluded as an important source of variation in large populations of fungi reproducing asexually. The efficient dispersal of many fungi suggests that migration too should never be excluded as a source of apparently new genotypes, especially in the case of aerially dispersed species. However, since the majority of fungi do possess some form of recombination, attention will now be directed to the regulation of the processes involved.

GENETIC SYSTEMS

Those factors which regulate the rate and nature of recombination in an organism can be described as its genetic system.

Recombination, be it meiotic, mitotic, or heterokaryotic, determines the potential amount of variation which can be produced.

Mating systems determine the potential ability of different kinds of nuclei, or the mycelia which carry them, to fuse and undergo fertilization. Hence they control the balance between inbreeding and outbreeding. They also define the minimum size of the mating group.

The probable or actual amount of variation produced, or the probability that nuclei, within the limits of a mating system, will, in fact, fuse is affected by many other factors. Many of these will reflect the biology of the species but one of general importance is the size and genetic heterogeneity of the *effective breeding population.* This will usually be larger than the minimal size of a mating group but not as large as the actual population.

Recombination and its regulation

The amount of recombination in diploid nuclei will be determined partly by the segregation of whole chromosomes and partly by the frequency of chiasmata. As a measure of this Darlington (1937) suggested the use of a recombination index, namely, the sum of the haploid chromosome number plus the number of chiasmata per nucleus. Pontecorvo (1958) modified this index and also adapted it for application to fungi which exhibit parasexual recombination. His indices are, respectively:

$$I_{mei} = (C + n - 1)m$$

and

$$I_{mi} = [E + (n - 1)h]d$$

where

I_{mei} = recombination index with meiosis;
I_{mi} = recombination index with mitosis;
C = chiasmata/nucleus;
E = no. of exchanges;
m = proportion of nuclei in meiosis;
h = proportion of diploid nuclei undergoing haploidization;
n = number of pairs of chromosomes;
d = proportion of diploid nuclei/colony.

These indices provide an estimate of the potential variation a fungus could release. It is clear, however, that their calculation must frequently involve approximations which may not be valid, especially in the case of I_{mi}. Three values have been calculated for the fungi *A. nidulans* and *A. niger* (Pontecorvo, 1959; Lhoas, 1967). They are:

A. nidulans
$$I_{mei} = (13 + 8 - 1)10^{-3} = c.\ 2 \times 10^{-2}$$
$$I_{mi} = [10^{-2} + (8 - 1)10^{-3}]10^{-3} = c.\ 2 \times 10^{-5}$$
A. niger
$$I_{mi} = [2 \times 10^{-1} + (6 - 1)0\cdot8 \times 10^{-2}]10^{-2} = c.\ 2 \times 10^{-3}$$

Points which emerge from this very limited comparison are:

(i) that mitotic recombination is less effective, as assessed by the recombination index, than meiotic recombination,
(ii) in an exclusively asexual form, like *A. niger*, the frequency of mitotic recombination is greatly enhanced.

Unfortunately, the relevance of these measurements to natural populations of fungi is not known. For example, there is neither data concerning the frequency of sexual reproduction in *A. nidulans* nor any indication of how many cycles of asexual generations are interspersed between sexual events. It may be that a combination of mutational and mitotic-recombinational events during a succession of asexual phases generate variability as effectively as successive, but less frequent, sexual events. It may be inferred that mitotic recombination has been selected in the exclusively asexual *A. niger* but it is not really known how typical, in this respect, this species is amongst Fungi Imperfecti. It has not proved difficult to find mitotic recombinants in several species from this group e.g. *F. oxysporum* f. *pisi* (Buxton, 1956); *Cephalosporium mycophilum* (Tuveson and Coy, 1961); *F. oxysporum* f. *callistephi* (Hoffman, 1966) and *Verticillium albo-atrum* (Hastie, 1970) and this suggests that *A. niger* is probably characteristic of the Fungi Imperfecti. They may well possess the ability to generate variation with efficiencies approaching those of sexually reproducing organisms by a combination of parasexual recombination and very large asexually reproducing populations. The collection of new and improved field and laboratory data are necessary before a reliable assessment can be made.

It is clear, however, that the regulation of recombination frequency is both important and amenable to selection. There is a good deal of additional data, albeit entirely from two species—*N. crassa* and *S. commune*, to indicate that genetic regulation of recombination frequency occurs in fungi. In *N. crassa*, Teas (1947) was the first to demonstrate that substantial differences in centromere–locus distances occurred between strains of different origin. Barratt (1954) summarized data which had accumulated in many crosses. He showed that heterogeneity of recombination data occurred in almost every linkage group and was associated with the strains employed. Finally, Frost (1961) was able to show that such heterogeneity could be eliminated if crosses were grouped according to their wild-type ancestry and that the differences were genetic in origin. Subsequently, Catcheside and his colleagues (Catcheside and others, 1964; Jessop and Catcheside, 1965; Smith, 1965, 1966; Catcheside, 1966, 1968) showed that genes occurred which affected the frequency of recombination between alleles at certain other specific loci. Thus *rec-1*[+] reduced recombination between alleles at the *his-1* (histidine) locus by a factor of about 10 compared with *rec-1*. Another pair, *rec-3*[+] and *rec-3*, affected the *am* locus (amination deficient), while *rec-2*[+] and *rec-2* affected not only the frequency of recombination between alleles at the *his-5* locus but also that between the non-allelic linked loci *pyr-3* (pyridoxinless) and *leu-2* (leucineless). In each case *rec-2*[+] reduced the frequency by about half. The *rec-1, 2* and *3* loci are not allelic. Landner (1971) has also shown that selection for low or high recombination frequency in the genotype as a whole is effective in *N. crassa*. In general, alleles which promote high recombination are recessive. A similar situation has been found in *S. commune*. Here it was shown that the recombination frequencies between the two loci determining each mating-type factor, the *A* and *B* loci (see p. 163), were genetically determined; low recombination being dominant to high recombination (Simchen, 1967; Stamberg, 1968, 1969; Stamberg and Simchen, 1970). Genes determining low recombination were found more frequently in nature than those for high recombination. Simchen and Stamberg (1969) also showed that in some cases genetic control of recombination was exerted on limited regions only, and different controls affected different regions. In other cases control could apparently be exerted on more than one short region of the genome.

The particular relevance of these findings to the mating-type factors of *S. commune* will be considered later (p. 164); here their general significance will be considered. It is quite clear in these fungi, and probably in all fungi, that meiotic recombination frequency is under genetic control and susceptible to selection. It can also be affected by inbreeding (Stadler and Towe, 1962; Simchen and Connolly, 1968). In both *Neurospora* and *Schizophyllum* the dominant alleles, which reduce the frequency of recombination, are the most frequent in the populations sampled. Situations can be conceived, however, where the recessive alleles would be at an advantage. For example, in small or isolated populations an increase of such alleles, whether by mutation or selection, would enable a more variable population to be produced. Thus a new balance between constancy and variability could be achieved. Such a situation has not been observed. It could usefully be looked for in populations established after long-range dispersal or after population decimation, e.g. in the case of an obligate pathogen, by the reduction of available hosts.

The study of the regulation of recombination frequency in fungi is in its infancy. Most of the data come from three or four species and very little of it reflects the situation in natural populations. By contrast, far more is known about fungal mating systems and these will now be considered.

Mating systems

Self-fertilization (homomixis) is probably the commonest mode of sexual reproduction in the fungi as a whole but a variety of mating systems exist which promote outbreeding. These latter can be described as dimictic and diaphoromictic systems respectively. In the former, two, and only two, kinds of complementary nuclei control mating. Their inheritance can be formally represented as being unifactorial with two alleles. In diaphoromictic systems several kinds of nuclei occur and the necessary condition for a successful mating is that the two nuclei concerned must be complementary. In these systems inheritance is either unifactorial, bifactorial or trifactorial, usually with multiple specificities in each case. Many examples are also known where outbreeding species have had some form of inbreeding system secondarily imposed upon them; these may be termed homodimictic or homodiaphoromictic respectively.

Table 8.1 (p. 158) summarizes some data on these mating systems including their distribution amongst different groups of fungi.

TABLE 8.1 The mating system of fungi and their taxonomic distribution.

Taxonomic group	Amixis Amictic	Homomixis Homomictic	Dimictic 2 m.t.f. $(A/a; +/-; \male/\female)$	Homo-dimictic 2 m.t.f.	Heteromixis and homoheteromixis Diaphoromictic — unifactorial 10's of m.t.f. $(A_x A_y)$	bifactorial 100's of m.t.f. $(A_x A_y B_x B_y)$	trifactorial ? $(A_x A_y B_x / B_y C_x C_y)$	Homo-diaphoromictic 10's–100's of m.t.f.
Phycomycetes								
Hyphochytridiomycetes and Chytridiomycetes	?	+ Commonest?	Some show phenotypic determination	−	−	−	−	−
Oomycetes ($2n$)	+ ?	+	+ Complex, range from $\male$ to $\female$ and +/−	−	−	−	−	−
Zygomycetes	+	+	+ Always +/−	−	−	−	−	−
Ascomycetes								
Hemiascomycetes, Plectomycetes, Pyrenomycetes, Discomycetes	+ Both obligate and facultative	+ Very common	+ Usually A/a; one group $\male/\female$ (Laboulbeniales)	+	−	−	−	−
Basidiomycetes								
Uredinales	?	+ Less common	+ Always +/− common	?	−	−	−	−
Ustilaginales	?	+ Uncommon	+ Always +/−	Facultatively ?	−	One sp. 2A with many B factors	−	−
Tremellales	?	+ ?	?	?	?	One sp. 2A with many B factors	−	−
Agaricales	+	+	−	−	+ Less common	+ Very common	One sp. known	+
Aphyllophorales	? Commonest	++	−	−	One sp. known	++	?	++
Gasteromycetales	?	?	−	−		++	?	

Note: Nomenclature of mating system varies greatly; some equivalents include:
1. Mating systems = incompatibility systems = homogenic incompatibility.
2. Amixis = apomixis.
3. Homomixis = homothallism = haplomonoecious; heteromixis = heterothallism; homoheteromixis = secondary homothallism = Amphithallism.
4. Dimixis = haplodioecious = morphological and physiological 2-allelomorph heterothallism = bithallic = bipolar sexuality.
5. Diaphoromixis = multipolar or multiple allelomorphic physiological heterothallism; unifactorial diaphoromixis = bipolar heterothallism; bifactorial diaphoromixis = tetrapolar heterothallism.
(See Whitehouse, 1949; Burnett, 1956; Esser, 1971.)

Homomixis

Potentially self-fertile fungi are not always *necessarily* homozygous and a variety of situations and rather imprecise regulating systems can result in heterozygosity.

Firstly, the multinucleate hyphal organization of fungi, with or without the occurrence of hyphal fusion, results in the widespread occurrence of heterokaryosis. Consequently even two adjacent nuclei which fuse within the mycelium of a self-fertile species may be genetically dissimilar. This can be easily seen in a species such as *Sordaria fimicola*. A single, haploid ascospore will give rise to a mycelium bearing fertile perithecia, nuclear association occurring entirely internally. In a heterokaryotic mycelium, however, where some nuclei carry the gene for normal, black-coloured ascospores and others the mutant allele for grey ascospores, perithecia develop with wholly black-spored, wholly grey-spored or segregating asci. The expectation, if nuclear fusion is at random, is a ratio of $1:1:2$ perithecia, i.e. 50% homozygous and 50% heterozygous in respect of ascospore colour. Surprisingly, Olive (1954) has reported an excess of perithecia heterozygous for ascospore colour on such a heterokaryon. A similar situation has been described in the self-fertile *A. nidulans* (Hemmons and others, 1953). Sixteen out of 26 heterokaryons involving different genetic markers gave rise to more than 50% segregating asci and in 10 of these 16 all the perithecia were heterozygous. The basis of this phenomenon is unknown. There are two possible explanations. One is that there is preferential fusion of genetically dissimilar nuclei. The other is that fusion is at random but either heterokaryotic hyphae or heterozygous diploid nuclei are at an advantage compared with homokaryotic hyphae or homozygous fusion nuclei.

How widespread this phenomenon will prove to be in heterokaryotic, homomictic Ascomycetes and Basidiomycetes is yet to be determined. It is less probable in Phycomycetes since here heterokaryosis is probably less common, originating as it does through mutation alone. Nevertheless in the Phycomycetes, especially the lower aquatic forms, outcrossing may be promoted by environmentally induced sexual dimorphism. Emerson (1950) has noted that a single swarmer of *Blastocladiella variabilis* can give rise on germination either to 'male' or 'female' thalli. Although the gametes produced are isogamous there is reason to suppose that normally gametes derived from phenotypically different, though possibly genotypically identical, thalli fuse. Clearly the juxtaposition of thalli and the motility and dispersal of the gametes could result in chance outbreeding. Juxtaposition is, of course, a factor in promoting outcrossing in any self-fertile fungal species. In the Oomycetes the mating behaviour is not fully understood but a case such as that illustrated in Figure 8.2 (p. 160) indicates the possibilities for selfing or outcrossing in a population.

It is not known, apart from the cases of *S. fimicola* and *A. nidulans* already cited, whether heterozygosity is favoured in self-fertile fungi. There is now

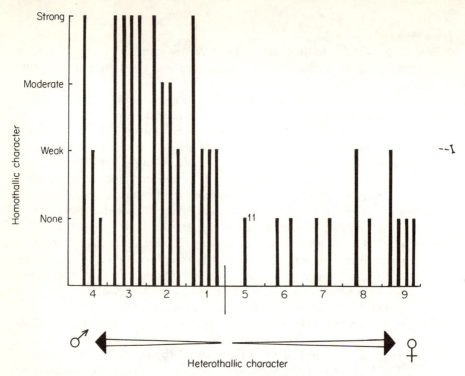

FIGURE 8.2 A diagrammatic representation of the mating system of *Pythium sylvaticum*. The behaviour of 36 isolates is shown. Each line represents the behaviour of a single isolate save that marked *11* which represents eleven. The so-called homothallic character is assessed by oospore production *in single culture*. The so-called heterothallic character is a measure of the intensity of sexual response, measured by the intensity of oogonium production, *in crosses*. (From Pratt and Green, 1973.)

much evidence for this in flowering plants (e.g. Allard and others, 1968) and it would not be surprising if a similar situation obtained in fungi.

In many Phycomycetes and some Ascomycetes, morphological differentiation either of gametes or of gametangia occurs although the species is homomictic. Loss mutations which could convert such species into potentially outbreeding, bisexual forms have been found. An example is the basically homomictic Ascomycete *Glomerella cingulata* (Wheeler and McGahen, 1952). Here the behaviour of natural isolates through mutation comes to resemble 'male', 'female', 'hermaphrodite', etc. but their expression is not clear-cut (Table 8.2). It is further complicated by the occurrence of diffusable agents which enhance or reduce perithecial formation (Markert, 1949).

Basically self-fertile fungi can evidently acquire additional genetical attributes which result, in various ways and to different degrees, in the promotion of heterozygosity. Indeed, it is a small step from some of these

TABLE 8.2 Morphology, behaviour and fertility of isolates of *Glomerella cingulata* which differ at their *A* and *B* loci. (Note: The *A* locus determines the production of protoperithecia, the *B* locus the degree of self-fertility. They are linked with a recombination frequency of 44·7%.) (Based on Wheeler and McGahen, 1952.)

Genotypes of cultures	Perithecial production							
	A+B+	*A2B+*	*A+B1*	*A+B2*	*A2B1*	*A2B2*	*A1B+*	*A1B1*
A+B+								
A2B+	*1*							
A+B1	2	X						
A+B2	2	X	X					
A2B1	2	2	*1*	*1*				
A2B2	2	2	*1*	X	0			
A1B+	1	*1*	2	2	2	2		
A1B1	3	3	X	X	X	X	0	
A1B2	3	3	X	X	X	X	0	0

Note:
1. Key: 3, very heavy ridge of perithecia; 2, heavy ridge of perithecia; 1, light but distinct ridge of perithecia; X, weak or irregular response; 0, no reaction.
2. Reactions in italics indicate *cross-fertility*.

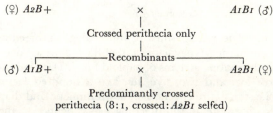

Reactions of two selected types and their recombinants

(♀) *A2B+*   ×   *A1B1* (♂)

Crossed perithecia only

──────── Recombinants ────────

(♂) *A1B+*   ×   *A2B1* (♀)

Predominantly crossed perithecia (8:1, crossed: *A2B1* selfed)

situations to regular outbreeding. El-Ani and Olive (1962) obtained two exceedingly closely linked mutants of *S. fimicola*, *a-3* and *st-59*, with partial and absolute requirements for arginine, respectively. Grown alone their asci or ascospores abort or are inviable. Grown together they complement each other and produce asci with viable ascospores. In 504 asci analysed, only the two parental genotypes segregated. Thus the mutants caused *S. fimicola* to resemble a dimictic fungus in its segregation of two kinds of 'compatible' partner.

Dimixis and diaphoromixis

These mating systems prevent the selfing of a single haploid but not that of mating between the products of a single zygote. On the other hand they increase enormously the opportunity for outcrossing (Mather, 1942). Figure 8.3 shows these relationships clearly. In dimictic systems, a haploid individual can never breed with more than 50% of the whole population but it can always mate with half the progeny of the same zygote. In diaphoromictic systems a haploid individual can mate with either half, one

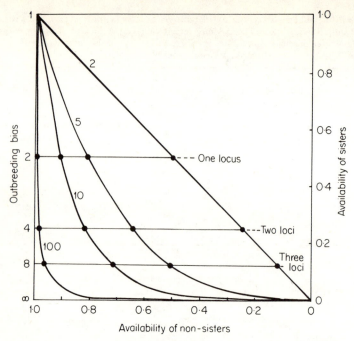

FIGURE 8.3 The relationships of inbreeding, outbreeding, the numbers of mating-type factors and of the loci which determine them, in the mating systems of haploid fungi. Data for one, two and three loci and two, five, ten and a hundred factors are shown, i.e. the figure covers the known dimictic and diaphoromictic mating systems. Outbreeding bias is defined as the ratio of the availability of non-sisters to the availability of sisters for a successful mating. Intermediate values between those shown can be obtained by projecting a line from the origin through the point of intersection corresponding with the numbers of loci and factors. The bias can be determined where this line cuts the outbreeding bias axis. (After Darlington and Mather, 1949.)

quarter, or one eighth of the progeny of the same zygote. But, because of the occurrence of multiple specificities, it can breed potentially with more than 50% of the population as a whole, the upper limit being proportional to the number of specificities in the population (Mather, 1942; Whitehouse, 1949).

The degree of potential outbreeding in unifactorial and bifactorial fungi in relation to the genetic structure of the mating-type factors has been considered by Koltin and others (1972) and by Stamberg and Koltin (1973).

The genetic structure of the mating-type factors has been investigated in very few diaphoromictic fungi, so that generalization is hardly possible. Raper's (1966) admirable book summarizes the work of his own pioneering group and that of others. In the most thoroughly studied bifactorial, diaphoromictic fungi each of the two unlinked factors, A and B, is determined by two linked loci, α and β, at each of which there is a number of alleles. In *S. commune*, for example, the number of alleles at each locus has been estimated as:

$$A\alpha \quad 9 \qquad B\alpha \quad 9$$
$$A\beta \quad 32 \qquad B\beta \quad 9$$

so that there are potentially 288 kinds of A factor and 81 kinds of B factor (Stamberg and Koltin, 1973). This is because every combination of an α-allele and a β-allele gives a unique specificity, e.g. $A\alpha_1\beta_2$ differs from $A\alpha_2\beta_1$, $A\alpha_1\beta_1$ or $A\alpha_2\beta_2$. This structure permits variation to occur in the number, symmetry and frequency of recombination of the alleles at each locus and, in respect of the first two of these, between factors.

Unifactorial diaphoromictic fungi have not been studied so extensively but there is no evidence yet for component loci of the mating-type factors.

TABLE 8.3 The effects of the number and symmetry of mating-type alleles on in- and outbreeding in fungi. (Data from Koltin, Stamberg and Lemke, 1972; Stamberg and Koltin, 1973.)

System	Component loci	Number and symmetry of alleles				Inbreeding potential %	Outbreeding potential %
		$A\alpha$	$A\beta$	$B\alpha$	$B\beta$		
Unifactorial 1 locus	$A\alpha$	2				50	50
		4				50	75
		10				50	90
		20				50	95
Unifactorial 2 loci	$A\alpha A\beta$	2	2			50–75	75
		5	5			50–75	96
		10	10			50–75	99
Bifactorial 2 linked loci	$A\alpha B\alpha$	2		2		25	25
		5		5		25	64
		10		10		25	81
		20		20		75	90
Bifactorial 2 factors each with two loci	$A\alpha A\beta B\alpha B\beta$	2	2	2	2	25–56	56
		3	2	3	2	25–56	69
		5	5	5	5	25–56	92
		8	2	8	2	25–56	88
		8	8	2	2	25–56	74
		14	2	2	2	25–56	72

Note:
1. When two loci specify a factor the inbreeding potential will vary with their frequency of recombination.
2. Inbreeding potential estimated from either $50\% + P_A - P_A^2$, where P_A is the frequency of recombination between $A\alpha$ and $A\beta$, in a unifactorial system, or $\frac{1}{4}(1 + 2P_A - 2P_A^2)(1 + 2P_B - 2P_B^2)$ where P_A and P_B are the frequencies of recombination between $A\alpha$ and $A\beta$, $B\alpha$ and $B\beta$, respectively, in a bifactorial system.
3. Outbreeding potential is estimated from $(n_A - 1)/n_A$ or $(n_{A\alpha}n_{A\beta} - 1)/n_{A\alpha}n_{A\beta}$ for a unifactorial system, or from $(n_A n_B - n_A - n_B + 1)/n_A n_B$ or $(n_{A\alpha}n_{A\beta}n_{B\alpha}n_{B\beta} - n_{A\alpha}n_{A\beta} - n_{B\alpha}n_{B\beta} + 1)/n_{A\alpha}n_{A\beta}n_{B\alpha}n_{B\beta}$ for a bifactorial system, where n is the number of alleles at each locus as indicated by the subscript.

In the only example of a trifactorial species, *Psathyrella coprobia*, no information is available concerning multiple specificities of the factors (Jurand and Kemp, 1973).

Table 8.3 shows how the number and symmetry of the alleles affects the in- and outbreeding potential of uni- and bifactorial, diaphoromictic fungi.

It can be seen in a unifactorial diaphoromictic fungus that the number of specificities determines the potential outbreeding and that if a two-locus structure occurred it would increase inbreeding beyond 50% without markedly increasing the outbreeding potential. By contrast, the two-locus structure in bifactorial diaphoromictic fungi makes possible a high outbreeding potential with a relatively small number of alleles, especially if these are symmetrically disposed amongst the four loci. A departure from symmetry either between two loci of one factor, or between factors, reduces the outbreeding potential, most markedly in the latter case. Since linkage between the loci may be genetically controlled, e.g. in *S. commune*, the inbreeding potential can also be regulated. In natural populations the dominant alleles for low recombination between the α and β loci are the most frequent, so that the inbreeding potential is about 25%, but where recessive alleles were the most frequent, this could rise to 56%, i.e. comparable with dimictic species.

Thus diaphoromixis provides an enormous potential for the control both of out- and inbreeding and hence the balance between them. It is not very clear how this is exercised in natural populations. Table 8.4 summarizes

TABLE 8.4 Out- and inbreeding potentials of different diaphoromictic fungi in natural populations of different sizes. (Data calculated from Anderson, unpublished; Roshal, 1950; Eggertson, 1953; Miles and others, 1966; Burnett, 1968; Puhalla, 1970; Silva, 1972.)

Organism	Sample size	Area or distance between limits of population	Outbreeding potential (%)	Inbreeding potential (%)
Bifactorial species				
Pleurotus ostreatus	35	16×32 km^2	96·2	25
	25	64 km^2	93·9	25
P. sapidus	7	64 km^2	82·5	25
	13	20 km^2	74·3	25
	18	5 km (1 tree)	66·7	25
	8	*c.* 0·1 m^2	65·0	25
Polyporus obtusus	24		96·9	25
Polystictus versicolor	37	*c.* 6·5 km	91·6	25
P. versicolor	12	50 cm^2 (1 stump)	80·5	25
Schizophyllum commune	12	15·8 ha	98·5	25
S. commune	15	11·7 ha	96·8	25
Ustilago maydis	23	1 small field	38·5	25
U. maydis	10	1 field	31·4	25
Unifactorial species				
Polyporus betulinus	100·5	Br. Isles	96·4	50
P. betulinus	33	3·8 ha	78·3	50
P. betulinus	29	0·76 ha	76·4	50

data which have been obtained from a limited number of samples of natural populations. In none of these cases was the genetic structure of the loci ascertained so that the balance between potential in- and outbreeding can only be assessed approximately. Nevertheless some differences can be demonstrated.

Perhaps the most notable feature is the high potential for outbreeding, rarely less than 75%, which exists in quite small natural populations save for the populations of *U. maydis*. Even in these two cases, however, the outbreeding potential is almost as high as is possible in such a mating system, i.e. two specificities at one locus and several at the other. It is twice as high as it would be if there were only two specificities at both the *A* and *B* factors, and the inbreeding potential is 25%, only half that of most other Ustilaginales, which possess a dimictic system.

It should be noted that although the role of mating-type factors in this discussion has been considered in relation to the mating system, this is not necessarily their only function. For example, the *B* locus in *U. maydis* determines the stability in growth and infectivity of the dikaryon for which both factors must differ, i.e. $A \neq B \neq$ (Rowell, 1954). In bifactorial diaphoromictic fungi the *A* factor regulates the formation of clamp connexions, while the *B* factor regulates nuclear migration as first shown by Fulton (1950) in *Cyathus stercoreus*. In the only trifactorial species studied, the *A* factor also regulates the formation of clamps but the *B* factor is concerned with the initiation of the basidiocarp primordium. The role of the *C* factor is unknown but heterozygosity at all three loci is necessary to ensure nuclear migration, the final development of the basidiocarp, fertilization and meiosis. Selection may, therefore, operate in nature on other, cellular functions of the alleles and so only indirectly come to regulate in- or outbreeding.

Other mechanisms also exist which affect the breeding potential of the mating systems; some are sporadic in their action, others impose a regulated restriction on outbreeding, converting the fungus to a homodimictic or homodiaphoromictic mating system.

Homodimixis and homodiaphoromixis

Either of these conditions can arise sporadically and accidentally. For example, sudden environmental changes or errors in cellular processes at or about the time of ascus formation may result in ascospores developing with nuclei of both mating types in them. This perpetuates the parental combination if a heterokaryon develops from such an ascospore. A similar error occurs in smut fungi and another not uncommon aberration is the presence of three-spored basidia in many agarics, e.g. *Coprinus* spp. (Lange, 1952); thus one basidiospore may carry both mating types. Even in four-spored basidial forms this may occur as the consequence of an additional mitosis after meiosis so that an 8-nucleate basidium develops. The basidiospores will, therefore, be homo- or heterokaryotic for mating type, depending on

166

which nuclei enter the basidiospores. In *Mycocalia denudata*, a microscopic Gasteromycete, precocious mitotic division in the basidium is determined by *Pd*; its allele *pd* defers this division to the basidiospore nucleus. In both cases nuclear migration is a random process so that from $(Pd + Pd)$ or $(Pd + pd)$ dikaryons the basidiospores are, on average, 50% homokaryotic and 50% heterokaryotic in respect of mating type but from $(pd + pd)$ dikaryons all the basidiospores are homokaryotic for mating type (Figure 8.4).

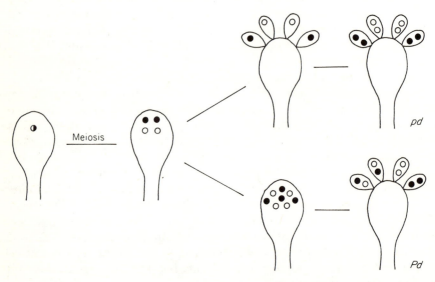

FIGURE 8.4 The action of the alleles *Pd/pd* on the post-meiotic mitosis in the basidium of *Mycocalia denudata*. Strains carrying *Pd* show a precocious mitotic division so that 8-nucleate basidia arise and the basidiospores are, on average, 50% homokaryotic and 50% heterokaryotic in respect of alleles represented by ● and ○. (From Burnett and Boulter, 1963.)

In the very limited sample studied the allele *Pd* occurred in 7/11 populations; clearly its frequency will impose some degree of inbreeding on the normal, unifactorial diaphoromictic system.

In other fungi mutation may result in genetically determined self-fertility. In dimictic *S. cerevisiae* the mutation of *a* to α or vice versa, resulting in self-fertility, has frequently been claimed (Lindegren and Lindegren, 1943, 1944; Ahmad, 1953, 1965). Winge and Roberts (1949) located a gene *D* in *S. chevalieri* (a species capable of hybridization with *S. cerevisiae*) which apparently results in diploidization. This is achieved through its action as a mutator gene of the mating-type allele during the first few divisions of haploid cells derived from a *D*-carrying ascospore. The compatible haploid cells so formed fuse and give rise to diploid clones, heterozygous, *a/α*, for mating type (Oeser, 1962; Hawthorne, 1963a, b). In other yeasts similar but usually more complex situations occur. For example in *S. oviformis* HO_α induces $\alpha \rightarrow a$ but the reverse mutation requires the additional presence of the modifier gene *HM* (Takano and Oshima, 1967). These specific mutator genes are not known in other dimictic species and may only be effective because the mating-type loci in yeasts are complex (Takahashi, 1964). The best evidence for this comes from the work of Leupold (1950, 1958) with *Schizosaccharomyces pombe* var. *liquefaciens*. Here there are two kinds of mating types designated h^+ and h^- but, in some crosses, 0·4% of the progeny gave rise to self-fertile progeny, designated h^{90}, almost 90% of which were capable of ascospore formation. Leupold suggested two possible explanations. One was that the *h* locus had undergone duplication and subsequent mutation to h^0. Thus an h^+ strain was genotypically either h^+h^+ or h^+h^0, an h^- strain h^-h^0, and the self-fertile strain, h^-h^+, could arise through recombination in h^+h^+/h^-h^0 diploids. The alternative explanation, perhaps more plausible when the situation in all yeasts is considered, was that there is a closely linked specific modifier s^h of h^- which converts h^- cells to self-fertility. Thus phenotypically h^+ cells are either h^+s or h^+s^h and h^- cells are h^-s. So, self-fertility could arise by recombination between the h^+s^h loci viz.:

$$\text{Parental diploid } \frac{h^+s^h}{h^-s} \longrightarrow \begin{array}{l} h^+s^h{:}h^+s{:}h^-s^h{:}h^-s \quad \text{Genotypes} \\ h^+ \;\; : h^+ \;\; : h^{90} \text{ self}{:}h^- \quad \text{Phenotypes} \\ \qquad\qquad\quad \text{fertile} \end{array}$$

Further studies with this fungus have not resolved the alternative explanations but have demonstrated the occurrence of many other loci which can interfere with the normal expression of the mating reaction. They may affect the onset or completion of either meiosis or sporulation but do not affect the action of the mating-type factors although phenotypically 'pseudo h^+' and 'pseudo h^-' cells could be produced (Bresch and others, 1968; Egel, 1973a, b).

Yeasts are perhaps exceptional amongst dimictic fungi since self-fertile mutants have not been detected elsewhere. A recent search for recombination of the mating-type locus of *N. crassa* revealed no evidence of complex structure (Newmeyer and others, 1973). The only other fungus where self-fertile mutants have been found is the diaphoromictic *S. commune* and here the commonest attribute of induced mutants is the loss of mating discrimination so that, for example an *A mut B mut* homokaryon '. . . aside from a tendency to form pseudoclamps instead of true clamp connections, is very similar to monosporous progeny of homothallic species in respect to general dikaryotic habit, nuclear pairing, conjugate division and fruiting' (Raper, 1966, p. 250). These mutants were only obtained after intensive laboratory investigations. In contrast, using data from natural populations Burnett and Evans (1966) inferred that either mating-type factors were very stable, or mutants had been selected against in the diaphoromictic *Marasmius oreades*, over tens, if not hundreds, of years. This estimate was possible because the age of the characteristic 'fairy rings' of this fungus could be estimated and the basidiocarps from any one ring always gave rise to the same mating-type factors. Evidence for mutation to self-fertility in natural populations of normally outbreeding fungi is, therefore, lacking.

However, in many species inbreeding has clearly been secondarily and indirectly imposed on outbreeding as a consequence of morphological or cytological changes. The genetical basis of these changes is rarely known. These fungi are the strictly homodimictic or homodiaphoromictic species.

The best known examples of homodimixis are the 4-spored Ascomycetes such as *N. tetrasperma* or *Podospora anserina*. In both of these species the haploid nuclei are so orientated in the ascus after the meiotic divisions and subsequent mitosis, that nuclei of complementary mating type are included in each ascospore (Figure 8.5a and b). The causes of this condition are unknown. The asci of *N. tetrasperma* are broader and shorter than those of the related 8-spored dimictic, *N. crassa* or *N. sitophila* and this could account for changed spindle orientations in the first compared with the last two (Dodge and others, 1950). Crosses between the 4- and 8-spored species suggested that the differences were explicable on a single-gene basis, the 8-spored condition being dominant to the 4-spored (Dodge, 1927). Other genes, e.g. *peak*, are known in *N. crassa* which affect the morphology and spindle orientation of the ascus (Pincheira and Srb, 1969). So, in *N. tetrasperma* homodimixis could have arisen as the secondary consequence of a mutation in a gene affecting ascus morphology.

In diaphoromictic fungi the commonest cause of homodiaphoromixis is the adoption of a 2-spored basidial habit (e.g. Sass, 1929; Smith, 1934; Lange, 1952). Both unifactorial and bifactorial homodiaphoromictic species occur, e.g. *A. bisporus* and *Coprinus sassi*, respectively (Raper and others, 1972; Kemp. 1974). So far as is known, in most of these species meiosis is normal, yet the majority of basidiospores are binucleate and heterokaryotic for mating type. For example, 94% of all the basidiospores of *Galera tenera* f. *bispora* give rise to dikaryons but 6% give rise to approximately equal numbers of monokaryons of different, complementary mating type, i.e. 94% ($A1 + A2$):3% $A1$:3% $A2$. So there must be some mechanism which ensures that nuclei of complementary mating types usually enter the same basidiospore. Evans (1959) has described different spindle orientations in the basidium of the cultivated mushroom, *Agaricus bisporus*, which would account for the characteristic production of about 80% of the spores capable of forming basidiocarps and 20% incapable. The former are postulated as being heterokaryotic for mating type, the latter homokaryotic (Figure 8.5c). It is now known that *A. bisporus* is indeed secondarily inbreeding, this having been imposed on a basically unifactorial, diaphoromictic system (Raper and others, 1972). However, although Evan's observations are suggestive they have not been confirmed either in *A. bisporus* or any other 2-spored Basidomycete. Other explanations are possible for the homokaryotic spores. For example, 4-spored basidia do arise irregularly in *A. bisporus* and here normal segregation could occur. Yet the problem of how

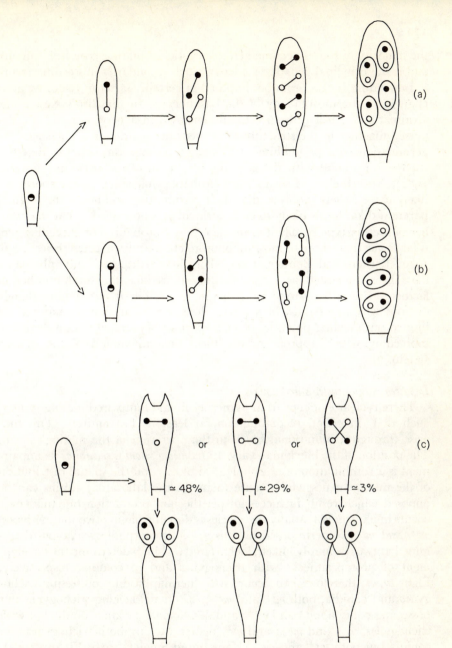

FIGURE 8.5 The origin of homodimictic and homodiaphoromictic conditions in *Neurospora, Podospora* and *Agaricus*. (a), (b) The origin of the homodimictic condition in *N. tetrasperma* and *P. anserina* respectively as a consequence of successive spindle orientations and spore delimitations in the ascus. Note first division segregation of mating-type factors (indicated by ○ and ●) in the former, and second division in the latter species. (c) In *Agaricus bisporus*, a unifactorial homodiaphoromictic fungus, a variety of spindle orientations can give rise to 80% basidiospores heterokaryotic for mating type. If both second-division spindles are parallel to the long axis then the spores are homokaryotic. (Based on Dodge, 1927; Franke, 1962; Evans, 1959.)

the majority of spores become heterokaryotic for mating type still remains, and is the principal genetic problem of such secondarily inbreeding forms.

Inbreeding is also probably imposed in certain Gasteromycetes by gross changes in the morphology of the basidiocarp. In the Bird's Nest Fungi (Nidulariales) the basidia are totally enclosed in a peridium and are dispersed together in the structure so formed, the peridiolum. The successful germination of a peridiolum may involve passage through the digestive tract of an animal with the possible dissolution of the peridium (Brodie, 1951). Nevertheless, it seems most probable that sister basidiospores are likely to be closely juxtaposed at their germination and hence the original parental dikaryon is likely to be re-formed. It is noteworthy that the number of mating-type specificities in these fungi, probably not exceeding ten, is far fewer than those of most diaphoromictic Basidiomycetes (Fries, 1936, 1940; Burnett and Boulter, 1963). Poor spore dispersal and inbreeding have been suggested as causes tending to reduce the total number of factors (Fries, 1940, 1943; Whitehouse, 1949) but it is not immediately obvious why this should be so. It is, however, clear that, in nature, the limited outbreeding potential of such species is associated with a degree of inbreeding which approaches the theoretical maximum of the mating system.

Impaired mating systems and amixis

There is clear evidence from numerous fungi of impaired mating systems such that, in effect, recombination, at least, is lost entirely. This may arise through a mutational block to the operation of the mating system. The position of the block may vary. In *Fusarium solani* f. *cucurbitae* its impairment is a combination of morphological blocks and the spatial distribution of the mutant genes which cause them. In the laboratory strains can be opposed which result in successful perithecium production but this never occurs in nature. Potentially the species is dimictic having two mating types, mt^+ and mt^-. Superimposed on this are four morphological types determined at two, probably linked, loci, M and C, which determine the development of protoperithecia with trichogynes and of conidia, respectively. Eight types, therefore, can occur: MC, 'hermaphrodite' possessing trichogynes and conidia, both mt^+ and mt^-; mC, 'male' lacking trichogynes but possessing conidia, mt^+ and mt^-; and Mc, 'female' lacking conidia but with trichogynes, mt^+ and mt^-; and mc, 'neuter' lacking both trichogynes and conidia but both mt^+ and mt^- (Hansen and Snyder, 1946; El-Ani, 1954). Clearly, differences of gene frequency at the M, C or mt loci could reduce fertility. Schippers and Snyder (1967) have found just such differences in Race 1 of the fungus. In California all isolates are $Mcmt^+$, in the Netherlands $MCmt^-$ and in Australia $MCmt^+$. Thus nowhere, in nature, do two morphologically and physiologically compatible types of this species occur together. Their reproduction is wholly asexual. In other cases the situation is simpler. In *Helminthosporium* the mating system is dimictic but two kinds of mutant impair its efficiency. In 79 isolates from 31 genera of grasses, 35 were mating type *A*, 33 mating type *a* but 11 would not react with either mating type (Nelson and Kline, 1964). This kind of sterility probably occurs in many fungi. It was early recorded in mucoraceous fungi by Blakeslee and others (1927) where it can account for 5–10% of all isolates. In *H.* (*Cochliobolus*) *carbonum* a second kind of impairment is due to a mutant gene *i*. If both compatible types carry this gene, i.e. *Ai* and *ai*, fertilization occurs but the development of the perithecia is inhibited. Out of 148 strains tested, 7/75 *A* and 9/73 *a* carried the mutant gene, so that 153/5475 possible crosses were rendered sterile.

Other species of fungi apparently possess all the morphological attributes of sexuality but meiosis and fertilization are wholly absent; they can be described as amictic. Amixis is difficult to detect but there is reasonably good evidence for its occurrence in the three main groups of fungi. *Szygites grandis*, although apparently a homomictic mucoraceous species, is said to lack meiosis and nuclear fusion both by Keene (1914) and Cutter (1942). *Podospora arizonensis* differs from many 8-spored species by possessing eight large, black and four small, hyaline ascospores. Early development is apparently normal and the asci develop from binucleate, penultimate cells of the ascogenous hyphae. But instead of fusing, the nuclei undergo two successive mitotic divisions to give the eight haploid ascospores. The determination of the two kinds of spore is obscure (Mainwaring and Wilson, 1968). Similar nuclear behaviour is shown by an ultraviolet-induced mutant, *sps-1*, of *Sordaria macrospora*; it behaves as a single-gene mutant (Heslot, 1958). No example is known with certainty from the Basidiomycetes but the curious genus *Nyctalis* includes two parasitic species in which the basidiospores are replaced by chlamydospores. Otherwise the basidiocarp is normal (Ingold, 1940).

In most cases it seems probable that the impaired mating system or amictic condition is derived from a preceding sexual condition. The selective situation where such loss mutations might be of advantage are not immediately obvious. Nevertheless apomixis, the analogous condition in flowering plants, is sporadic but widespread. In fungi the advantages of dormancy and spore dispersal from the reproductive structure are retained but, so far as is known, genetical versatility is wholly lost. There appears to be no correlation in fungi between amixis and polyploidy but there is little evidence. *P. arizonensis* with *n* = 2 has a lower chromosome number than other *Podospora* species where *n* = 7 is common (Heslot, 1958; Beckett and Wilson, 1968).

The adoption of a mating system

It is not known what factors determine the kind of mating system adopted by a fungus. Selection must be involved in some way, but whether it operates directly or indirectly is not known. Since, in many fungi, mating competence is the final expression of a series of events which affect the growth and behaviour of the mycelium and its cellular contents as a whole, selection could operate at various points in the growth sequence. It will be recalled that the B factors in *U. maydis* determine the stability and pathogenicity of the dikaryon as well as determining mating behaviour. Therefore, to look only at situations where a change in outbreeding efficiency, or the adoption of inbreeding, for example, might be advantageous, is to look at only one kind of adaptive response. Such a response could occur. Simchen (1967) has argued that selection for high recombination between the two loci of the A and B factors in *S. commune* would occur, 'only under special circumstances such as isolation or colonization of a new habitat'. There is no direct evidence concerning this suggestion. However, in such circumstances inbreeding is likely to occur. In fact, when inbreeding was imposed experimentally for 8–15 generations on seven lines derived from a single wild dikaryon of *S. commune*, increased recombination between $A\alpha$ and $A\beta$ occurred in 4 of them; there were no other changes of significance (Simchen and Connolly, 1968). This experiment provides some support for Simchen's suggestion and demonstrates unequivocally that an imposed breeding regime can result in direct adjustment of the mating system. There is, indeed, clear evidence of such changes from the very existence of homodimictic and homodiaphoromictic species where inbreeding has been imposed upon an originally outbreeding system.

The indirect adoption of a mating system could also occur as a consequence of a change elsewhere in the fungus. No natural examples are known but Pateman's selection experiment for increased ascospore size in *N. crassa* (see pp. 143–145) resulted eventually in a 4-spored homodimictic form comparable with the naturally occurring species *N. tetrasperma*. It will be recalled also that the shape of the ascus of *N. tetrasperma* differs from that of *N. crassa* so that the selection of mutants affecting ascus shape could impose a change upon the mating system. In general, changes in the mating system are most easily conceived when they result in the adoption of inbreeding in outbreeding forms; the opposite situation is more difficult to envisage. Nevertheless, this must presumably occur and in the case of *S. fimicola* described earlier (see p. 161) the selective agency behind the change was the acquisition of protrophy and fertility. Indirect selection may, therefore, play some role in the origin of new mating systems.

Several cases are known in fungi where apparently closely related taxa differ in their mating systems (Table 8.5). This suggests another way of investigating the origin and consequences of adopting a new mating system.

A careful examination of the comparative biology, ecology and distribution of such taxa might well indicate correlations between their mating systems and some aspect of the biological success or lack of success of the taxa. Such correlations have been detected in flowering plants (e.g. Baker, 1953; Grant, 1958) but at present little comparable data exist in fungi. The genus *Mycocalia* provides an example. *M. duriaeana* is homomictic, morphologically uniform and restricted to maritime or submaritime habitats.

M. denudata, in contrast, is more variable, occurs on a wider range of substrates and is more widely distributed. It is diaphoromictic. It is not possible from such a comparison to determine cause and effect and it should certainly not be concluded that outbreeding is the cause of the success of *M. denudata* compared with that of *M. duriaeana*. Indeed, the opposite conclusion could be drawn from other fungi. For example one of the most

TABLE 8.5 Some examples of related species which differ in their mating systems.

	Fungus species	Mating system
	Phycomycetes	
	Rhizopus nigricans	Dimictic
	Rhizopus sexualis	Homomictic
	Ascomycetes	
	Neurospora sitophila	Dimictic
	Neurospora crassa	
	Neurospora tetrasperma	Homodimictic
	Neurospora galapagosensis	Homomictic
	Basidiomycetes	
	Coprinus ephemerus	Diaphoromictic-unifactorial
	Coprinus congregatus	
Agaricales	*Coprinus sassi*	Homodiaphoromictic-bifactorial
	Coprinus bisporus	Homomictic
	Coprinus stellatus	
	Sistotrema brinkmanni I	Homomictic
Aphyllophorales	*Sistotrema brinkmanni* II	Diaphoromictic-unifactorial
	Sistotrema brinkmanni III	Diaphoromictic-bifactorial
Gasteromycetales	*Mycocalia duriaeana*	Homomictic
	Mycocalia denudata	Diaphoromictic-unifactorial and facultatively homodiaphoromictic

successful of the 65 coprophilous species of *Podospora* is *P. anserina*, a homodimictic form, while in the 50+ species comprising the genus *Aspergillus* there is but a single obligatory outbreeder, the dimictic *A. heterothallicus* (Raper and Fennell, 1965; Kwon and Raper, 1967a). Although the discovery of correlations between the mating systems of a fungus and its biology are worthwhile it is evident that they will not be easy to interpret. That some such correlation does exist in a very broad way is apparent from the distribution of mating systems within the fungi (see Table 8.1, p. 158). The most advanced fungi, in an evolutionary sense, have the most effective and flexible outbreeding systems known whereas the more primitive Phycomycetes are either homomictic or, at best, dimictic.

Mating systems are important elements of the recombination system but they are only a component of the genetic system upon which selection acts. Population size also affects the situation and this will now be considered.

POPULATION SIZE

It has been realized for many years that the frequency of selectively neutral alleles in a large, stable, breeding population given by the Hardy–Weinberg formula, describes their mean values. In any particular generation the actual values will fluctuate around the means by chance but, if the frequency of an allele reaches 100% or 0%, then it is either fixed permanently or lost altogether. Wright (1931, 1948) noted that such deviations from the mean were inversely correlated with the number of breeding individuals in the population, which he termed the population number and symbolized by N. This non-selective loss or fixation of an allele he termed *random drift*.

If, however, one allele has selective advantage (s) over another, then the loss of the former through random drift may be prevented or retarded, that of the latter hastened; but the precise consequences will be affected by the magnitude of the population number. Wright has given formulae which express the interrelationship between N and s as well as the mutation pressure, u, and gene flow by migration, m. In effect if,

$$N \geqslant 1/2s\text{—selection predominates}$$
but if
$$N \leqslant 1/4s\text{—drift predominates.}$$

Between N values of $1/4s$ and $1/2s$ both processes affect gene frequencies. Similar relationships apply in relation to u and m, i.e. drift is of significance when $N \geqslant 1/2u$ or $1/2m$. Figure 8.6 summarizes the situation in respect of population number and selective value.

It can be seen that for an allele possessing selective values of 0·01, or 0·001 respectively, the corresponding population numbers where drift will either be of predominating importance, or play some significant role are 0–25 and 0–250.

Very little is known of the selective value of particular alleles in any fungus and there are no reliable estimates of population numbers. No experiments have been reported on the relationship of population number to gene frequency in fungi; this area, therefore, awaits investigation. Pontecorvo and Gemmell (1944) have drawn attention to the use of mixed colonies of strains differing in pigmentation but of identical growth rates 'as models of the accidental multiplication or elimination of genes, and its dependence on the size of populations'. Here the sampling procedure operates on differently pigmented hyphal tips within a small circumferential arc where such colour mutants are supposedly selectively neutral. Accidents of growth may result in all being of one kind and hence a sector of a single colour arises. 'To grasp fully the analogy between the Sewall Wright effect in sexually reproducing organisms and our example, one has to substitute a hyphal tip for a gene, and a small arc of a colony for a small population' (N.B. Sewall Wright effect = random drift). Despite the example of this analogy, no quantitative data were published on the model.

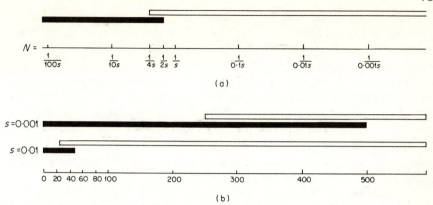

FIGURE 8.6 The interactions of drift, selection and population size. (a) The importance of drift and selection for different proportions of N, the effective population size and s, the coefficent of selection. Note a narrow overlap where both drift and selection may operate simultaneously. (b) The importance of drift and selection for different numbers of breeding individuals at two levels of selection. Note the very different ranges of overlap of the two processes.
Key: Solid black line—drift; open line—selection. (After Grant, 1963.)

In his original paper Wright (1931) drew attention to the fact that the population number could often be better replaced by the effective population number N_e. This concept is of particular value in these species where there exist either gross discrepancies in the numbers of kinds of parent or where populations fluctuate from generation to generation. In these circumstances, the value of N_e is closer to the minimal figure than the maximal. In the former case, N_e is twice the harmonic mean of the numbers of the two parental types; in the latter, it is the harmonic mean of the numbers in each generation (Wright, 1931, 1940).

Both these situations can occur in fungi and fluctuations in population size have often been recorded. This latter is manifested not only by differences in spore samples in successive generations but can also be seen, for example, in the differing numbers of basidiocarps of agarics produced in successive autumnal flushes (Parker-Rhodes, 1951; Hora, 1959). The causes of such fluctuations are numerous. The close relationships between environmental triggers of various kinds—temperature, photoperiod, presence of free amino acids or glucose—and growth and reproduction are important and are frequently highly localized causes. Amongst pathogens the introduction of new resistant host genotypes can drastically reduce population size of the pathogen. Whether, however, population size is ever reduced to a sufficiently low level for random drift to occur is an open matter. In this context it should be recalled that the effective dispersal mechanisms of many fungi must render an increase in effective population number through migration probable.

Before leaving the topic of population size, it may be observed that two features of fungal biology have an especially important bearing on effective population number. The first is the widespread occurrence of extensive, clonal reproduction by spores. So, for example, a barley plant may carry several hundred distinct mycelia of *Erysiphe graminis* of identical genotype and, indeed, all of the same mating type. There is quite a high probability in such circumstances that there will be big departures from equality of the ratio of $+ : -$ mating types so reducing N_e, e.g. for a 500:5 ratio of $+ : -$, $N_e \simeq 20$ as compared with $N_e \simeq 505$ for a ratio of 250:255. The second feature of importance is the widespread occurrence of anastomoses and, subsequently, nuclear migration. In contrast to the previous example, a single mycelium of *Polystictus versicolor* derived through numerous anastomoses can represent an N_e of as large as 18. This is the number of mono-karyons which could have been involved, via fusions, in making up this particular genetical mosaic (Burnett and Partington, 1957).

Examples such as these reinforce the requirement, stated earlier, that N or N_e must be determined by investigation and make clear that the effective population number rarely equals the actual number of individuals in the population. However, in most populations selection will be operative and frequently the dominant influence. Its consequences will be considered in the next two chapters.

CHAPTER 9

Selection: experimental

The operation of natural selection in fungal populations has often been inferred but rarely studied. Experimental studies on the mode of action of selection on fungi and its consequences have not often been attempted, despite a relatively early study by La Rue (1922). The reasons for this comparative neglect are twofold. Firstly, early studies, such as those of La Rue (1922) and Hanna (1926), were concerned with selection for change in spore dimensions to which they appeared not to be susceptible. This failure to achieve results probably deterred others. Such failure is all the more surprising since selection for ascospore size has, more recently, proved extremely effective (see p. 143 and p. 191). Secondly, and of more importance, it is now clear that selection can act in a variety of ways depending upon the biology of the fungus concerned. Until the different causes of variation in fungi were reasonably well understood, interpretation of the mode of operation of selection was hardly possible. It is now known that selection may result not only in change of gene frequencies, both of oligogenes and polygenes, as in all organisms, but also in frequencies of extra-chromosomal elements, of different nuclei in heterokaryons, or in any combination of these. It is evident, therefore, that great care is necessary in the design of selection experiments with fungi as well as in their interpretation. Extrapolation to the action and consequences of natural selection is an even more hazardous procedure.

In this account a number of illustrative examples from experimental situations will be described first and some examples of natural selection will be considered in the next chapter.

Selection in heteroplasmons

It seems likely, as stated in Chapter 6, that many changes observed in fungi are due to extrachromosomal elements. If such elements are numerous or capable of replication it should be possible to accumulate them by appropriate selective processes. This is, indeed, the case and Jinks (1957)

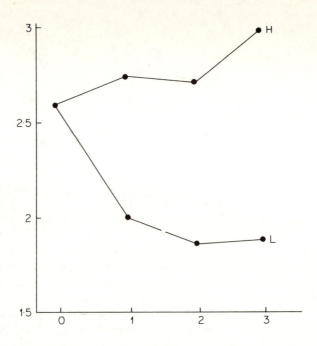

FIGURE 9.1 The consequences of selecting for high (H) and low (L) growth rate (mm/day) in three successive conidial transfers of a homokaryotic clone of *Aspergillus glaucus*. (From Jinks, 1957.)

has demonstrated significant differences in growth rate within a homokaryotic clone of *Aspergillus glaucus* selected for high and low rates (Figure 9.1).

Changes result from altered equilibria in the hyphal content of extra-chromosomal elements, the nature of which is unfortunately unknown. Similar striking changes have been demonstrated for perithecial density in *A. nidulans* (Croft in Jinks, 1966). The fact that almost any character may be affected by extrachromosomal elements indicates that whenever selection experiments are carried out an appropriate test should be made to determine whether or not the changes are chromosomal.

In some cases extrachromosomal mutants are at an advantage compared with chromosomal ones. Although Ephrussi and Hottinguer (1950) were able to show that cytoplasmic *petites* in *S. cerevisiae* arose directly by the mutagenic action of euflavine and were retained by recurrent mutation rather than by selection, they are at an advantage in cobalt sulphate solutions, being some three times more resistant than normal cells (Horn and Wilkie, 1966). Thus, on media containing 2 mM $CoSO_4$ the normal cells are inhibited and the mutants form colonies.

Heavy-metal and fungicide resistance often appears to have an extrachromosomal basis. It is often possible to select clones showing increasing tolerance to the toxic agent by transferring growing survivors to media with increased concentrations of the agent. In this way very high tolerances indeed can be achieved, e.g. to organic mercury compounds in *Pyrenophora avenae* (Greenaway and Cowan, 1970), but in general this is unusual and tolerance exhibited both to inorganic and organic fungicides is not high (Ashida, 1965; Georgopoulos and Zaracovitis, 1967). In the majority of cases, moreover, resistance is lost rapidly on transfer to a medium free of the toxicant. The genetic basis for such changes is not known but some kind of extrachromosomal selection seems most plausible.

Selection in heterokaryons

Dodge (1942) first demonstrated that a heterokaryon could have an advantage over its component homokaryons. He combined two genetically distinct, slow-growing strains of *Neurospora tetrasperma*, *dwarf-16* and *C-4*, and observed that the growth rate of the heterokaryon equalled that of the wild type. In general, unless a heterokaryon has some advantage over its component homokaryons, it is unlikely to be stable (see Chapter 5, pp. 92–94). To analyse changes in a heterokaryon an important requirement is the estimation of its nuclear constitution; this was first achieved by Beadle and Coonradt (1944) with *N. crassa*. Their technique consisted of isolating hyphal tips after selection and assaying their nuclear content either by seeing whether they complemented genetically marked homokaryons, or by allowing the tips to grow and conidiate and then carrying out the test with the conidia.

A number of other sampling techniques are available:

(a) In species with uninucleate conidia, the conidia can be plated and distinguished genetically by an appropriate technique. There are a number of snags to this simple procedure. It has to be assumed that nuclei are distributed at random in the mycelium and that they are not differentially incorporated into conidia. The only way in which these possibilities can be dealt with in practice is to sample at a number of points on the mycelium and to determine the frequencies of homo-karyotic versus heterokaryotic conidial heads or masses, where the fungus morphology enables this to be done, e.g. *Aspergillus fumigatus* (Raper and Fennell, 1953).

(b) In species with multinucleate conidia, the conidia can be classified as either homokaryotic or heterokaryotic. On the supposition that nuclei are distributed at random within the mycelium, formulae can be established for estimating their ratio by observations of the ratio of conidia homokaryotic for one or the other nuclear type, and those heterokaryotic for both types. In fact, although this is a plausible supposition, published data suggest that there is some tendency for like nuclei to occur together (e.g. Prout and others, 1953). This could reflect nuclear divisions at the tip of the conidiophore. Atwood and Mukai (1955) have developed formulae which allow for this effect. If the proportion of one type of nucleus in a heterokaryon is p and of the other $(1 - p)$ then,

$$p \simeq \frac{r(1 - r) + a(\bar{n} - 2r)}{\bar{n}(1 - r)} \tag{9.1}$$

and $$(1 - p) \simeq \frac{r(1 - r) + b(\bar{n} - 2r)}{\bar{n}(1 - r)} \tag{9.2}$$

where r = proportion of heterokaryotic conidia; a, b = proportion of conidia homokaryotic for type 1 and type 2 nuclei respectively; $\bar{n}$ = average number of nuclei per conidium. These formulae assume that $\bar{n}$ does not differ between homokaryotic and heterokaryotic conidia; this is so for *N. crassa* at least. It is probably applicable to all fungi with multinucleate spores, apart from those cases where initially uninucleate spores become multinucleate by subsequent divisions, e.g. *Fusarium oxysporum* (Buxton, 1960). Species like this are treated exactly as uninucleate spores.

(c) Maceration and plating-out of the fragments can be employed with sporing fungi and this is the only technique applicable to non-sporing species, e.g. most Basidiomycete dikaryons. The fragments are classified and a similar treatment applied to them as to multinucleate spores. A limitation imposed on this method is the minimal size of a viable mycelial fragment and its nuclear content; if this is relatively small, say not more than 10 nuclei, then the method is effective. It has been applied by Snider (1963b) to heterokaryons of *Schizophyllum commune*. The method can also be used to compare nuclear ratios in mycelia with ratios determined from spore plating techniques.

The application of these techniques to heterokaryons exposed to selection has demonstrated a number of interesting, and sometimes contradictory, situations. Most of the cases studied have been two-component heterokaryons including one or more auxotrophic components. How far such systems reflect the behaviour of naturally occurring systems is not known.

In many heterokaryons of *N. crassa* wild-type growth rates, with or without an essential metabolite in the medium, are associated with a wide range of nuclear ratios. For example, Beadle and Coonradt (1944) found wild-type growth rates with ratios ranging from 1:18 to 0·2:1. They interpreted these situations by supposing that the dominant genes present could support the appropriate function even if the number of nuclei of either type fell below 50% of the total. In the contrasting situation, where weakly dominant genes occurred, at least half of the total nuclei had to carry the dominant gene for it to be fully functional. In such cases it seems likely that normal growth can only be sustained in a narrower range of nuclear ratios (Pontecorvo, 1947).

Some cases are known where a wide range of nuclear ratios is associated with normal growth and only when a limiting value is reached does growth become proportional to the nuclear ratio, e.g. *pan-1*$^+$/*pan-1* heterokaryons already mentioned (pp. 93–94). Such observations could reflect limits beyond which intranuclear physiological compensation is no longer effective and this notion seems to fit those cases where growth is determined by maintaining a balance between the component nuclei, through their cyclical fluctuations around a mean nuclear ratio, e.g. the *pan*/*pan-m* and the *sfo pab*/*sfo pab*$^+$ heterokaryons of *N. crassa* (Emerson, 1948; Davis, 1952, 1960). The first example has already been mentioned. In the second case normal growth occurs only in the presence of sulphonamide and when the *sfo pab*:*sfo pab*$^+$ nuclear ratio is about 2:1. This represents a physiological balance point between inhibitory over-production of *p*-aminobenzoic acid by the *sfo pab*$^+$ nuclei and a growth-limiting under-production by the *sfo pab* nuclei. Alteration of exogenous sulphonamide, *p*-aminobenzoic acid or, indeed, threonine or methionine can all affect this balance. Emerson has provided evidence for nuclear fluctuations around the 2:1 ratio during heterokaryon growth, which tends to be erratic even in constant conditions.

In other cases the selective response of a heterokaryon appears to be reflected by a fixed change in the nuclear equilibrium but the nature of this change is not always predictable. One of the earliest cases described involved a leucine-requiring mutant *leu-1* and a reverted wild type from it *leu-1*$^+$ (Ryan and Lederberg, 1946; Ryan, 1947). The behaviour of the reverted wild type, the mutant and their heterokaryon in the absence, or in the presence, of a limiting amount (0·01588 mg/l) of leucine is illustrated in Figure 9.2 (p. 182).

It can be seen that on minimal medium the heterokaryon virtually equals the wild type in growth rate. This can be attributed to the presence in the heterokaryons of *leu-1*$^+$ nuclei. In the presence of limiting leucine, however, the heterokaryon grows no faster than the *leu-1* homokaryon. It might be concluded that in the former case *leu-1*$^+$ nuclei had been selected, in the latter that *leu-1* nuclei were at an advantage. Confirmation of this hypothesis was obtained by further labelling each nuclear type with a different albino mutant, so giving *leu-1*$^+$ *al-2* and *leu-1 al-1* nuclei. Because *al-1* and *al-2* nuclei complement each other the heterokaryon was pigmented but after a period of growth the hyphal margin became unpigmented and, on minimal medium, a sample of conidia were all *leu-1*$^+$ *al-2*, whereas on limiting-leucine medium a similar sample were all *leu-1 al-1*.

Ryan (1947) pointed out that these consequences of selection were not always completely predictable. Sometimes, on minimal medium, hetero-karyon growth would slow down or cease completely, while on limiting-leucine medium it might suddenly speed up, perhaps in a sector, and adopt a wild-type growth rate. He attributed these growth responses to fluctuations in the *leu-1*$^+$/*leu-1* nuclear ratios; but they were not measured.

This experiment illustrates clearly that selection can alter nuclear ratios

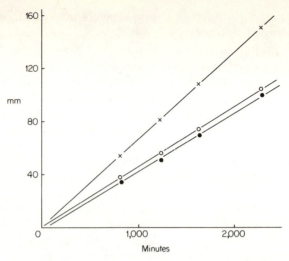

FIGURE 9.2 Growth rates of strains of *Neurospora crassa* auxotrophic for leucine (○ : *leu-1*), protrophic for leucine (× : *leu-1* $^+$) and the heterokaryon between them (●) on medium containing a limiting amount of leucine for growth. Note that the protrophic strain appears to be at an advantage. (From Ryan, 1947.)

in a heterokaryon and that its behaviour reflects that of the component nuclei. On the other hand, the selection of *leu-1* nuclei in the presence of limiting leucine suggests that physiological compensation is not always fully effective even by wild-type, *leu-1* $^+$, nuclei.

There is a good deal of evidence which suggests that nuclear ratios in *N. crassa*, once established, are remarkably invariant. Pittenger and Atwood (1956) made up heterokaryons containing different nuclear ratios ranging from 1:1 to 1:50 and 1:50 to 1:500. They employed a wide range of mutant alleles in two-member systems and allowed them to grow along 'Ryan' growth tubes (or 'race tubes'). Nuclear ratios were measured at the outset and again when the mycelium reached the end of the tube. In some cases tubes 20 mm in diameter and 1000 mm long were employed and conidial samples were taken every 100 mm. The mean difference between the initial and final nuclear ratios was about 2–3% on a range of media from initial ratios of 1:1 to 1:500. Normal growth occurred in ratios up to 1:50 but beyond this it was submaximal and often irregular. This fixity of behaviour contrasted with the diversity of twenty-five 1 mm hyphal apices, each containing about 100 nuclei, isolated from abnormal heterokaryons. Eight grew comparably with the parent heterokaryon, six failed to grow, and in eleven cases maximal growth was restored. Those which failed to grow were interpreted as being homokaryotic, those that grew normally as having nuclear ratios of 1:50 or less. Thus the gross behaviour of the mycelium did not reflect that of the nuclear ratios at its hyphal tips, where growth actually occurs. Physiological compensation occurs between tips and appears to be able to over-ride the apparent advantage that certain tips possess once isolated from the whole. This suggests that selection operates at the time of establishment of the heterokaryon rather than throughout its growth. Some confirmation was obtained from the long growth-tube experiments. Discrepancies sometimes occurred between the initial ratio and the first sample, after 100 mm, but from then onwards to 1000 mm the ratio usually remained constant. Pittenger and others (1955) supported this suggestion from experiments where they varied the ratio of the two components from 20:1 to 5000:1 employing three, two-member combinations (Table 9.1).

TABLE 9.1 Changes in the ratios of nuclei carrying different auxotrophic mutants during the growth of two-component heterokaryons of *Neurospora crassa*. (Data of Pittenger, Kimball and Atwood, 1955.)

Initial ratio of major:minor components	Final ratios of components		
	pan/al-1 *nic-2/al-2*	*pan/al-1* *arg-6*	*pan/al-1* *lys-3*
20:1	25:1	47:1	19:1
100:1	54:1	57:1	56:1
500:1	67:1	80:1	25:1
1000:1	52:1	24:1	30:1
5000:1	15:1	26:1	6:1

Note: The initial ratios were prepared by mixing appropriate numbers of conidia carrying the different nuclear types. The mycelium grew along a growth tube and was sampled between 100 and 1000 mm from the start.

It can be seen that the final ratios differed from the initial ratios and that the differences were greater as the initial ratios became more unequal. Moreover, each heterokaryon behaved differently. They found that the established nuclear ratio Y could be expressed as a function of the initial ratio X by the expression:

$$Y = (\lambda_b/\lambda_a)X^m$$

λ_b/λ_a reflect differential rates of nuclear division and m is a geometric restriction on the contribution of the majority type to the initial heterokaryon. In general, λ_b/λ_a only differed slightly from 1, which suggests that m played the dominant role in establishing differences. They envisaged the selective process as one in which, when a mass of conidia were plated together, they would virtually all become connected by anastomoses of their germ tubes. Thus heterokaryotic hyphal tips, so formed, and in a favourable position to grow out from the anastomosed mass, would have an enormous advantage over both homokaryotic tips and other heterokaryotic regions not able to exploit their growth potential. Thus the nuclear ratios of such tips would effectively determine the ratio of the growing mycelium. This initial phase of growth is, therefore, one of intense selection in which the establishment of a heterokaryon and spatial position play the principal roles.

In striking contrast to these results are the observations of Flentje and Stretton (1964) on natural heterokaryotic isolates of *Rhizoctonia solani* and the experiments of Jinks (1952) with heterokaryons of *Penicillium cyclopium*.

In *R. solani* hyphal-tip cultures give rise to colonies which resemble their parental colonies. However, it is possible to isolate short cells which have arisen by secondary septation of the multinucleate cells of the older hyphae. Those give rise to colonies which differ from the parental colony, from hyphal-tip isolates and from each other in a variety of characters. It was inferred that heterokaryons carried several kinds of nuclei and these observations suggest, in contrast to heterokaryons of *N. crassa*, that the nuclei are irregularly disposed save at the apices. A similar situation may occur in *Fusarium* heterokaryons where the apical cell is multinucleate, the remainder uninucleate (Buxton, 1954).

Penicillium cyclopium resembles *N. crassa* in some respects but has a much slower growth rate. Moreover, Jinks' experiments differed in a number of features from those using *N. crassa*. The heterokaryon employed had been obtained as a single point infection and its components were isolated subsequently as single spores. Both were protrophic but one, *4A*, grew faster on a minimal (Czapek) medium. The heterokaryon sectored spontaneously on minimal medium in plate-culture to give homokaryotic sectors of the *4A* type. If apple pulp was added to the medium the heterokaryon could be resynthesized. Nevertheless, there is some evidence that failure to establish the heterokaryon on solid minimal medium could have been due to competition, since, in liquid minimal medium, the heterokaryon was established. In liquid culture, spatial factors are less likely to be limiting and competition between mycelia during the initial phase of growth is less intense than on solid media. Nevertheless, Jinks prepared different mixtures of solid minimal medium and 10% apple pulp. The heterokaryon and each homokaryon were

TABLE 9.2 Changes in the nuclear ratio in heterokaryons of *Penicillium cyclopium* when grown on medium including different amounts of apple pulp and the comparative growth rates of the component homokaryons. (Data from Jinks, 1952.)

Medium		Number of colonies scored	Percentage of type *4A* nuclei in heterokaryon	Comparative growth rates of homokaryons *4A:4B* (*4B* = 1)
10% apple pulp (%)	minimal medium (%)			
100	0	324	8·55	0·47
80	20	381	7·75	0·53
60	40	167	11·11	0·54
40	60	830	12·66	0·67
20	80	831	13·51	1·00
0	100	632	51·81	1·56

Note: The growth rates were measured over the period 2–8 days while the nuclear ratios were sampled by plating-out conidia. On 100% minimal medium the heterokaryon showed a tendency to sector into its component homokaryons.

grown on these mixtures, their growth rates measured and the nuclear ratios of the heterokaryon determined. Typical data are included in Table 9.2.

From data such as these Jinks was able to conclude

'First, the ratio of the two kinds of nucleus found in the heterokaryon varies with the medium on which it is growing and is, indeed, characteristic of the medium. Second, the nuclear ratio on any medium is related to the comparative growth rates of the two homokaryons on that medium, in the sense that the proportion of the *4A* type nucleus in the heterokaryon rises as the comparative growth rate of homokaryon *4A* rises. In the extreme case when the *4A* homokaryon grows as well as the heterokaryon, the latter breaks down to give sectors pure for the *4A* nucleus.'

Other examples of similar behaviour are now known, e.g. *A. oryzeae* (Kiritani, 1959); *Fusarium oxysporum* f. *pisi* (Tuveson and Garber, 1961).

It appears that selection can change the nuclear equilibrium in an adaptive manner in some heterokaryons so that their behaviour matches their substrate. It is not clear how the behaviour is altered. A number of possibilities exist:

(i) Adaptive nuclear ratios could be maintained throughout growth by a combination of stabilizing and directional selection acting on the mycelium as a whole, or on the apical hyphae. Rates of nuclear division could be differentially affected by external metabolites and this would bring about the adaptive response especially if internal, physiological compensation was less effective than in *Neurospora*.

(ii) Beadle and Coonradt originally suggested that growth was determined by competition between hyphal apices. Those apices which, by chance, had the fastest growth rates would be selected. There would be no further alteration in these successful nuclear ratios after the initial selection since division is usually synchronized (Robinow in Dubos, 1947; Rees and Jinks, 1952; Flentje and others, 1963). A change in exogenous metabolite could disrupt this synchrony when selection would again operate between apices.

The second alternative is not so very dissimilar from that proposed by Pittenger and others (1955), described earlier. Nevertheless, there does seem to be a discrepancy between the great stability of nuclear ratios once established in *Neurospora* and the changes described in other experiments. The fact is that until there is clear proof of differential rates of nuclear division, of hyphal, or of apical selection, it is not possible to account for the results.

It is, however, significant that heterokaryons will often persist despite selection. Davis (1959) investigated the rate of loss of one nuclear component of a heterokaryon when specifically selected against. Successive transfers were made of conidia from a *pan-1 al-1/nic-2 al-2* heterokaryon to media supplemented either with pantothenate or nicotinic acid. The transfers were

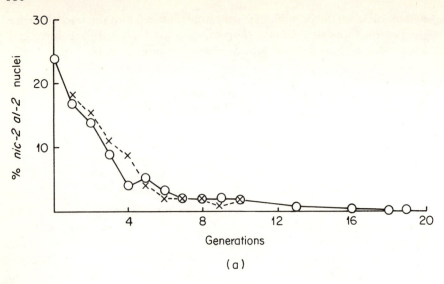

(a)

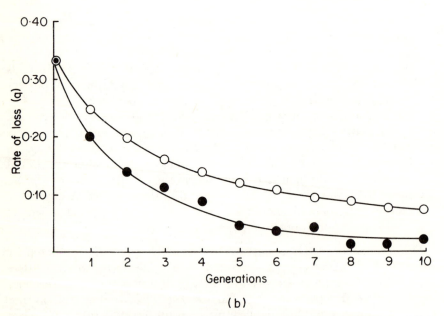

(b)

FIGURE 9.3 (a) The rate of loss of *nic-2 al-2* nuclei over successive conidial transfers of a *pan-1 al-1/nic-2 al-2* heterokaryon of *Neurospora crassa* on pantothenate medium, i.e. *nic-2 al-2* nuclei were selected against. Key: —o— observed and —×— expected rates of loss. (b) A comparison between the rates of loss of a 'lethal' nucleus over successive conidial transfers of a heterokaryon (such as shown in (a)) and the rate of loss of a lethal gene over successive generations of a randomly breeding, diploid population. Key: Rate of loss of nucleus, ● and of a lethal gene, o. (From Davis, 1959.)

between 100–600 conidia, a relatively small sample. On the pantothenate-supplemented medium the proportions of *nic-2 al-2* nuclei were reduced rapidly at first but persisted for as many as 20–30 transfers before finally being lost (Figure 9.3a). In Figure 9.3b the rate of loss of *nic-2 al-2* nuclei from a heterokaryon is compared with the rate of loss of a recessive lethal in a diploid organism. It can be seen that there is considerable similarity between these situations.

In the other selection experiment both nuclei persisted so that apparently homokaryotic conidia were as successful as heterokaryotic ones in initiating growth. Davis was able to show that the nuclear frequency could be estimated by adopting the formulae developed by Atwood and Mukai for nuclear ratios in a heterokaryon (see p. 180).

If it is assumed that homokaryotic conidia are selected against then the proportion of nuclei in the heterokaryotic conidia will be, from equations (9.1) and (9.2)

$$p = \frac{(1 - r) + a(\bar{n} - 2)}{\bar{n}(1 - r)} \tag{9.3a}$$

and
$$(1 - p) = \frac{(1 - r) + b(\bar{n} - 2)}{\bar{n}(1 - r)} \tag{9.3b}$$

If, however, homokaryotic conidia are not selected against completely and so contribute a proportion, x, to the nuclei of the next generation then,

$$p_{t+1} = \frac{r(1 - r) + a(\bar{n}r - 2r) + \bar{n}a(1 - r)x}{\bar{n}(1 - r)(r + ax)} \tag{9.4a}$$

Similarly if the other homokaryotic conidia contribute a proportion, y,

$$(1 - p)_{t+1} = \frac{r(1 - r) + b(\bar{n}r - 2r) + \bar{n}b(1 - r)y}{\bar{n}(1 - r)(r + by)} \tag{9.4b}$$

The generation t is that for which a, r and $\bar{n}$ are measured.

A combination of this persistence effect and the advantage of a hetero-karyon over its component homokaryons could well account for the accumulation of spontaneous mutants in stock cultures. An admirable example has been provided by M. B. Mitchell (quoted in Emerson, 1952). Many lysineless mutants of *N. crassa* are also inhibited by arginine but Mitchell noticed that some stocks had become less sensitive to such inhibition. They proved to be heterokaryons carrying both the unchanged mutant gene for the lysine requirement plus additional spontaneous mutant genes which reduced the conidial sensitivity to arginine inhibition. Even if a component confers no apparent advantage on a heterokaryon it can apparently persist. Atwood and Mukai (1953) demonstrated that of some 26 spontaneous mutants which arose in a synthetic heterokaryon, some 24 when isolated were lethal. It seems probable that many stocks of fungi may accumulate such lethal mutants and so become heterokaryotic for them. Selection experiments on such stocks have not been performed but they would clearly be of relevance to the maintenance of fungal stocks. The population structure (to be described in the next section) of *Penicillium chrysogenum* Q 167 may well have arisen and persisted in this way, although here there is no evidence for heterokaryosis. The effects of selection on multicomponent heterokaryons have not yet been described.

Since many natural heterokaryons probably arise through mutation the selection of mutants will now be considered.

Selection within a population

Mutant selection

Fungal geneticists seeking to employ new alleles both induce and select for appropriate mutants as described in Chapter 2. In such procedures the selective situation is so adjusted that the mutants possess an enormous advantage over other types. Such experiments, therefore, give little information on conditions where the selective value between two genotypes differs only slightly. Moreover, mutant selection is normally carried out in circumstances where novel genotypes can be readily expressed and detected. This contrasts with situations already described, where mutants arise during continued growth of a homokaryon and then persist undetected in the resulting heterokaryon unless some kind of segregation occurs or is induced. This kind of situation frequently arises in culture collections, during industrial fermentations and in nature. Algebraic treatments of the balance between mutation and selection are difficult to apply to fungi because the quantities involved, u, the forward mutation rate, v, the back mutation rate, s, the coefficient of selection and N, the number of generations are either exceedingly difficult to measure during mycelial growth, or are rather meaningless. For example, N is virtually meaningless in a hyphally propagated organism and in prolonged culture, changes in the medium are likely to result in differing selective pressures and hence s is not constant. However, it is useful at times to express the appropriate formulae in terms of numbers of subcultures rather than generations. Then, the expected frequency of mutant phenotypes p is:

$$p_N = u \frac{(1 - (1 - s)^N)}{s} \tag{9.5}$$

where the culture is free of mutants initially and where v is negligible compared with s.

If the mutant is selectively neutral and $s \to 0$, mutants will accumulate in a linear manner and the expression reduces to:

$$p_N = u \frac{1 - (1 - Ns)}{s} = uN \tag{9.6}$$

Moreover, the rate of increase of frequency of mutants per generation (Δ_p) will be:

$$\Delta_p = u - sp \qquad (9.7)$$

But, if p is large compared with u, because mutants have accumulated, then $u \to 0$ and so:

$$\Delta_p = -sp \quad \text{or} \quad s = \frac{\Delta_p}{p} \qquad (9.8)$$

In practice, u and s can be estimated by determining the mutation rates, firstly after a single subculture of a mutant-free strain and, secondly, from a reconstructed culture containing a known number of mutants after it has been subcultured (Example 9.1).

MUTATION RATES: EXAMPLE

Example 9.1 Estimation of frequency of mutants by reconstruction experiments

A population was sampled after one subculture from a mutant-free strain and found to have 10^{-5} mutants.

An artificial population was constructed containing 1 mutant in 10 and after one subculture had 1·08 mutants.

The spontaneous mutation rate $\mu = 10^{-5}$

Selection coefficient, $\quad s = \dfrac{0 \cdot 08 - 0 \cdot 1}{0 \cdot 1} = 0 \cdot 2$

Hence after N generations the expected frequency of mutations p_N is given by:

$$p_N = \frac{\mu(1 - (1 - s)^N)}{s}$$

So for 10 generations

$$p_{10} = 10^{-5} \frac{1 - (1 - 0 \cdot 2)^{10}}{0 \cdot 2} = 4 \cdot 5 \times 10^{-5}$$

or, for a large number of generations, i.e. when $(1 - 0 \cdot 2)^N \to 0$,

$$p_N = 10^{-5} \frac{1}{0 \cdot 2} = 5 \times 10^{-5}$$

(Data of Sermonti, 1969.)

An alternative procedure is to modify the expressions developed by Davis for application to heterokaryons (see Equations 9.3a, b, p. 187).

Populations comprising a constant mixture of genotypes in equilibrium may become established in various situations. One such has been described for the Wisconsin strain Q176 of *Penicillium chrysogenum* (Backus and Stauffer, 1955). When plated on honey-peptone agar the uninucleate conidia gave rise to five constant morphological types which differed in their radial growth rates and ability to sporulate. These probably represent distinct genotypes. Their equilibrium frequencies were:

Type	U	D	C	B	A
% Frequency	65	20	8	3	4

Now, in general, at equilibrium:

$$uq - vp - sp = 0 \quad \text{where } q = (1 - p) \qquad (9.9)$$

but, since v is exceedingly low,

$$uq = sp \qquad (9.10)$$

This formula can be applied to situations such as that described for Q176 and so selective values can be attributed to each mutant type, provided that u, and, if significant, v, is known for each mutant. The interpretation of such situations is difficult. Firstly, it is not clear whether in such situations the population really does consist of genotypically constant and distinct strains or whether, in effect, the types isolated reflect segregation of a multicomponent heterokaryon. Secondly, even if different strains occur, it is highly probable that the values of s reflect only averages. Studies with bacterial populations have shown how a variety of synergistic or antagonistic interactions can occur between different genotypes. These interactions will alter not only as the genotype frequencies alter but also at constant frequencies during the growth cycle or in different metabolic conditions (e.g. Braun, 1965). Similar behaviour is extremely probable in fungi and examples have been given earlier of interactions between wild type and mutant, or between mutants of this general type (see Chapter 2, p. 40). Nevertheless, so long as only one or two descriptive accounts of this phenomenon exist it is not possible to see whether worthwhile generalizations can be made. The importance of understanding such phenomena in connexion with 'strain degeneration', whether expressed morphologically in stock cultures or as reduced yield in continuous fermentations, can hardly be underestimated. Data relevant to such situations are likely to be obtained from competition experiments between genotypically different strains and from selection experiments for quantitative characters. These will now be considered.

Selection for quantitative characters

The most informative experiments with fungi have been concerned with directional selection for quantitative characters within a breeding population. However, the first reported selection experiment, for changed conidium length and number of appendages in the imperfect fungus *Pestalozzia guepinii,* is worthy of mention. La Rue (1922) was concerned to investigate whether prolonged selection within a pure line could be effective; he demonstrated the ineffectiveness of selection in this case. The care he took to reduce uncontrolled variation is noteworthy; for example, conidia of the same age possessing features enabling their maturity to be assessed precisely; sufficiently large samples for statistical analysis, and prolonged experiments, in some cases 25 transfers extending over a calendar year. He also commenced each experiment with a single conidium but although each was 2–4 celled, they were evidently homokaryotic.

Two other experiments have been reported on selection for alteration in spore size, that by Hanna (1926) on basidiospores of *C. sterquilinus,* and those initiated by Pateman on ascospore size in *N. crassa* (Pateman, 1955, 1959a, b; Lee and Pateman, 1959, 1961; Lee, 1962). No response was detected in *C. sterquilinus* but selection was highly effective in *N. crassa.* In addition, selection for increased or reduced growth rate has been effective in *N. crassa* (Papa, Srb and Federer, 1966, 1967) and in monokaryons of *Schizophyllum commune* (Simchen, 1966a; Connolly and Simchen, 1968). All these successful experiments showed a number of common features; these are set out below and some features are illustrated in Figure 9.4.

(i)   Selection, both plus or minus, is usually effective for several generations. The response is gradual but reaches a plateau. When both types of

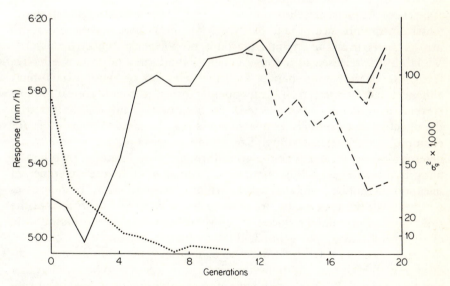

FIGURE 9.4 Response to selection for fast growth rate in a cross between two strains of *Neurospora crassa.* Note positive responses both to selection and to reverse selection and also the rapid fall in genetic variance by generation 10. Key: — Mean of two replications to generation 10; – – reverse selection at generations 11 and 17; . . . . . genetic variance. (From Papa, Srb and Federer, 1967.)

       selection were practised on common starting material the response was usually asymetrical.

(ii) During selection some lines frequently became infertile and this could only be overcome by relaxing selection in some way. Nevertheless, there was usually at least one line where such problems were not encountered.

(iii) Frequently genetic variance decreased with selection although the extent was variable. Lack of response to selection was sometimes, but not necessarily, correlated with low genetic variance.

(iv) Reverse selection from 'plateaued' lines usually showed some response indicating that potential genetic variability was still available. Crosses between comparable selected lines of different origins also restored response to selection.

(v) Response to selection was effected through polygenes. The numbers of effective factors (linked groups of polygenes) was usually quite small, less than 10, and their chromosomal positions were determined in some cases.

These common features are similar to those which have been found in selection experiments with diploid organisms. They are amenable to the same type of explanation as has been given for diploids (e.g. Mather, 1943, 1953; Mather and Harrison, 1949), allowance being made for the differences between diploids and haploids.

Selective response is believed to be due to the release of stored variability which, in the absence of heterokaryosis or heterozygosis (for the organisms are haploid) must be in linked groups of polygenes (effective factors). Variability is released as a consequence of the formation of new effective factors through recombination and the selection of those which most enhance the trait selected for. Consequently, other traits may appear to show correlated responses to the selected trait because they are closely linked to the effective factors. Such correlated traits may be disadvantageous as, for example, a decline in fertility. Relaxation of selection will permit genes determining such traits to be replaced, through recombination, by others of natural advantage to the organism. Thereafter, response to renewed selection will not be correlated with the disadvantageous traits. In some cases selection will result in fixation of some genes or effective factors so that the genetic variance will be reduced. If, however, new linkage relationships are established between polygenes within an effective factor, e.g. from $+ -$, $- +$ to $+ +$, $- -$, there need be no change in genetic variance and, indeed, it could increase. Response to selection is expected to be rapid at first and then to level off as a consequence of the establishment of new, balanced, stable, polygenic complexes. Thereafter, further response will only occur with relaxation of selection pressure (when a new equilibrium will be achieved), the introduction of new genetic material from outside the selection line, or by very rare recombinant events within it.

There is a good deal of confirmatory evidence for this kind of explanation from the experiments summarized above.

The steady response to selection achieved in every case and the segregation data suggest strongly that the traits were controlled by systems of polygenes. In some cases it proved possible to locate some of the effective factors and to distinguish them from major genes which affected the characters selected. A correlated response to selection, reduced fertility, was not uncommon and when it occurred it could be by-passed by relaxing selection —as expected from hypothesis. Indeed, when selection was relaxed, or reversed selection practised, genetic change resulted, even though the selection lines had by then become more or less stable. Moreover, in *N. crassa* when selection had resulted in a plateau associated with greatly reduced genetic variability in two lines of different origin, further selection was effective with the progeny of their cross. Growth rate increased from 5·8 mm/h to 6·2 mm/h in seventeen generations of selection and did not appear to have reached its limit (Papa, 1971). This is in accordance with expectation for there was reason to suppose that there were considerable genetic differences between the strains. Thus, even if genetic fixation had occurred in the initial selections, the crossed progeny introduced new genetic material between which novel recombinants could be expected to arise and be available for selection.

A puzzling feature of the results both for ascospore-size selection in *N. crassa* and for growth rate in *S. commune*, was the asymmetrical response to plus and minus selection. In both cases selection for a reduced manifestation of the character resulted in the lines becoming less stable than either the plus-selection lines, or the parental strains. This was especially clearly shown by *S. commune* where the environmental component of the variances (V_E) was much higher in the low-selection lines than in the high-selection lines. Most of the causes of asymmetrical response to selection applicable to diploid organisms are not applicable to haploids save for non-additive interactions, genotype–environment interactions and cytoplasmic effects. The instability associated with the low-selection lines could possibly have arisen by indirect selection of genes for instability through their linkage to the effective factors determining reduced growth rate. This would be a form of non-additive interaction. [In this context it should be noted that, both in *S. commune* and *A. nidulans*, non-additive interactions occur between polygenic complexes originating from different parental strains. So such interactions are known from fungi (Jinks and others, 1966; Simchen, 1967).] There was also inconclusive evidence in the selection lines that the basis of the instability was either cytoplasmic or could have arisen from the persistent consequences of genotype–environment interactions at an early stage of mycelial development. It was not possible to discriminate between these possibilities and the phenomenon needs to be studied further.

In addition to these experiments where adequate genetic analysis has proved more or less possible, many have been reported where the genetics are not clear. This is particularly true of selection experiments concerned with tolerance of toxic agents and selection for increased pathogenicity, usually with obligate pathogens. These experiments are of such importance for the biology of fungi that they must be considered though briefly. In general, they all share the property of illustrating the efficacy of selection for the desired objective. In many cases, this is achieved by gradual steps, sometimes the response finally reaches a plateau and relaxation of selection is associated with fairly rapid loss of the selected characteristic. In other cases selection is associated with rare, irregular but definitive changes which may continue to be exhibited if selection is relaxed. It is tempting to account for the former pattern as being due to multiple genetic, possibly polygenic, changes and the latter pattern as reflecting the selection of rare chance mutants of major genes. Unhappily, in these cases plausibility is not usually supported by proof.

Resistance to fungicides and toxic agents has been mentioned before (Chapter 2, pp. 30, 36; and this chapter, p. 179). In a number of fungi there is unequivocal evidence for clear-out gene mutations to resistance, e.g. five loci for resistance to aromatic hydrocarbons have been identified in *Nectria haematococca* (Georgopoulos and Panopoulos, 1966). In others the position is uncertain; either adaptive enzyme development, selection of nuclei or extrachromosomal elements, polygenic response or some combination of these could have occurred.

Selection in relation to pathogenicity has proved difficult to interpret not only because of inadequate genetic information but also because pathogenicity can be expressed in different ways and is itself a complex character. A convenient simplification is to distinguish between 'aggressiveness' and 'virulence', although both are still complex characters. The former is defined here as the ability to infect and cause symptoms to varying degrees in the host, the latter as the ability to infect genetically different hosts differentially. So the former is a measure of the intensity of pathogenicity, the latter refers to the range of specificity towards hosts. In each case, of course, the response is a unique host–parasite reaction and thus involves the genetics of both organisms. This alone increases the complexities of interpretation.

In some cases changes both in virulence and in aggressiveness are involved. For example, Watson and his colleagues have induced mutants in Race 21-2, 3, 4, 5, 7 of *Puccinia graminis tritici* which are then capable of infecting wheats carrying the resistance gene *Sr8*, i.e. they have changed from avirulent to virulent. Such mutants, however, differ from their parent biotype and each other in characters such as growth rate and uredosoral-pustule type, e.g. normal, small or non-erumpent pustules. Selection can subsequently be exercised on such characters, which affect the competitive ability and survival value of the mutants, so that they are capable of competing effectively with other biotypes (Watson, 1970).

Most experimental studies have been concerned with selection for aggressiveness either within or between populations. There is some support from such experiments that aggressiveness can be increased by repeated passage through a resistant host, provided infection can occur. For example, Cherewick (1958) increased the percentage of infected oat plants of *cv*. Monarch to *Ustilago avenae*, isolate 47-4, from about 10% to about 90% in 8 successive generations. The successive selections were tested at each generation on some 11 other oat cultivars but no selection was practised on any of them. A number, all initially more susceptible than *cv*. Monarch, showed an increased response to the pathogen although less than *cv*. Monarch; others, initially resistant, showed only irregular and sporadic infections. Since each fungal generation represented a sexual cycle, this experiment suggests that directional selection for genotypes better adapted to the particular host environment represented by *cv*. Monarch had occurred. This selection evidently improved adaptation to certain other cultivars, which presumably represented a similar environment to that of *cv*. Monarch. Those to which no changed response occurred presumably represented different host environments (Person, 1968). It may be plausibly suggested that this gradual specific response is likely to have been multigenic, probably polygenic, but genetic markers were not available in the pathogen, nor were appropriate crosses made. Other examples are even more confusing. Reddick and Mills (1938) claimed that passage (via zoospores) through the leaves of potatoes showing progressively increased resistance, increased the pathogenicity of isolates of *Phytophthora infestans*. Other workers could not confirm these results (e.g. Thurston and Eide, 1952). Whether or not selection was effective, complete confusion reigns so far as possible explanations are concerned. Firstly, it is not clear whether changes in aggressiveness or virulence are involved; secondly, it is not clear if selection is operative on mutants, heterokaryons, heteroplasmons or is acting on major genes or polygenes; thirdly, uncertainty as to whether the species is haploid or diploid bedevils any explanation. All these phenomena and conditions have been claimed to occur by one worker or another (summarized in Gallegly, 1970).

Even if there were more certainty concerning genetic mechanisms involved, and even if appropriate crosses were made, one problem would still remain. This is the availability of suitable genetic markers. It is unfortunate that the simplest kind of genetic markers to employ and to obtain in saprophytic forms are either very difficult or impossible to identify, in obligate pathogens, since they may impair pathogenicity. This was first demonstrated clearly with the apple-scab fungus *Venturia inaequalis* (Keitt and others, 1959). Some induced mutants were completely pathogenic, e.g. inositol-less, but others only induced leaf flecking, e.g. *rib* riboflavinless and *gua* guanineless. Provided leaves were exposed to exogenous riboflavin solutions during incubation prior to infection, normal symptoms could be obtained with the *rib* mutant but comparable treatment with guanine was ineffective for the *gua* mutant. On the other hand, in suitable cases colour markers appear to behave comparably to wild-type forms. Brown and Sharp (1970) could detect no differences in behaviour between a spontaneous white uredospore mutant of *P. striiformis* and the normal, and Harding (1973) was able to recover an albino spore mutant of *Helminthosporium sativum* from soil some 15 years after it had been introduced experimentally. Thus, direct experiments to discover how selection for increased aggressiveness affects the genotype may prove difficult, but not impossible, both to design and to carry out. An alternative type of experiment, which can also give information on the selective value of changes in virulence, is to study competition between biotypes.

Selection between biotypes of pathogenic species

Experiments of this nature can be of increasing complexity. The simplest are those where the biotypes of the pathogen are genotypically similar in virulence, as determined by host range, and the host employed is equally susceptible to the biotypes. The selective response to competition can be further analysed by recording features involved in the complex character 'aggressiveness' and the genetics further explored. More complex investigations involve competition between biotypes differing in their virulence, either on a host equally susceptible to all, or on hosts differing in their susceptibility. In fact, most experiments to date have been rather simple and the partitioning of features involved in .aggressiveness has not often been achieved. Some examples will illustrate the types of experiment carried out.

Figure 9.5a, b illustrates a simple experiment carried out by Loegering (1951) on the survival of different pairs of races of wheat-stem rust when propagated solely by (asexual) uredospores. Infections were made on equally susceptible seedlings with mixtures of various proportions of each race. Care had been taken to ensure that sporulation was comparable on each host in its time of inception and that uredospore viability was at least 95%. It can be seen that Race 17 survived better than Race 19 on all the test cultivars after seven successive transfers regardless of the initial proportions of the mixture. The results were not those expected from field observations, where Race 19 was predominantly isolated from *cv*. Mindum, although in the experiment Race 17 appeared to have a slight advantage over Race 19 on this wheat. Loegering commented 'both from observational and from experimental evidence, that combinations of relatively minor ecological features, when operating together, affect the success or failure of physiologic races in nature'.

Katsuya and Green (1967) carried out similar investigations but attempted to identify the components, morphological, physiological and ecological, which were related to the selective response. Figure 9.5c shows their results in outline. However, they employed a range of temperatures, 15°, 20° and 25°C, and investigated the response of nine different features. In six of these features the response differed between the two biotypes, Races 56 and 15-B1. Uredospores were more infective in Race 56; the incubation period to

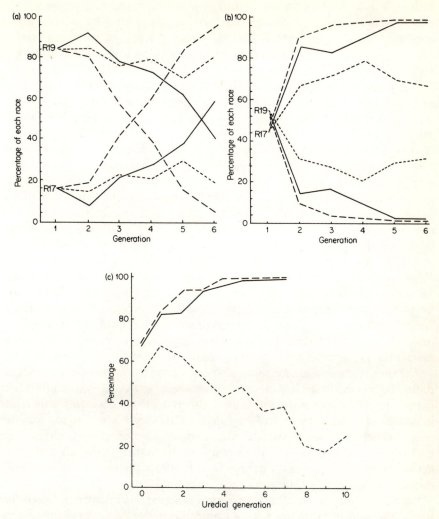

FIGURE 9.5 Selection between different races of *Puccinia graminis tritici*. (a), (b) Percentages of Races 17 and 19 in mixtures of uredospores inoculated on three susceptible cultivars. Uredospores harvested and mixed for subsequent inoculation over six generations. Note changes in successive generations and relationship with initial composition of inoculum in each case. Key:—Little Club; — — Fulcaster; --- Mindum. (c) Percentage changes over successive generations of uredospores of Race 56 from an initial mixture of Races 15-B1 and 56 on the susceptible *cv.* Little Club at three different temperatures. Key: — — 25°C; —— 20°C; --- 15°C. ((a), (b) From Loegering, 1951; (c) from Katsuya and Green, 1967.)

sporulation was shorter; its sporulating pustules grew more rapidly at first, although those of Race 15-B1 eventually overtook them; and its uredospore production per pustule was greater. As infection density on the leaves increased Race 15-B1 predominated, whereas Race 56 predominated in sparse infections. There were no differences in either uredospore viability or their germination.

Although this analysis showed those features of host–parasite interaction in which the superiority of one parasite's genotype was expressed over the other, it provided no further information. The nature of the genetic determination of all these characters is unknown. Moreover, it will be apparent that it is not likely to be simple even if some characters are correlated, some represent pleiotropic expression of particular genes, or some their interactions.

In earlier studies Watson and Singh (1952) investigated competition between pairs of races of *P. graminis tritici*, differing in the number of virulence genes which they carried, over five uredospore passages through *cv.* Federation, a wheat cultivar susceptible to all the races used. In each pair one race carried an additional gene compared with the other, e.g. Race 126-6 versus 126-1, 6, and in every case this race was at a disadvantage. In contrast Brown and Sharp (1970) made up mixtures of different proportions of uredospores of two races of *Puccinia striiformis*, B1 and FB1, the latter carrying more virulence genes than the former. Regardless of the initial proportion of Race FB1, during 8 transfers it increased and became predominant, representing 93–100% after 7 transfers.

Another type of behaviour was also demonstrated in this kind of experiment. In several fungi it was shown that there was competitive inhibition between races virulent on the same cultivar so that there was either less infection in a mixture, e.g. *P. infestans* (Thurston, 1961) or less disease response, e.g. *Puccinia graminis avenae* (Leonard, 1969), than if either race had been inoculated alone. In *Diplodia* it seems as though this kind of interaction stems from the production of inhibitory metabolites presumably under genetic control (Hoppe, 1936).

Survival ability is clearly not necessarily related to virulence. Conversely, the acquisition of virulence without an appropriate genotype determining aggressiveness, competitive or survival ability, will be ineffective until these have been acquired also.

This discussion of experimental studies on selection for increased pathogenicity has begun to indicate the complexity of the situation, both physiological and genetical. It should be recalled that, in addition to the several features bearing on aggressiveness and the ability to survive and compete, there must be included different attributes of virulence, e.g. *P. graminis tritici* Race 36 probably possesses at least 9 loci determining virulence/avirulence to different host resistance genes (Loegering and Powers, 1962). It would not seem improbable to suggest that 10–20 loci could be involved in determining pathogenicity in a fungus, plus at least as many loci carrying polygenes. Moreover, not only are there opportunities for these loci to interact in the fungal genotype but interactions can also occur in the host genotype and in the joint expression of both genotypes. The 'aegricorpus', as Loegering (1966) has termed the unique genetic and biochemical system established by a host–parasite complex, is the background against which the effects of natural selection on pathogens have to be interpreted.

CHAPTER 10

Natural selection

It is not possible to be as precise in discussing natural selection in fungal populations as it was in discussing experimental selection. This is because natural selection is usually detected by its consequences, rather than when it is actually operating and by then it may be too late to determine its mode of action. Moreover, in experimental situations, the design is usually such that directional selection is exercised whereas, in nature, stabilizing selection is likely to predominate. It is for this reason that directional selection is most clearly exhibited in pathogens as a consequence of man's agricultural activities such as changes in agricultural practice, or the introduction of new pesticides, or of disease-resistant crop plants. So striking, indeed, is this last effect, resulting as it does in the exertion of massive selection pressures on fungal populations, that Johnson (1961) has published an article entitled 'Man-guided evolution in plant rusts'. First, however, the action of stabilizing selection will be considered.

Stabilizing selection

The action of stabilizing selection in nature can be inferred from observations on many fungi. Soil-inhabiting fungi afford several instances. For example, when soil Fusaria are isolated, even as single uninucleate conidia, from a particular soil they are remarkably constant and recognizable in their characters. This continues if they are subcultured in soil tubes but if grown on nutrient agars they behave very differently. Morphologically abnormal areas, flattened, with fluffy aerial hyphae, plectenchymatous masses, or sclerotia, develop rapidly. Miller, (1945, 1946a, b) called these 'patch mutants' and it seems reasonable to suppose that they do, indeed, arise by mutation since heterokaryosis can be ruled out in microconidial isolates. In the soil it would seem that natural selection (environmental factors in this case) weed out such mutants but selective pressures are evidently relaxed on agar plates. In nature, stabilizing selection is likely to be more stringent in its effects on mutants of haploid organisms which lack buffering systems such as heterokaryosis, dikaryosis or diploidy. There is good evidence that when present such systems act in this way in fungi.

Heterokaryosis, dikaryosis and diploidy as agents of stabilizing selection

It is not surprising to find that in the heterokaryotic *Thanetophorus cucumeris* only a 'very limited range of cultural types of any one pathogenic strain...have so far been isolated from any one soil' (Flentje and others, 1970). Indeed, in this fungus the characteristic resting structures, the sclerotia, germinate to give rise to colonies resembling the parent mycelium despite the fact that cells isolated by secondary septation in older parts of the mycelium give variable colonies. Here, therefore, variability is masked by heterokaryosis. Stability is further ensured partly by the adaptive behaviour of the sclerotia and partly by the effective selection which must be exercised against cells cut off by secondary septation and capable of expressing the masked variation. The occurrence of purple, red and white strains, differing in gibberellin production as 500:250:5, derived from single uninucleate conidia of *Fusarium fujikuroi*, is even more remarkable (Ming, Lin and Yu, 1966). This pathogen of rice was only isolated from the wild as a heterokaryon including two, sometimes all three, of the component strains described. Since pathogenicity was correlated with gibberellin production there would clearly be strong selection against the white component, yet it had apparently persisted in a heterokaryotic condition.

The persistence of lethals or semi-lethals in heterokaryons is not improbable. In particular, lethals have been isolated through segregation from many dikaryotic fungi. *Ustilago bullata* was found to carry five haplo-lethals, four of which were linked to the mating-type factor, in isolations from wild grasses (Fischer, 1940). Monokaryotic basidial segregants underwent progressive lysis but, if such isolates were incorporated into a dikaryon through anastomoses before lysis was complete, then they persisted at least until the next segregational event.

Stabilizing selection of an entirely different kind has been proposed in the dikaryotic *Marasmius oreades*. Burnett and Evans (1966) provided presumptive evidence for the stability of the mating-type factors in the terrestrial mycelium over a period of at least 100 years. Day (1970) has noted that in experimental culture old dikaryons of *S. commune* accumulate recessive haplo-lethals which would become exposed if the mating-type factors mutated to permit either homokaryosis or haploid fruiting. He suggests, therefore, that stability in *M. oreades* reflects such selection against mutations which would permit haploid fruiting.

The occurrence and persistence of rare but stable diploid strains of some fungi can also be regarded as a special case of stabilizing selection. The best documented case is *Verticillium dahliae* var. *longisporum*, isolated by Stark (1961) together with the type species in the Hamburg Botanic Garden from diseased horse-radish. This has been shown by Ingram (1968) to be an extremely stable diploid. Synthetic diploids and heterokaryons of *Verticillium* are, in contrast, rather unstable save in those cases where one component of the diploid is derived from *V. dahliae* var. *longisporum* (Hastie, 1970). In this species, therefore, stabilizing selection clearly favours the diploid over all other possible forms.

Host specialization

A very common example of stabilizing selection in many pathogenic fungi, and possibly amongst saprophytes also, is host, i.e. substrate, specificity. This is especially well shown by the *formae specialis* of the rust fungi, each of which is confined to a limited number of host species (see Table 13.1, p. 260).

Some of these forms, e.g. *P. graminis* f. spp. *tritici* and *secalis*, have *Hordeum vulgare* as a common host, but they never infect this as vigorously as they each infect their specialized host (Johnson and others, 1932). Similarly, even though some of the *formae* can hybridize, their progeny are nearly always less pathogenic to both hosts than is either parent to its particular host, e.g. f. spp. *avenae* and *tritici* (Johnson and Newton, 1933). The origins of such specialization must lie in disruptive selection, and will be considered later, but its outcome is maintained by stabilizing selection. It has never proved possible to adapt one *forma* to a novel host despite Ward's original claims that this could be done (Ward, 1903). Indeed, in a reinvestigation of Ward's claims, Bean and others (1954) employing various techniques over 14 years, showed that no changes in host specificity could be achieved.

Such host specialization occurs in other fungi but is not so clear-cut, e.g. the *Polyporus abietinus* complex. The poroid form (*Hirschioporus abietinus*) grows most frequently on *Picea* and *Pinus* in Japan, as it does in N. America, but there it also grows on other conifers, and rarely on dicotyledonous genera such as *Betula* and *Prunus*. In Europe it is more common on *Pinus* but occurs on *Picea* and, less frequently, on other conifers (Macrae, 1967). Since these coniferous genera are often growing in close proximity the host preferences must reflect the effects of stabilizing selection.

Directional selection

Directional selection in fungi is especially well shown by responses of pathogens to man's activities. Three examples will illustrate this.

Continuous cereal cultivation, both of wheat and barley, has been an important change in agricultural practice in Britain since the second world war. An important root pathogen of cereals is *Ophiobolus* (*Gaeumannomyces*) *graminis*. Under conditions of monoculture disease intensity declines—the so-called 'Take-all decline'. It has been claimed that inoculum density in the soil declines and the morphology of hyphae in the cereal's rhizosphere is modified with subsequently reduced pathogenicity (Shipton, 1973). However, Asher (unpublished) has demonstrated that isolates from barley monocultures are, in general, far more aggressive than those from wheat monocultures grown in comparable conditions. This change is genetic and barley is known to be more resistant than wheat. Here, therefore, is a clear example of directional selection in response to changed environmental conditions arising from altered agricultural practice.

A more predictable example of directional selection is the origin and increase of fungal strains resistant to fungicides. Particularly striking examples have arisen with the recent introduction of systemic fungicides, notably, benomyl:

Benomyl = methyl (1-butylcarbamoyl)-2-benzimidazole carbamate

This was employed as a 0·15% spray to control *Botrytis cinerea* on cyclamen in Dutch greenhouse culture. After several weeks however, it appeared to be less efficacious and a fungal isolate was shown to be resistant to 1000 p.p.m. benomyl; the wild type was sensitive to 0·5 p.p.m. benomyl. Resistant strains of various fungi have now been isolated from the field and, in addition, genetically characterized mutants have been induced experimentally in others (Table 10.1, p. 204).

It is unfortunate that the genetics of the natural resistant isolates have not been studied. However similar u.v.-mutants of *A. nidulans* demonstrated the occurrence of two, unlinked, recessive, single-gene mutants, *ben-1* and *ben-2*; they did not show additive effects.

In some cases resistant strains have arisen sporadically in different localities but, after withdrawal of the fungicides, they have continued to spread. Dimethirimol is a classic example. It was introduced to control cucumber mildew, *Sphaerotheca fulginea*, in Holland in 1968 and was used on a large scale in 1969. Results became less satisfactory in the autumn of 1969 and even less so in spring, 1970. Moreover, by this time tests in glasshouses where the fungicide had never been used showed that their fungal populations were somewhat tolerant. The fungicide rapidly fell into disuse but dimethirimol-tolerant mildew was still widespread in Holland in May, 1971 and was detected elsewhere in N.W. Europe. So far, such resistant strains have not been detected in field-grown crops. The nature of the adaptive change is not known but it is clearly heritable. The acquired tolerance evidently did not impair competitive ability of the fungus in fungicide-free glasshouse conditions. Hence tolerant strains were dispersed and became established. Dispersal could have been effected principally by man but it is difficult not to suppose that some air-borne dispersal also occurred. In the field environment, therefore, it must be supposed that tolerance impairs fitness in some way so that such strains are at a disadvantage compared with susceptible strains. A genetic analysis of the situation would be invaluable and this should be possible either in natural conditions or simulated ones. Sensitive strains still occur in glasshouses and the fungus has a readily inducible sexual stage.

TABLE 10.1 Resistance to various systemic fungicides either detected in nature or induced in the laboratory.

Fungus	Fungicide	Reference
Spontaneously-occurring resistance		
Botrytis cinerea	B, T	Bollen and Scholten, 1971
Erysiphe cichoracearum	B	Netzer and Dishon, 1970
E. graminis tritici	E	Wolfe, 1971
Penicillium brevicompactum	B, T	Bollen, 1971
Penicillium corymbiferum	B, T	Bollen, 1971
Sphaerotheca fuliginea	B	Schroeder and Provvidenti, 1969
	D	Bent and others, 1971
Induced resistance		
Aspergillus nidulans	B	Hastie and Georgopoulos, 1971
	Di	Dekker (unpub.) in Dekker, 1972
Cladosporium cladosporioides	B	Bollen (unpub.) in Dekker, 1972
C. cucumerinum	B	Dekker (unpub.) in Dekker, 1972
Erysiphe graminis hordei	D	Leeming (unpub.)
Fusarium oxysporum lycopersici	B	Thanassoulopoulos and others, 1971
F. oxysporum melonis	B	Bartels-Schooley and McNeil, 1971
Neurospora crassa	B, O	Sisler, 1971
Pyricularia oryzeae	B/S	Nakamura and Sakurai, 1962
	Ka	Ohmori, 1967
	Ki	de Waard (unpub.) in Dekker, 1972
Rhizoctonia sp.	O	Grover and Chopra, 1970
Sclerotinia sclerotiorum	Tr	Paster and Dinoor (unpub.) in Dekker, 1972
Ustilago hordei	B, O	Ben-Yefet and Dinoor (unpub.) in Dekker, 1972
U. maydis	Ch	Tillman and Sisler, 1971
	O	Georgopoulos and Sisler, 1970

Note: Key to fungicides; B, benomyl and related compounds; B/S, blastocidin-S; Ch, chloroneb; D, dimethirimol; Di, dichlorozoline; E, ethirimol; Ka, kasugamycin; Ki, kitazin; O, oxathiins; T, thiophanates; Tr, triarimol.

The last example to be considered has probably attracted most attention and is the most spectacular and best documented of adaptive responses by fungi in nature. It is the successive development of new races of pathogenic fungi capable of infecting successions of crop plants bred for resistance to them. The universality of this man-directed phenomenon is illustrated in Figure 10.1 for wheat-stem rust and wheat cultivars in N. America, Australia and E. Africa.

The basis of this phenomenon is the gene-for-gene relationship between host and parasite first enunciated by Flor (1942, 1955, 1956, 1971) for the flax rust *Melampsora lini* and its host *Linum usitatissimum*. Rust resistance is inherited as a dominant character at several loci and virulence is recessive to avirulence at corresponding loci in the pathogen. For each gene conditioning resistance in the host there is a specific gene conditioning pathogenicity in the parasite (Chapter 13, pp. 262ff).

Changes in a pathogen's virulence have been recorded in several situations. One clearly established sequence of changes is that recorded by

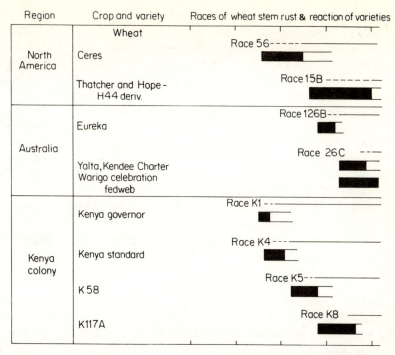

FIGURE 10.1 The relationship between the prevalence of certain physiological races of *Puccinia graminis tritici* and the resistance and susceptibility of wheat cultivars in different parts of the world. Key: Races scarce----; races widespread ——; cultivars when resistant are shown solid, susceptible open, and they end when generally withdrawn. (Based on Person, 1967.)

Bailey (1950) in Ontario, Canada, for *Cladosporium fulvum*, the cause of leaf mould on tomato:

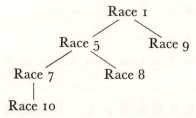

A mutational origin for the probable changes involved was made more probable by Day's (1957) success in inducing the virulence gene a_3 by X-irradiation from its avirulent allele, A_3, i.e. Race 5 from Race 1.

The occurrence of single-step changes in pathogens, related to the introduction of new resistance genes in the host is now documented for many host–pathogen combinations and is implicit in all those where a gene-for-gene relationship occurs (Table 13.7, p. 268).

The change from avirulence to virulence in the pathogen is not necessarily a single-step change. Watson and Luig (1968) have shown that full virulence to the resistance gene *Sr11* in wheat requires a two-step change in *P. graminis tritici*. New virulent races may also arise through recombinational events. Genes for virulence may occur in populations even when the corresponding resistance genes do not occur in the host population, e.g. in New Zealand where no host cultivars carry *Sr5*, several strains of *P. graminis tritici* carrying the virulence gene to this host gene are quite frequent. In such circumstances the genes for virulence must either be selectively neutral, and the pathogens carrying them must be selected for other attributes, or selectively advantageous as a result of their pleiotropic action. These attributes are presumably ones which determine aggressiveness or general fitness (cf. pp. 196–199). An example is Race 126-6,7 which was probably introduced into Australia about 1926. It represented almost 100% of the prevalent races by 1929 and so persisted until 1941. Up to 1938 no wheat cultivars resistant to this race were grown in Australia so that its rise to predominance must have been due to its superior aggressiveness over the formerly prevalent races. In 1938, *cv.* Eureka was released. It carried the resistance gene *Sr6* and was highly and specifically resistant to Race 126-6,7 which thereafter declined to 9% of all prevalent races by 1950. As it declined, Race 126-1,6,7 possessing virulence to *Sr6* and hence to *cv.* Eureka,

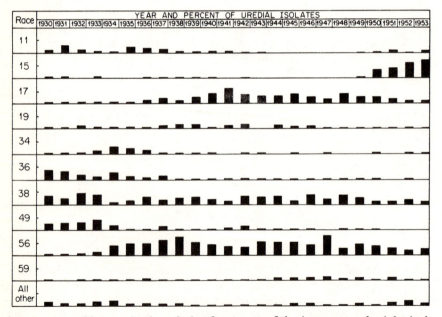

FIGURE 10.2 Changes in the relative frequency of the important physiological races of *Puccinia graminis tritici* in the U.S.A. over the period 1930–1953. Key: Each block represents a potential frequency of 100%. (Based on Stakman and Harrar, 1957.)

arose and increased to 67% by 1947 (Watson, 1970). Here, therefore, selection initially favoured increased aggressiveness alone and, later, virulence combined with aggressiveness. Similar data are available for N. America where Race 56 resembled Race 126-6,7 in its history. Figure 10.2 illustrates detailed changes in the frequency of rust races in N. America over several decades.

Most of these changes are 'man-directed' since they can be related to the introduction of cultivars carrying specific resistance genes, but their origins could be various, both mutational and recombinational. A somewhat different situation obtained for *Helminthosporium maydis* whose Race T caused the serious epidemic of Southern leaf blight of maize in 1970. In this year most maize grown in the U.S.A. carried a cytoplasmic factor, *Tms* for male sterility. This rendered the maize especially susceptible to *H. maydis* Race T which had been known from 1955. There is evidence that this race had increased its aggressiveness between 1955 and 1970. Comparisons of initial and epidemic isolates showed in the latter a 15% increased capacity for sporulation, more rapid colonization of susceptible tissue and a wider climatic tolerance (Nelson, 1972).

An interesting feature of data concerning fluctuations in the frequency of races is that a race rarely disappears totally. Various reasons may account for this. In composite data for a large area like N. America (Figure 10.2) results from different cultivars are lumped together. Thus a race appearing in the national data in very low frequency may be confined to a small region where a susceptible cultivar is grown permitting it to occur locally in very high frequency. However, apart from spurious effects such as this it is possible that different frequencies of races could reflect a polymorphic situation in either stable or unstable equilibrium. Indeed the changes described by Knott (1972) in the frequencies of races of *P. graminis tritici* as they move from south to north along the 'Puccinia path' in the U.S.A. demonstrate clearly the effectiveness of selection, resulting in different balances in different regions (Figure 10.3, p. 208).

Person (1966) has considered ways in which a stable polymorphic situation could arise and this is implicit also in the mathematical models of Mode (1957, 1958) and Cook (1971).

Person points out that as new *R* genes are introduced the corresponding *a* genes in the pathogen will increase from a minimal frequency to a maximum one. Such *R*-carrying hosts will then decline through disease and, as a consequence, the corresponding *a* genes will also decline. If new *R* genes arise in the hosts a pattern could become established in which the frequencies of host and pathogen genes for resistance and virulence would show correlated cyclical changes (Figure 10.4, p. 209). Their equilibrium at any time would reflect the *R* and *a* gene interactions.

Mode's and Cook's treatments are rather different. Mode (1958) showed that where a gene-for-gene relationship obtained, a stable equilibrium could be achieved. Where, however, the hosts consisted of a mixture of cultivars

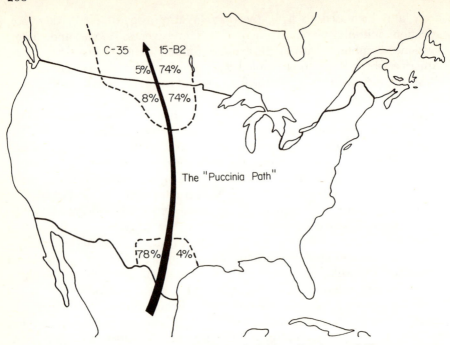

C-35   15-B2
5%   74%
8%   74%

The "Puccinia Path"

78%   4%

FIGURE 10.3 Changes in the frequency of field collections of Races 32–113 (C-35) and 15B-2 of *Puccinia graminis tritici* as the fungus spreads along the 'Puccinia path' in N. America. (Data from Knott, 1972.)

rather than a monoculture, a stable equilibrium could never be reached and any variation in the frequencies of cultivars or races resulted in a drift away from the equilibrium position. Mode (1957) only comments on this situation with reference to a composite mixture of cereal cultivars and rust races, yet it seems not improbable that this may also reflect the situation in truly natural populations. In practice, however, it might be difficult to distinguish such a situation from the equilibrium arising from the cyclical fluctuations described by Person. Since there is virtually no data on host–pathogen genetics in situations other than those of cultivated crops, it has not been possible to assess natural situations. They might well repay investigation. In most cultivated plants the situation is still open-ended since in them new synthetic combinations of resistance genes are still possible. Hence Person's cyclical system is, in effect, distorted before completion.

Cook (1971) considered only the situation where both host and pathogen are haploid and reproduce asexually. He concluded from his model that such a situation would result in an unstable equilibrium. Examples are known of haploid mycoparasitism (e.g. Barnett, 1964) but their population genetics have never been investigated, so that the applicability of Cook's treatment cannot yet be tested in fungi.

Although these considerations of the polymorphic situations which arise

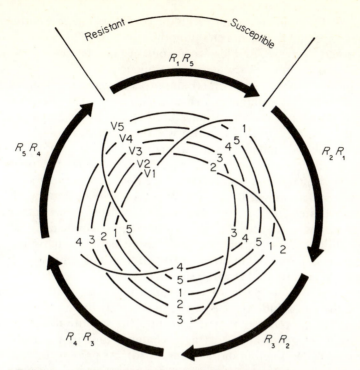

FIGURE 10.4 Diagram to illustrate the successive interactions between virulence genes in a pathogen and resistance genes in a host where a gene-for-gene relationship obtains. Note how the introduction of a novel resistance gene is followed by the rise of the corresponding virulence gene so maintaining a cyclical polymorphism. (From Person, 1966.)

from the gene-for-gene relationships of hosts and pathogens are valuable as possible models, they are not sufficiently sophisticated to account for crop situations. For example, it has been remarked earlier that the fitness of a race cannot be equated solely to its attributes in respect of virulence. But this is the only measure of fitness considered in Person's and Mode's treatments. A particular example will illustrate the point. In situations, such as those described, where a pathogen successively accumulates additional virulence genes, it will eventually be carrying a number of genes of little or no selective value. For instance, if a host carrying R_1 becomes susceptible, due to the acquisition of a_1 by the pathogen, it will tend to be withdrawn and replaced by a new host carrying R_2. The pathogen in due course acquires a_2—a mutation at a different locus from the A_1/a_1 locus. Thus it now carries in its genotype two genes a_1 and a_2, the former having no selective advantage, the latter being of high selective value. Flor (1953) found, through an analysis of epidemiological data of *M. lini* collected over 20 years, that races possessing the least number of virulence genes necessary for survival predominated. This is in accordance with the considerations just outlined, for it would be expected that genes enabling the rust to infect cultivars no longer grown would decrease. As Flor (1956) puts it 'The tendency has been towards the loss of unnecessary genes for virulence.'

This principle has been adopted almost as an axiom by van der Plank (1968), who has cited numerous cases which support it. Nevertheless, an increasing number of exceptions have been found. For example, from 1962 to 1969 *P. graminis tritici* Race 21-1,2 was the main strain in Australia but was then replaced, after a severe drought, by Race 21-2,3,4,7, i.e. one carrying 4 virulence genes instead of 2. This supports the points made earlier that virulence and aggressiveness are both necessary for survival and that fitness is determined by the genotype as a whole. Presumably, in the example just cited, the 'unnecessary' genes for virulence were carried in a genotype strongly selected for its aggressiveness, competitive ability or survival ability. These attributes, as discussed earlier (pp. 196–199), are likely to be determined multi-genically, if not polygenically. It is not impossible, therefore, that it will be found that some 'unnecessary' virulence genes are linked with genes, or polygenic complexes, that determine the fitness of the race. Indeed, one of the general attributes of genetic polymorphism (Ford, 1965) is that it promotes the evolution of super-genes, i.e. 'a group of genes acting as a mechanical unit in particular allelic combinations' (Darlington and Mather, 1949). At present, there is insufficient data on the position and linkage relationships of virulence genes in any pathogenic fungus for this hypothesis to be examined adequately, although multiple allelic series have been claimed in *Melampsora lini* (Flor, 1946; Lawrence, 1973) and in *U. hordei* (Ebba and Person, 1973). An alternative hypothesis, mentioned earlier (p. 206), that virulence genes are pleiotropic in their effects, could of course be reconciled with a similar super-gene concept; loci determining fitness being very tightly linked to the virulence locus itself.

There is very little data concerning this phenomenon in truly natural populations of wild plants and their pathogens. However, there is some evidence that the position may be quite different in wild plants and a very clear example has been described for *Ustilago avenae*, the cause of loose smut in oats. This fungus not only infects cultivated oats but also the wild oats, *Avena fatua* and *A. sterilis*. Collections from cultivated and wild oats were assessed to determine their virulence patterns on a range of oat cultivars carrying different resistance genes. The isolates from the cultivated oats could be grouped into 10 virulence patterns, usually showing high virulence to at least 4 or 5 of the 7 tester oat cultivars. In contrast, 13 strains from wild oats included 10 genes avirulent or only weakly virulent to no more than 2 of the tester oat cultivars (Table 10.2). The three strains showing wider and greater virulence were all collected close to oat cultivar experimental plots and it seems probable that they owed their pathogenicity to strains derived from cultivated oats, i.e. here the virulent fungal strains had spread *from the*

TABLE 10.2 Infection patterns exhibited by isolates from cultivated and wild oats of *Ustilago avenae* when inoculated on to 7 differential cultivars of oats. (Based on data from Holton, 1967.)

Origin of isolate	Infection level (% total panicles/row with smut) on differential cultivars						
	Aln	Ath	BD	Vtry	Gl	Mrh	Flg
Avena sativa	95	0	0	0	0	0	0
4 common patterns of infection including	95	80	32	24	0	0	75
most and least frequent	90	92	97	76	50	94	0
	93	0	92	0	80	97	93
Avena fatua	18	0	0	0	0	0	0
highest, intermediate and least infected	0	7	0	0	0	0	0
	0	0	0	0	0	0	0
Avena fatua	98	90	38	35	15	85	0
near to oat *cv.* plots	0	43	36	35	7	46	0
Avena sterilis	0	0	0	0	0	0	0
different populations	34	10	0	0	0	0	0
Avena sterilis	4	0	67	3	66	0	0
near to oat *cv.* plots							

Note: Key to oat cultivars: Aln, Atlantic; Ath, Anthony; BD, Black Diamond; Vtry, Victory; Gl, Gothland; Mrh, Monarch; Flg, Fulghum.

domesticated to the wild-oat populations (Holton, 1967). Selection pressures in wild populations are likely to differ from those imposed by diverse, man-constructed host-resistance complexes.

Evidently, as yet, a detailed understanding of how selection operates for enhanced pathogenicity is neither possible nor predictable for various technical reasons. The differential selection of pathogenic races, nevertheless, provides one of the best documented examples of genetic polymorphism in fungi and is clear evidence of the operation of powerful selective forces. Other polymorphic situations exist in fungi and they will be considered briefly.

Genetic polymorphism in fungi

The two best known and widespread examples of genetic polymorphism in fungi are the situation just discussed, i.e. different pathogenic races within a species, and the occurrence of multiple mating types in diaphoromictic fungi. Ford (1965) has written

'Genetic polymorphism...proved to be a distinct form of diversity, recognizable and endowed with predictable properties. These, for the present purpose are three in number: (1) the phases, which are necessarily discontinuous, are relatively common compared with those maintained merely by mutation. (2) Polymorphism promotes the evolution of supergenes. (3) It is essentially associated with selection, often of a powerful kind.'

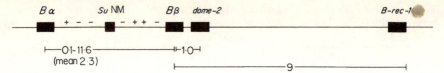

FIGURE 10.5 The *B* mating-type locus of *Schizophyllum commune*. Key: *Bα*, *Bβ*, loci determining mating-type specificity; *B-rec-1*, determining recombination between *Bα* and *Bβ*, hence range of map distances between these loci; +, − etc., recognition sites, number unknown perhaps polygenes, of *B-rec-1* action; *Su*, suppressor of gene which disrupts nuclear migration determined by *B* factors; *dome-2*, a morphological mutant, recombination between *Bβ* and *dome-2* is not affected by *B-rec-1*.

Multiple mating types possess all these properties. Indeed, they illustrate the second property more clearly than do the virulence genes of pathogens just discussed. The current status of the linkage relationships of the components of the *B* factor of *S. commune* are set out above (Figure 10.5).

There is close association between the α and β loci, whose interaction determines the specificity of the factor, so switching mating competence. Between them lies at least one 'recognition site' which responds to the alleles controlling recombination between the loci, so affecting the probability of producing a new mating-type factor by recombination.

No departures in the frequency of mating-type factors from random have been observed but neither the samples available nor the situations from which they have been obtained were conducive to the detection of such differences. Indeed, little is known of the selective advantage of one mating-type factor over another, although it may be inferred that any new factor is likely to be at an advantage compared with the rest in a panmictic population. The distribution of mating-type factors may, therefore, represent a stable polymorphic situation.

Other polymorphic situations doubtless occur in fungi. Although little is known of their genetics, different colour forms of various fungi are not uncommon, e.g. the purple and brown forms of the agaric *Laccaria laccata*, while forms showing constant morphological or physiological differences are also known, e.g. hermaphrodite, self-fertile, and female self-sterile forms of *Ceratocystis fimbriata*. One of the most remarkable examples is the luminescent and non-luminescent forms of *Panus stypticus*. The former is confined to N. America, whereas the latter is known only in Europe and yet the difference is due to a single allele (Macrae in Buller, 1941). This type of polymorphism may not be uncommon in Basidiomycetes but it has not been studied systematically. By analogy with other organisms there is no doubt that cryptic polymorphisms also occur.

There is good reason to expect that the study of fungal polymorphisms would provide information on the way in which selection acts on fungi in nature.

CHAPTER 11

Isolating mechanisms

Selection can result in the establishment of recognizable populations which differ in their gene frequencies. So long as they are not reproductively isolated they may be swamped or replaced by crossing followed by altered selection. They cannot go their separate evolutionary ways completely. This is not to say that recognizably differently populations cannot persist for long periods in a dynamic equilibrium, determined by relatively constant selection pressures. In fungi admirable examples are provided by the physiological races of pathogens, examples of whose persistence have already been cited (see p. 207) and, probably, by yeasts. The former can cross and their offspring, through segregation and recombination, include a wider range of types than the parents. If selection by the prevalent hosts is unchanged, most of the segregants will be less well fitted than those resembling the parents and so will be selected against. For example, Race 111 × 36 of *P. graminis tritici* gives rise to 9 other recognizable physiological races on the differential hosts employed in the greenhouse, but only those resembling parental types persisted in the field (Loegering and Powers, 1962). Of course, where selection is weak, relaxed or changed a new equilibrium can become established. Yeasts maintain their populations in separate ecological niches which rarely overlap.

If reproductive isolation develops, potential genetic flexibility is lost. In compensation, each population can now become even more precisely adapted to its niche without the danger of genetic dilution from other populations. An example is the *formae specialis* of *P. graminis* already mentioned (pp. 202) between which there is a greater or lesser degree of reproductive isolation. Thus there has resulted the characteristic, adaptive host range of each *forma* which is remarkably immutable. Genetic flexibility has been replaced by increased fitness.

Those processes which result in reproductive isolation are called *isolating mechanisms*. Grant (1963) has suggested that three main components can be recognized in them, namely 'spatial distance, the nature of the environment, and the reproductive characteristics of the organisms'. The first of these has frequently been stressed as of initial importance in animal populations (e.g. Mayr, 1954) and botanists have tended to emphasize the second factor for green plants (e.g. Stebbins, 1950). The little evidence available suggests that neither of these is so important in fungi and genetic isolation appears to be capable of developing even in the absence of spatial or environmental differences. Fungi differ in another way: their isolating mechanisms are predominantly pre-zygotic in their expression. So, parental populations frequently fail to initiate reproduction at all, whereas in other organisms the F_1 is formed but may then be inviable, sterile or break down reproductively. A consequence of the frequent inability to form a hybrid is that the genetic analysis of isolating mechanisms in fungi is not often possible.

Esser (1959, 1962, 1971) has investigated one case, *Podospora anserina*, where genetic analysis was possible. As a consequence he has coined the term 'heterogenic incompatibility', i.e. the inhibition of zygote formation between partners due to the heterogeneity of their incompatibility loci. This contrasts with 'homogenic incompatibility', i.e. those phenomena described as mating systems in Chapter 8 (see Table 8.1, p. 158).

Heterogenic incompatibility is exhibited between different strains of the same species and the number of loci which determine it are relatively few. Since the phenomenon has only been fully analysed in *P. anserina* the term will be restricted here to that fungus. Nevertheless, Esser believes that the phenomenon is widespread and has reviewed possible cases in fungi and other organisms (Esser and Blaich, 1973). Plausible and attractive as this hypothesis is, it will not be adopted here because of the lack of evidence. Since, however, it is an admirable model to which other examples of pre-zygotic isolating mechanisms in fungi can be related, it will be fully described.

Heterogenic incompatibility in Podospora anserina

P. anserina has a homodimictic mating system (cf. Figure 8.5b, p. 169) but errors in ascospore formation can result in uninucleate ascospores of + or − mating type. When such compatible isolates derived from strains collected from different areas were paired in all combinations, four kinds of reaction were observed (Figure 11.1):

(a) Neither heterokaryons nor perithecia developed.
(b) Heterokaryons not formed but some perithecia.
(c) Heterokaryons not formed but abundant perithecia.
(d) Both heterokaryons and abundant perithecia developed.

Esser was able to account for these results by supposing that four unlinked loci, *a*, *b*, *c* and *v*, were involved. These loci exhibit both allelic and non-allelic interactions regulating both heterokaryon formation and fertilization, which is normally effected by microconidia and trichogynes. Provided both partners carry identical alleles at all four loci, heterokaryons develop, fertilization is effective and perithecia are formed, usually in two parallel rows representing reciprocal fertilization between the opposed strains. However, not all the resulting new strains are viable. Strains homo-karyotic for $a1b$ or $c1v$ are inviable. They commence growth but their nuclei progressively degenerate and the mycelia die out. This interaction extends to heterokaryons where $a1$ and b or $c1$ and v may be in different nuclei. In

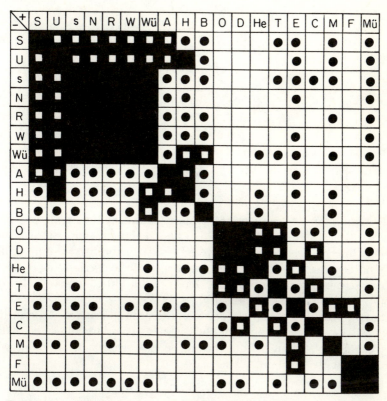

FIGURE 11.1 The mating reactions between strains of *Podospora anserina* from different localities in France and Germany. Key: $+/-$, mating type: ■, sexually and heterokaryon compatible; ◪, sexually compatible, heterokaryon incompatible; ● sexually compatible but with reduced perithecium production, heterokaryon incompatible; open squares, sexually and heterokaryon incompatible. (From Esser and Blaich, 1973.)

this case the $a1$- or $c1$-carrying nuclei die out and the heterokaryon becomes homokaryotic. A further expression of these non-allelic interactions is the development between opposed strains of 'barrage' zones of non-pigmented cells, which may break down, and where no perithecia form. These arise whenever different alleles are carried at any one of the four loci, i.e. a-$a1$, b-$b1$, c-$c1$ or v-$v1$.

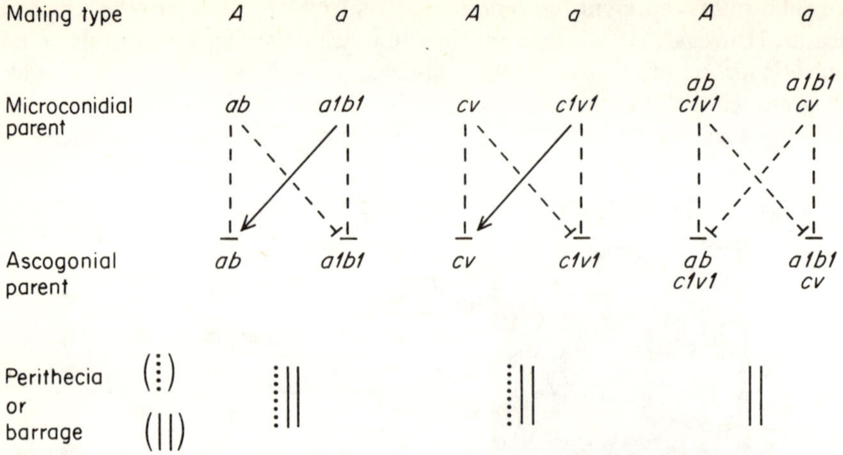

Mating type	A	a	A	a	A	a

FIGURE 11.2 *Podospora anserina*. Three basic interactions, save full fertility, of the alleles *a/a1*, *b/b1*, *c/c1* and *v/v1*, in various combinations. (From data of Esser, 1956.)

The regulation of fertilization is more complex and is set out above (Figure 11.2). Esser has termed these two reactions 'semi-compatibility', where a line of perithecia are formed, and 'reciprocal incompatibility', where no perithecia develop as a consequence of the overlapping of the two kinds of semi-compatible reaction. The possible interactions of all viable genotypes are set out in Figure 11.3.

The four reactions described between the natural isolates can clearly be accounted for by the actions of these genes. Their frequency in nature, however, differs strikingly from that to be expected when all viable genotypes are equally frequent, viz.:

	Observed (Figure 11.1)	Expected (Figure 11.3)
Neither heterokaryons nor perithecia	39·5	2·2
No heterokaryons, few perithecia	30·5	33·3
No heterokaryons, abundant perithecia	13·2	15·5
Both heterokaryosis and abundant perithecia	16·8	48·8

In nature, almost 40% of the confrontations were incapable of developing perithecia and only 30·0% were fully effective as compared with an expectation of 64·3%. As judged by the 19 natural isolates tested, heterogenic incompatibility is evidently an effective mechanism resulting in reproductive isolation in *Podospora anserina*. Since this species is homodimictic it is already, predominantly, an inbreeding organism. The detection of a similar system in other strains of *P. anserina* by Bernet suggests that heterogenic incompatibility is both important and widespread in this species (Bernet and others, 1960; Bernet, 1963, 1965).

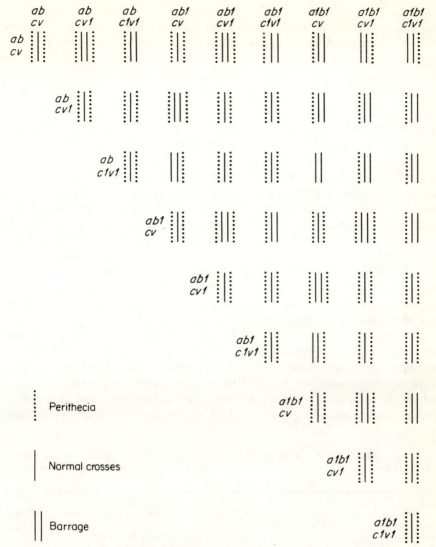

FIGURE 11.3 *Podospora anserina*. The range of interactions between all possible genotypes involving the alleles a/a_1, b/b_1, c/c_1 and v/v_1. Recall (p. 215) that $a_1 b$ and $c_1 v$ are inviable. (Based on Esser, 1956, 1959.)

Features of heterogenic incompatibility in other fungi

The heterogenic reaction is compounded of various processes, namely, heterokaryon incompatibility, barrage zones, failure of one or both gametangial regions to fuse and reduction in the numbers of zygotes formed. Several of these phenomena have been detected in other fungi. Evidently genetic determinants for such processes are widespread and have been selected for in various situations. To date, no case has been detected where all of these processes have operated in concert, determined by a genetic system comparable with that in *Podospora*. Nevertheless, since alone or together they may result in some sort of genetic isolation, they will now be briefly reviewed.

Heterokaryon incompatibility

This has already been considered in some detail in Chapter 5 (pp. 88–90). It will be recalled that in some fungi e.g. *Neurospora* and *Aspergillus* spp., it is determined by a relatively small number of loci, as in *Podospora*. On the other hand, in *Rhizoctonia solani* AS group IV, the genetic basis of heterokaryon formation is entirely different. In all these cases, however, there was no clear evidence that genetic determinants for heterokaryon incompatibility also determined reproductive isolation. The case most resembling *P. anserina* is that of *Aspergillus nidulans*. Although complete reproductive isolation between heterokaryon-incompatible groups was not detected, in some crosses the numbers of cleistothecia were reduced and the progeny of the cross were said to be less vigorous. Unfortunately the precise mechanism of sexual reproduction in *A. nidulans* is still obscure and until this has been resolved a detailed comparison with *Podospora* is not possible.

In those species, mostly Basidiomycetes, where the only route to sexual reproduction is via hyphal anastomoses, heterokaryon incompatibility is a significant cause of reproductive isolation. Several cases have been recorded although the details vary. Broadly speaking, two kinds of situation have been found. In some species dikaryon formation has been recorded as a rare or infrequent event between morphologically indistinguishable populations,

TABLE 11.1 Examples of fungi in which heterokaryon (or dikaryon) incompatibility groups occur between fungi similar in morphology, ecology and habitat preferences, i.e. sibling species.

Fungal Species	Groups	Features
Ascomycetes and Fungi Imperfecti		
Neurospora crassa	2	Incompatible fusion cell dies
Podospora anserina	2 + ?	Heterogenic incompatibility
Aspergillus nidulans	5 +	Mechanism not known
Basidiomycetes		
Agaricales		
Coprinus micaceus	2	Mechanism not known
C. callinus	2	Mechanism not known
Aphyllophorales		
Polyporus betulinus	2	No fusions occur
Fomes pinicola	2	? No fusions occur
Sistotrema brinkmanni I	5 +	Mechanism not known
Thanetophorus cucumeris	Many	Determined by multiple alleles at two loci. Fusion cells die
Gasteromycetales		
Mycocalia denudata	2	No fusions occur

which also differed in their ecological or geographical range; these will be considered later (pp. 221–226). Others show no such differences although in some there may be minute morphological differences (Table 11.1).

Heterokaryon incompatibility in these cases must usually be equated to inability of monokaryons to form dikaryons, since in most cases that was all that was examined. In addition it is not usually known whether the rare dikaryons, when formed, were capable of reproducing sexually since this is often technically difficult to test, even in fully fertile crosses in the laboratory. The reaction is usually expressed in one of two ways. Either, as in *R. solani*, hyphal fusions are reduced and the fusion cell eventually autolyses, or fusions simply do not occur at all (Mounce and Macrae, 1938; Flentje and Stretton, 1964). In some cases the relationships based on the pattern of fusions are complex, e.g. *Merulius himantioides* (Harmsen, 1960) (Figure 11.4).

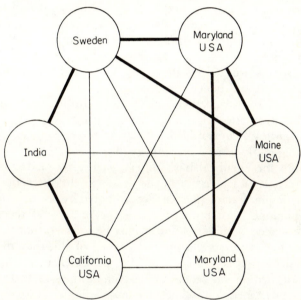

FIGURE 11.4 Interfertility of six isolates of *Merulius himantioides*. Key: fertile crosses, solid line; sterile crosses, thin line; no connecting line, cross not attempted. (Redrawn after L. Harmsen, 1960.)

The remarkable thing about this example is that isolation is neither correlated with geographical or ecological range, nor with morphological differences. Inability to fruit the fungus left the genetic situation unresolved.

Aversion, barrages and inability to fuse

In many species aversion regions, into which hyphae rarely penetrate, develop in plate culture, or in nature when different mycelia are opposed.

In other cases the hyphae become heaped up in opposed regions, show increased branching, pigmentation, chlamydospore formation, more septa per cell or exhibit any combination of these. Such regions are called barrages. In yet other species hyphae of opposed mycelia simply intermingle and fail to fuse, or, if gametangia are involved, they fail to develop or to fuse or both. The former is not uncommon in Basidiomycetes although it has also been recorded in *Sordaria fimicola* (Carr and Olive, 1959). In *Mycocalia denudata* and *Polyporus betulinus* the failure to anastomose is complete whether between two monokaryons, two dikaryons or between a monokaryon and a dikaryon. This inability to fuse occurs quite independently, so far as can be judged, of geographical or ecological differences (Saunders, 1956; Burnett and Boulter, 1963; Burnett, unpublished). The most striking case is that of strains 144 and 144N of *M. denudata*. These were originally obtained from what was thought to be a group of peridiola derived from a single individual. Clearly, however, the fruiting bodies of two individuals had become confluent. Comparisons were made of their growth rate and mycelial colour, size range and colour of the peridiola and the basidiospore size range, but no constant differences were found. Their ecological niche, the base of dead *Juncus effusus* shoots, was clearly the same. No hyphal fusions between them were ever detected and hence the genetic differences between them could not be analysed. Each had the same haploid chromosome number. How isolation arises in such circumstances is not known.

Amongst species of mucoraceous fungi or Saprolegniales, reproductive isolation is frequently associated with a failure of gametangia to develop or fuse. Although the chemical basis differs, each group exhibits a complex sequence of events which results in the development, attraction and fusion of their gametangia. Each sequence is coordinated by one or more diffusable hormones (gamones). In both groups 'imperfect sexual reactions' occur where the sequence is impaired and it may be inferred that this has arisen from a failure to coordinate the sequence. It is probable that, in each group, the hormonal basis is common to all the species (Raper, 1950; Gooday, 1973). Sex determination seems to be comparable, in the mucoraceous species, so that its expression has become modified quantitatively in different species. This strongly suggests interaction of major genes with different polygenic complexes or modifiers. Indirect selection for either of these last could well have occurred in different populations of fungi through correlated response, so resulting in the reproductive isolation observed. It may be noted that, in favourable material, hyphal anastomoses have been shown to be brought about by an apparently comparable sequence of events to those resulting in gametangial fusion in mucors. A similar explanation may, therefore, be offered for the behaviour already described in *M. denudata*. It is exasperating that genetic analysis has not been possible in either of these groups for, even when reproductive isolation is partial or lacking, it has rarely proved possible to germinate the zygotes (e.g. Burgeff, 1924; Gauger, 1965; Hocking, 1967).

The occurrence of heterogenic incompatibility

It is a most plausible hypothesis that heterogenic incompatibility of the *Podospora* type will be found to occur in other species (Esser, 1973). However, this brief assessment of a range of situations does seem to indicate that other mechanisms also occur which need not have the same genetic basis. It seems not improbable that the outward similarities in modes of reproductive isolation mask, as so often happens in genetics, genotypic differences. What is striking, and contrasts greatly with other organisms, is the existence of processes resulting in reproductive isolation in the apparent absence of prior isolation by distance, habitat or other external agencies such as mechanical or seasonal isolation. In other fungi, however, such differences do occur.

Isolation by distance and habitat

Examples in fungi of isolation by distance alone are not common. This may reflect the very efficient dispersal mechanisms which so many fungi possess and this is borne out by the apparently world-wide distribution of many taxa. Nevertheless, three examples suggest that distance is of importance in some situations. In order of increasing complexity these are the coprophilous agaric *Panaeolus*, the polypore *Peniophora pini* and the crust fungus *Sistotrema brinkmanni*.

Parker-Rhodes (1949, 1950) has studied populations of *Panaeolus campanulatus* and *P. papilionaceus* on the mainland of Pembrokeshire (Wales) and the off-shore islands, Skokholm, Gateholm and Skomer. By means of a complex statistical comparison and analysis of basidiospore dimensions, he concluded that island populations, when separated by at least 2 miles of open water from the mainland, were more inbred than mainland populations. He suggested that such a distance was an appreciable barrier to anemophilous spore dispersal for these species.

Peniophora pini (syn. '*Stereum pini*') was believed to occur both in N. America and in Europe. It has been shown to be an aggregate species comprising three morphologically very similar taxa, parasitic on *Pinus* spp. or other conifers. These are, *P. pini-pini*, *P. pini-duplex* and *P. pseudo-pini*; the first is confined to Europe, the last two being N. American. *P. pseudo-pini* is quite incapable of forming dikaryons with either of the other two, between which there is some, although greatly reduced, compatibility. So far as they are concerned, therefore, there seem to be two, morphologically just-recognizable subspecies differing in their geographic range and showing a considerable degree of potential reproductive isolation between them. It has not proved possible to fruit these subspecies or their cross, so that a more definitive analysis cannot be made. However, it should be noted that in N. America the other taxon, *P. pseudo-pini*, occupies a similar habitat to *P. pini-duplex* and differs only in small features from it but is totally isolated reproductively. Thus, in *P. pini*, distance is associated with partial isolation, sympatry with total isolation (Nobles, 1956; Weresub and Gibson, 1960).

One of the most complex cases to have been partially analysed is the aggregate species *S. brinkmanni* (syn. *Corticium coronilla*). The species is widely distributed throughout the world. Biggs (1937) recognized four groups which differed in their mating systems but could be partially separated by minute differences in their basidiocarps and cultural characters. All are capable of producing typical dikaryotic hyphae bearing clamps. Their interfertility has been studied by Biggs (1937), Lemke (1969) and Ullrich (1973) whose findings can be summarized as follows (Table 11.2).

TABLE 11.2 Intersterile groups within the *Sistotrema brinkmanni* aggregate species. (After Biggs, 1937; Boidin, 1958; Lemke, 1969; Ullrich, 1973.)

Minimum number of intersterile groups and their designation	Mating system and morphology	Distribution
Well-defined groups		
Group I	Homomictic	U.S.A., Canada, Germany, France,
2 intersterile groups	Catenulate bulbils on mycelium	Switzerland, Netherlands, Israel, conifers and hardwoods
Group II-1		U.S.A., Canada, conifers and hardwoods
Group II-2	Unifactorial diaphoromictic	U.S.A., hardwoods
Group II-3	Mycelium lacking bulbils	U.S.A., *Pinus strobus*
Group II-4		U.S.A., France, Germany, conifers or hardwoods
Group II-5		Australia, *Pinus radiata*
Group III-1	Bifactorial diaphoromictic	Canada, 'Europe', discarded paper
Group III-2	Mycelium with clustered bulbils	Canada, Australia, hardwoods
Less well-defined groups		
Biggs' Group IV	Heteromictic Conspicuous apical or catenulate spherical cells	Canada

Note:
Group I   = Biggs' Group II and Boidon's Type *A*.
Group II = Biggs' Group I and Boidon's Type *B*.
Group III = Biggs' Group III, Lemke and Ullrich's Group IV.

Group I is homomictic but by making auxotrophic mutants Lemke was able to demonstrate that amongst three isolates there were at least two intersterile subgroups. Paired auxotrophs from Nova Scotia and the Netherlands formed protrophic dikaryons but if either was paired with an auxotroph derived from a British Columbian collection, 'hyphae of slight vegetative vigor and essentially devoid of clamp-connections', developed.

Group II has a unifactorial diaphoromictic mating system and sympatric isolates were found to be cross-fertile by both Biggs and Lemke. Neither the latter nor Ullrich, however, were able to obtain dikaryons between isolates from widely different localities, e.g. Adelaide, Central Corsica, France and New York. There appeared to be some possibility of host specialization (II-3, II-5).

Group III comprised bifactorial diaphoromictic types, amongst which Biggs recognized three intersterile subgroups which had no geographical or habitat basis. Lemke showed that isolates from areas as wide apart as Ontario, Europe and New York were completely compatible.

Group IV was ill-defined by Biggs. It is certainly outbreeding but its mating system was not determined.

Biggs failed to demonstrate cross-compatibility between any of the 6 groups or subgroups she recognized (her Groups I, II, IIIA, B, C and IV). Lemke, however, found some exceptions in Group I × II confrontations, although detecting no such behaviour between confrontations of the I × III or II × III types. Lemke's success must be attributed to his employment of auxotrophic mutants which enabled him to detect apparently protrophic dikaryons, one of which developed a basidiocarp. From this cross, *A2 inositolless* II × *methionineless* I, 109 germinating basidiospores were isolated but only 8 mycelia were viable. Seven of these were recombinant but none protrophic. One was homomictic, six were mating-type *A2* and the eighth reacted both with parental *A1* and *A2* stocks.

In summary, in the *S. brinkmanni* complex, there are clear examples of reproductive isolation, sometimes associated with isolation by distance (Group I), at other times not (Group IIIA, B, C). There is also clear evidence that even populations from central Corsica and British Columbia, which differ in their mating systems (I and II) are not wholly isolated reproductively and can produce some, albeit few, fertile progeny.

Isolation by habitat is a much more frequent phenomenon. Some examples are given in Table 11.3.

A number of these examples appear to support conventional steps in the origin of isolation by divergent adaptation to different environments. In

TABLE 11.3 Basidiomycete species of similar morphology, separable by small habitat differences and showing different degrees of intersterility. (Data from Weresub and Gibson, 1960; Macrae, 1967; Duncan and Macdonald, 1969; Ullrich, 1973.)

Species	Nature of habitat differences
Hirschioporus fusco-violaceus	*Abies, Picea* and other conifers v. predominantly *Pinus*, rarely *Picea*.
	Slightly reduced compatibility
Peniophora mutata Groups I and II	Group I on deciduous trees and Group II only on *Populus*
	Reduced compatibility shown
Auricularia auricula-judae	Deciduous hosts v. coniferous hosts.
	High degree of intersterility
Sistostrema brinkmanni Group II	Certain isolated types are confined to *Pinus strobus* (II-3), some to hardwoods (II-2) and others unrestricted (II-1 and II-4).
	Intersterility complete

some cases, however, the occurrence of isolation in the same species complex without such divergence does raise the question whether the genetic or the ecological isolation was the initial cause. Two especially illustrative examples of this type of problem are the *Polyporus abietinus* complex and the genus *Peniophora*. The first of these has already been mentioned (Chapter 10, pp. 202). Macrae (1967) has suggested that the complex is best regarded as three separate species which differ somewhat in gross sporophore morphology, namely, *Hirschioporus abietinus*—a typical poroid form; *H. fusco-violaceus*—an irpicoid form and *H. larcinus*, the lamellate form. All possess a bifactorial diaphoromictic mating system.

Within *H. abietinus* three groups can be recognized on the basis of monokaryotic confrontations. These groups are not correlated with host preferences, which include several coniferous genera especially *Picea* and *Pinus*. Two of the groups are completely isolated and occur in N. America, the third occurs in Europe and shows partial compatibility with *both* N. American groups (Figure 11.5). Apart from mating behaviour no other constant morphological or host differences occur.

This behaviour contrasts with that in *H. fusco-violaceus* where N. American and European collections showed almost complete compatibility. However, the geographically separated groups show rather different host preferences.

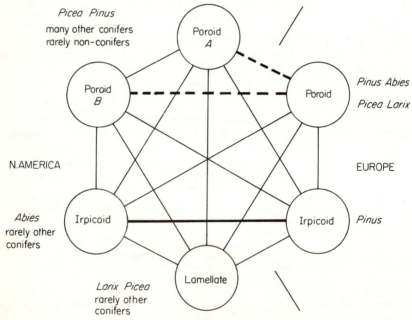

FIGURE 11.5 Interfertility of different isolates of different origins of the *Polyporus abietinus* complex; some indication of host preference is given. Key: fertile crosses, solid line; partially fertile crosses, broken line; sterile crosses, thin line. (Based on Macrae, 1967.)

In Japan the forms grow predominantly on *Abies* (74/74) and also on *Picea* (14/54); in N. America the predominance on *Abies* is even more marked (53/61: *Picea* 1, *Pinus* 2, *Pseudotsuga* 4, *Tsuga* 4) but in Europe it is almost exclusively confined to *Pinus*, rarely on *Picea* (Raestad, 1941; Macrae, 1967).

The lamellate species *H. larcinus* occurs on a range of coniferous genera in N. America, Japan and Europe but confrontations have only been made so far between N. American isolates: they were all compatible.

Confrontations between the three species very occasionally show some degree of hyphal fusion but it is usually either abnormal or results in some kind of cellular degeneration. The most striking behaviour is shown between dikaryons of *H. larcinus* and the other two, where a characteristic antagonism develops.

'When the tip of a hypha came near or touched a hypha of the other partner in the pairing, it wound around the hypha and sent out short branches to form a ball of intertwined hyphae at the point of contact. Either one of the partners in the pairing could produce the knot of hyphae and usually there was deterioration in the contents of the hyphae at the point of contact' (Macrae, 1967).

This example is especially instructive and reinforces the contrast between fungi and other organisms. Firstly, it shows clearly how reproductively isolated groups can arise within a completely sympatric population. (Groups A and B, *H. abietinus*.) Secondly, that isolation by distance need not be associated with reproductive isolation and, moreover, that the occurrence of genetic isolation between two segments of a population need not, of necessity, cut off either from the rest of the population outside them (Groups A, B and C). Thirdly, while the irpicoid form shows selection for different habitats (conifers v. *Pinus*) this need not result in genetical isolation although the slight reduction in compatibility detected may perhaps represent an incipient stage in the process. This is more likely since, apart from *Abies sibirica* in N. E. Russia, there is no native *Abies* in Europe. Thus the fungus may have been introduced to Europe and be adapting to *Pinus* as a substitute for *Abies*. Finally, the occurrence of three taxa occupying the same habitat—frequently, indeed, the same tree bears at least two of them—not only demonstrates sympatric isolation but illustrates how such isolation can be reinforced by antagonistic behaviour such as that of *H. larcinus* to the other two.

McKean (1952) has studied three similar species of *Peniophora, P. heterocystidea, P. populnea* and *P. mutata*; they are mutually incompatible. All grow on deciduous trees as hosts but the first two are ecologically isolated, *P. populnea* being confined exclusively to *Populus* spp., *P. heterocystidea* occurring on broad-leaved trees other than *Populus*. An identical distinction was detected within the third species, *P. mutata*, which can thus be divided into two ecologically distinct races. However, in this case the races still show partial compatibility between their monokaryons. Here then, in contrast to *H. fusco-violaceus*, a habitat preference is correlated with the occurrence of partial or complete reproductive isolation. At the same time it illustrates, yet again, the occurrence of isolation in sympatric populations on both broad-leaved trees and *Populus* spp., i.e. *P. heterocystidea* versus *P. mutata* Gp. I and *P. populnea* versus *P. mutata* Gp. II. Thus this species aggregate also shows a remarkable, parallel development of an allopatric isolating mechanism in which differentiation has proceeded somewhat further in one case than in the other.

Good examples of ecological isolation in fungi can be found outside the Basidiomycetes especially amongst pathogenic species, e.g. the *formae specialis* of *Erysiphe graminis*, or *Ophiobolus graminis* and *O. graminis avenae*. In many of these cases it has proved possible in experimental conditions to overcome this isolation, so that introgression can occur between different related taxa. It is not clear whether this occurs in nature. In other cases post-zygotic isolation appears to operate.

Post-zygotic isolation

When the zygote fails to develop or, if it develops, is sterile, or produces offspring in reduced numbers or with reduced viability, post-zygotic isolation is said to operate. Apart from the case, already described, of the cross between a Gp. I and a Gp. II isolate of *Sistotrema brinkmanni* whose f_1 basidiospores showed reduced viability, there are no other authenticated cases of post-zygotic isolation amongst diaphoromictic Basidiomycetes. It should be recalled that not many of these fungi have been fruited experimentally. It is not clear, therefore, whether such fungi characteristically develop pre-zygotic isolating mechanisms or, whether they possess post-zygotic isolating mechanisms still to be studied. In contrast, a number of cases of post-zygotic isolation are known in dimictic fungi.

Hybrid inviability as exemplified by the development of carpophores with reduced fertility is one of the commonest modes of isolation in Ascomycetes. This condition may often be overcome by selection in a few generations, e.g. *N. crassa* × *N. sitophila* (Fincham, 1951), but it is not clear whether this has ever occurred in nature. Frequently the status of the taxa crossed is uncertain. While the cross in the *Neurospora* case just cited is clearly interspecific, many of those described between 'species' of yeasts of the genus *Saccharomyces* are more probably intraspecific crosses, e.g. those of Winge (1941). Such mechanisms will be described, therefore, without regard to the status of the taxa crossed, since their operation is comparable.

Impaired viability of the zygote may be determined by a single gene acting as a sterility barrier or it may be expressed quantitatively and show a more or less continuous range of expression. This suggests multigenic or polygenic control. Excellent examples of these two modes occur in the genus *Helminthosporium*. In *H. carbonum* (syn. *Cochliobolus carbonum*), a pair of alleles is known in which the recessive causes sterility. If both parents of a compatible combination carry the recessive, i.e. $As \times as$, the cross is sterile, whereas other combinations, e.g. $AS \times aS$ or $As \times aS$, are fertile. The alleles are scattered through the population. Out of 148 isolates S was carried by 17 of 75 A mating-type isolates and 9 of 73 a mating-type isolates (Nelson, 1964a). Similar genes are known in $C.$ $heterostrophus$ and $C.$ $spicifer$ (Webster and Nelson, 1968). The latter species also carries a further pair of duplicate genes, either of whose recessive alleles, if present in an opposed compatible pair, can block the mating reaction, i.e. $i_1:i_1$, $i_2:i_2$, or $i_1i_2:i_1i_2$. The S allele in this fungus prevents ascus formation. An i-type allele has also been found in $C.$ $carbonum$ but it is only associated in natural isolates with a mating type. Isolates of $C.$ $carbonum$ lacking i are fertile in compatible crosses but not uniformly so (Nelson, 1970). Initially, he crossed all the 58 AI isolates with the 64 aI isolates and recorded with how many of the opposite mating type each would produce fertile perithecia. The fertility so assessed of AI isolates ranged from 22 to 98% and of aI isolates from 28 to 98%. He then compared 10 of each kind of isolate for their cross-fertility. Each sample of ten ranged from low to high fertility, and he assessed the relative fertility of the crosses on the frequency of perithecial production in 15 replicates. The results set out in Table 11.4 show that crosses between isolates of the lowest fertility gave no perithecia, those between isolates of the highest fertility frequently gave 100%, while isolates of intermediate fertility gave intermediate results.

Fertility was here behaving as a quantitative character showing a graded response and suggestive of a multiple-gene or polygenic basis.

TABLE 11.4 The results of crossing compatible isolates which differed in fertility when tested against the total sample of *Cochliobolus carbonum*. (Based on data of Nelson, 1970.)

		Mating-type A isolates of differing % fertility									
		28	35	43	50	57	64	75	87	92	96
		Percentage fertility of crosses									
Mating-type a isolates of differing % fertility	22	0	0	20	34	47	40	53	67	80	80
	33	13	27	27	34	47	47	47	13	87	80
	42	20	47	34	47	40	60	53	80	80	80
	50	40	47	47	60	60	60	73	80	87	87
	58	40	60	53	67	73	80	80	80	87	80
	64	47	53	53	67	67	73	80	87	80	87
	75	60	60	73	73	80	73	87	87	93	87
	86	73	73	73	80	80	93	93	100	100	100
	91	80	87	73	80	87	93	100	100	100	100
	98	80	80	73	93	87	100	100	100	100	100

Note:
1. The relative fertility of each isolate was determined by mating it with all available compatible isolates, i.e. each A isolate with the 64 a isolates and each a isolate with the 58 A isolates, and recording as a percentage the proportion with which each isolate was fertile.
2. The percentage fertility of each cross is derived from the number of fertile crosses out of 15 replicates of each cross.
3. All percentages have been rounded to the nearest whole percent.

It is of especial interest to notice, therefore, that in fungi post-zygotic isolation may be determined by single genes of major effect or, very probably, by genes having minor effects but cumulative in their action. This suggests that isolation may arise abruptly, as by the mutation of $S \to s$, or gradually by the accumulation of genes that impair fertility.

The precise ways in which fertility of the f_1 may be impaired has also been demonstrated in *Helminthosporium*. Nelson and Kline (1964) studied the fertility in crosses of all combinations of 79 isolates of Eu-helminthosporium from cereals or grasses throughout the American continent from Argentina to Alaska. Eleven isolates were completely sterile in all combinations but 40·2% of crosses between the remaining 68 were fertile. Amongst the sterile crosses, many developed perithecia but in some cases they lacked dikaryotic tissue within or only developed pseudo-parenchyma, never asci. In some the asci remained immature or aborted and in others the asci were sterile or the ascospores aborted. The most frequent situation was that where the asci were sterile (169/432 examined). The genetic basis of these aberrations was not investigated further but, in *C. carbonum* a genetic block preventing perithecium formation or impairing fertility could frequently be restored by supplying exogenous sterols or their precursors, e.g. squalene, β-sitosterol, ergosterol (Nelson and others, 1967). Reduced fertility in Eu-helminthosporium isolates could be related to their habitat or provenance. Ten of the 11 totally sterile isolates had been collected from wild grass species and of those that showed some fertility in crosses, only 14% had been isolated from wild grasses. This suggests that ecological isolation is associated with reproductive isolation in Eu-helminthosporium. There is also a geographical component. Cross-fertility between isolates from Mexico was only 24%, that between isolates from the U.S.A. was 50% and that between Mexican and U.S.A. isolates was 36%. This suggests that selection in Mexico for reproductive isolation is more intense than in the U.S.A. but the reason for this is unknown. It might reflect the rather different genotypes of the host cereals in Mexico from those in the U.S.A.

There is no other body of data as extensive as that of Nelson and his colleagues but similar data are known from British isolates of *A. nidulans*. As already mentioned reproductive isolation is not closely correlated with heterokaryon incompatibility. However, in some crosses cleistothecium production is reduced, while in others ascospore viability is either impaired or the growth of the progeny is less vigorous.

Hybrid sterility or impairment of subsequent generations has not been studied very often in unequivocally intraspecific crosses in fungi. One example is the smut fungus *Sphacelotheca sorghi*. Crosses between certain haploid isolates gave rise to infective dikaryons but their teleutospores frequently underwent lysis so that the 'hybrid' dikaryon was sterile. A similar phenomenon has been observed in a cross between *U. avenae* with echinulate spores and the form of *U. hordei* which occurs on oats called *U. kolleri*, with smooth spores. The hybrid is only viable on susceptible oats and in many crosses its teleutospores lyse so that it is apparently sterile (Holton, 1931). In culture experiments, Nielsen (1958) showed that *U. avenae*-derived nuclei were lost as the hybrid dikaryon broke down but that this could be prevented either by altering the temperature or in the presence of metabolites of a *Fusarium* sp. He is of the opinion that, in nature, lysis effects are not a barrier to crossing between the species. In other crosses, Holton (1932) was able to demonstrate a wide range of dikaryons, as assessed by host symptoms on oats, most of which were viable.

Further studies suggest that identical virulence/avirulence genes are carried by *U. avenae* and *U. kolleri*. The existence of genetic material in common between those species can be explained in two ways. It could be due to the two species having arisen from a common gene pool from which they are now diverging. Alternatively, it could be due to the transfer of genetic material from one taxon to the other through introgressive hybridization. This seems a more plausible explanation. It will be recalled that *U. kolleri* is a form of *U. hordei*, the cause of covered smut of barley, but also found on oats, whereas *U. avenae* is confined to oats. Thus if *U. kolleri* were initially just capable of infecting oats, and if reproductive isolation broke down, the chances are that the hybrid progeny would back-cross to *U. avenae*. Clearly there would be selection for characters which favoured pathogenicity to oats and hence for the virulence genes derived from *U. avenae*. In circumstances like those just hypothesized, it is clear that post-zygotic isolation may not be completely effective. It is not clear how often this breakdown of reproductive isolation occurs in fungi nor how significant it may be.

Hybridization: the failure of reproductive isolation

Hybridization can only occur because of the failure of post-zygotic reproductive isolation or when normally allopatric populations, lacking any reproductive isolation, are brought together. Broadly speaking, successful hybridizations have been achieved experimentally but there is little evidence for them in natural populations. The smuts are an exception. Because experimental hybridization has been achieved readily there has been sufficient knowledge of hybrid types for them to be recognized in the field. In this way, for example, Holton (1944, 1954) was able to recognize natural hybrids between *Tilletia caries*, *T. foetida* and *T. controversa*. There is only one case known from fungi where the converse process has been employed, i.e. the recognition of a putative hybrid followed by an experimental test of its origin. This is the case of the water mould *Allomyces javanicus*. This is in several respects an intermediate form between *A. arbuscula* and *A. javanicus* v. *macrogynus* (*A. macrogynus*). Emerson (1941) and Emerson and Wilson (1954) were able to show that hybrids between certain isolates of these species did, indeed, resemble *A. javanicus* and this was also supported by chromosome counts on the hybrids, their F_2 and F_3 generations and wild isolates of *A. javanicus*. They were also able to show that such hybrids were not always produced so that reproductive isolation did not necessarily break down. For example, in a number of cases when *A. arbuscula* was the female parent, there was evidence that the progeny arose parthenogenetically from an unfertilized female gamete.

In other species crosses, e.g. *N. crassa* × *N. sitophila* (Fincham, 1951), viability is extremely low in the f_1 but can be restored in two to seven generations. This is equally true with repeated back-crossing to one of the parental species and in this way introgression has been successfully achieved with *Neurospora*. There is no evidence, however, that this, or other crosses made experimentally, have occurred in nature.

Introgression can occur in two other genera as a consequence of rather different processes. Nelson (1964b) has studied crosses between different species of *Cochliobolus* (*Helminthosporium*) and has shown that multiple crosses often produce more viable progeny than do direct interspecific crosses. For example, no success was achieved in 782 attempted crosses between *C. cynodontis* × *C. sativum* but 3 out of 800 were successful when different strains of each species were crossed first and then their f_1s, i.e. (*C. cyn.* × *C. cyn*) × (*C. sat.* × *C. sat*). As with *Neurospora* further selection of such crosses and back-crosses resulted in full fertility in 2 to 7 generations.

Hastie (1973) has described a remarkable situation in the two species of Fungi Imperfecti, *Verticillium albo-atrum* and *V. dahliae*. By using auxotrophic haploid strains he was able to select diploid, protrophic crossed strains. These haploidized infrequently but gave an excess of haploid recombinants which resembled either parent. This restricted genetic recombination was believed to be due to a lack of homology between the chromosomes of the two species. Such behaviour, therefore, resulted in the production of viable, asexual haploids comparable to one of the parental species but showing some features of the other. This is the first demonstration of introgressive hybridization via the parasexual cycle in a Fungus Imperfectus.

These last three examples have been taken from experimental studies. There is not yet any unequivocal evidence that such crosses have occurred in nature. If introgression has occurred in nature it might well prove difficult to detect but it might pay to investigate different species which in some circumstances, at least, occupy the same ecological niches. In such situations it is likely that either isolation will be exhibited most strongly or that introgression will occur.

Polyploidy

Amongst green plants polyploidy, especially allopolyploidy, is widespread and, frequently, the polyploid derivatives are partially or wholly isolated from the parental taxa. The evidence for polyploidy as an important factor in fungal speciation is still not very convincing although several possible cases of polyploidy are now known (Burnett, 1968; Rogers, 1973). The most fully documented cases are in *Allomyces* where isolates of *A. arbuscula* had $n = 16$, 22–24 and 32 and *A. macrogynus* had $n = 14$, 28 and *c.* 56 (Wilson,

1952). Unfortunately, nothing is known concerning the results of crossing different intraspecific cytotypes although the absence of multivalents in the higher polyploid isolates suggests they were not autopolyploids. Claims have been made for the existence of segmental allopolyploids (i.e. an allopolyploid derived from a cross between two cytotypes descended from the same ancestral species) in fungi, on the basis of the occurrence of variable numbers of bivalents and multivalents, e.g. *Cyathus stercoreus* (Lu, 1964); *Xylaria curta* (Rogers, 1968). Here, too, there is no further evidence. An unequivocal demonstration of allopolyploidy in a fungus would indicate an initial breakdown of reproductive isolation. There is, as yet, no evidence of isolation between taxa of different ploidy in the fungi. Genera where such a search might be made on the basis of published chromosomes include *Allomyces* (Wilson, 1952) and *Puccinia* (McGinnis, 1953, 1956; Sansome, 1959).

This account has shown how very fragmentary is existing knowledge concerning isolating mechanisms in fungi. Apart from heterogenic incompatibility in *Podospora anserina*, genetic analysis has rarely been carried far enough for adequate understanding. Nevertheless, the evidence for widespread pre-zygotic isolating mechanisms is striking and suggests that this may be a characteristic feature of fungi. Equally striking are the numerous examples of sympatric isolation which contrast with its apparently less common occurrence in other organisms. This does not necessarily mean that post-zygotic isolating mechanisms are not of importance also. Here the data are so fragmentary that no valid generalizations can be made. This also applies to polyploidy and the breakdown of reproductive isolation; the significance of neither can be adequately assessed in fungi. Much more information is needed but it is already clear that fungi will be especially valuable to the student of evolution in the study of the earliest stages of reproductive isolation within a population occupying a single habitat.

Section 4

Applications of fungal genetics

The preceding sections have been concerned with the formal genetics of individual fungi and their population genetics. Here applications of such knowledge will be considered.

Fungi are widely employed in a variety of industrial fermentations for the production of chemicals, enzymes or antibiotics as well as in the older established industries of brewing and wine making. The application of genetics to their improvement is of increasing relevance if they are to be fully exploited.

Equally important is an adequate understanding of fungal genetics if pathologists and plant breeders are to stay one jump ahead of the activities of what von Haller once described as 'a mutable and treacherous tribe'. Without an adequate understanding of how pathogens vary, control can neither be exercised effectively, nor can plant breeders introduce resistance into crops in a rational manner.

Apart from these direct applications of fungal genetics to applied mycology and pathology, the subject can also contribute to general problems of genetics. The occurrence of linear tetrads on the one hand and the metabolic versatility of the fungi on the other makes them uniquely suitable for the elucidation of certain problems in eukaryotes. The most notable of these are the mechanism of recombination, gene action and regulation within the cell, and the genetic control of development.

CHAPTER 12

Industrial applications of fungal genetics

Introduction

The industrial uses of fungi, already considerable, are both increasing and diversifying. The traditional processes of alcoholic fermentation, baking and cheese making are still of enormous economic value and fungi are unlikely to be displaced. The use of fungi as food is increasing, not only in the mushroom industry, based on *Agaricus bisporus* in Europe and Asia and on the paddy-straw mushroom (*Volvariella* spp.) and shiitake (*Lentinus edodes*) in Asia, but new fungi are being brought into cultivation, notably truffles (*Tuber* spp.), boletes (*Boletus* spp.) and oyster mushrooms (*Pleurotus* spp.), for direct consumption. Filamentous fungi, including yeasts and *Aspergillus* spp., are increasingly being employed as novel sources of protein or, after processing, as synthetic meat substitutes. The rise in the twentieth century of the study of fungal metabolism and an appreciation of its diversity has revealed expanding horizons for the industrial production of primary and secondary metabolic products and the utilization of fungi to perform particular synthetic steps with high precision. Examples are now so numerous that only a few can be mentioned. Amongst primary metabolic products are organic acids such as citric, gallic, gluconic and itaconic; amino acids like lysine and threonine; and alcohols, notably glycerol. Secondary metabolic products include enzymes of the amylase, protease and pectinase types; vitamins, especially riboflavin; antibiotics, notably penicillins, cephalosporins and griseofulvin; alkaloids of the ergot type, gibberellins, carotenoids and, very recently, indole hallucinogens of the psilocybin and psilocin type. Fungi are increasingly being employed to carry out synthetic steps where normal chemical methods are both difficult and only result in low and mixed yields, e.g. stereospecific steps in steroid biosynthesis such as hydroxylation at the 11α position plus saturation at the 4,5 carbon, double-bond position by *Rhizopus nigricans*. These are but a few of the industrial uses to which fungi have been put and which are detailed elsewhere (e.g. Singer, 1961; Peterson, 1963; Hesseltine, 1965; Christensen, 1967; Hoffer and Osmond, 1967; Taber and Taber, 1967; Smith, 1969; Turner, 1971; Brian, 1972).

In many of these situations man has long exercised unconscious selection in his domestication of fungi, for example, in the selection of yeasts for brewing or of Penicillia in cheese making. Conscious selection and the breeding of fungi is increasing. Such methods are about a century old. They were introduced into the brewing industry about 1876 by Hansen at the Carlsberg Laboratories in Copenhagen and they were followed, some sixty years later, by the application of precise genetic methods by Winge in the same laboratory. In other cases genetic methods have been employed with greater or lesser success from the outset, e.g. penicillin production, where mutation production and selection is now beginning to be enhanced by the employment of the parasexual cycle.

Sufficient expertise has not yet been acquired in any fungus, or in any production, for general principles to have become established. Alikhanian (1973), writing of the microbial production of secondary metabolic products, has suggested five generalizations, namely:

(i)   The choice of an effective mutagen is highly specific for any particular organism.
(ii) Selection associated with the employment of a particular mutagen decreases progressively in effectiveness. A change in mutagen is then desirable.
(iii) The best yield of mutants for a product may not coincide with the optimum dose for producing morphological mutants or auxotrophs.
(iv) Product yield is not necessarily correlated with morphological change.
(v)   Multistep breeding, involving a gradual accumulation of slight increases over 10–20 steps, associated with the use of mutagens, is often the most effective selective procedure.

It will be apparent that these observations reflect the heavy dependence given to mutant selection in the improvement of many microorganisms. In the same article Alikhanian also considers 'hybridization', or breeding, and notes that:

(i)   Recombination and segregation enable particular genotypes to be obtained and biosynthetic sequences to be elucidated.
(ii) Breeding can disrupt a 'character' where selection has failed.
(iii) Breeding enables heterozygotes, which show hybrid vigour, or polyploids to be obtained.

He writes, 'However, in spite of the apparent importance of microbial hybridization from the point of view of a breeder, progress in the field is still insignificant.' This is, indeed, the case as will become apparent in the three processes now to be considered, namely, penicillin production, yeast improvement and mushroom culture.

Penicillin production

The antibiotic activity of *Penicillium notatum* (in the *P. chrysogenum*-series) against gram +ve bacteria was first detected in 1928 by Sir Alexander Fleming (1929) in plate culture as the result of an experimental contamination. Fleming retained his culture and attempts were made to isolate stable preparations of penicillin from it by Raistrick in 1932 (Clutterbuck and others, 1932) and finally and successfully by Lord Florey and his colleagues in 1939. The stability and efficacy of penicillin *in vivo* having been established (Chain and others, 1940; Abraham and others, 1941) programmes were developed to improve yield. These still continue and the programme has passed through three phases (a) selection of natural variants, (b) induction of mutants and their selection and (c) breeding by other means.

Selection of natural variants

Many wild strains were isolated and shown to possess a great range of variation in penicillin production, morphology and sporulation, both in liquid-surface culture and in submerged culture (Table 12.1).

TABLE 12.1 Variation in penicillin production of 241 wild isolates of the *Penicillium notatum—chrysogenum* group collected throughout the world. (Data from Raper, Alexander and Coghill, 1944.)

In surface culture

Penicillin, I.U./ml

Range	0	1–10	11–20	21–30	31–40	41–50	51–60	61–70	71–80	81–90	91–100	100+
Strains	24	31	23	45	29	13	18	32	16	4	4	2
% of total	10	13	10	19	12	5	7	13	7	2	2	1

In submerged culture

Penicillin, I.U./ml

Range	1–10	11–20	21–30	31–40	41–50	51–60	61–70	71–80	81–90	91–100	100+
Strains	3	9	8	7	3	6	3	3			
% of total	7	21	19	17	7	15	7	7			

Note: The data for submerged culture represent a sample of those selected from the strains shown in the first part of the table, 22 from those giving more than 60 I.U. and 20 from those giving less than 50 I.U. excluding those giving no yield at all.

Attempts were made to correlate morphological features with penicillin production but despite an early claim for an association between antibiotic yield and sporulation, they proved spurious (Raper and others, 1944). Progress was made with surface cultures initially but when submerged culture came into vogue different selections overtook the earlier ones. The first of the submerged cultures NRRL 832 (NRRL = Northern Regional Research Laboratory, Peoria, U.S.A.) was displaced by a culture from a single (uninucleate) spore of *P. chrysogenum* derived from a strain, NRRL 1951 found on a rotten cantaloupe. This culture, NRRL 1951 B25, gave a better yield than Fleming's original strain, 100–200 units/ml compared with 2 units/ml, but no better than the best surface culture, NRRL 1249 B2, which produced 220–280 units/ml. However, because NRRL 1951 B25 was effective in submerged culture, it was destined to be the progenitor, so far as it is known, of all subsequent industrial strains. Thus, modern penicillin production stemmed largely from the unbelievably narrow genetic basis of a single, haploid nucleus, even though it was subsequently modified by mutation.

By this stage several problems with a genetic basis had already arisen. These included strain degeneration, variability and the preservation of stock cultures. The last gave rise to the widespread adoption of three basic methods: soil culture, lyophilization or freeze drying and preservation under mineral oil. Initially, the first of these was the preferred method and is still employed. Basically, a tube of sterile soil is inoculated with a heavy spore suspension. The tubes are shaken, desiccated in a vacuum desiccator and stored in a cool, dry place. It will be recalled that in *Neurospora* there is evidence that the spontaneous mutation rate is reduced by low temperature (p. 32) and that selection appears to favour wild types of *Fusarium* over variants in soil tube culture (p. 200). In such cultures growth is slight and this is true of cultures stored under mineral oil where metabolism is also reduced. Lyophilization, freeze-drying and the more recent technique of silica-gel storage all maintain the cultures so treated in a state of suspended animation. Presumably under all these conditions the replication of DNA is minimal, or non-existent and the negligible metabolic activity reduces thermal agitation and hence the opportunity for loss or replacement of the bases in nucleic acid. In fact, no tests have been reported on mutation rates under these conditions.

Strain degeneration was shown by Whiffen and Savage (1947) to result in a loss of 50 to 140 units/ml after 5–7 subcultures of a surface-cultured strain. Such 'degenerate' cultures sporulated more rapidly and heavily than their parents and formed a more solid mycelial pad. It was shown that if the subculturing procedures rigorously excluded spores, the yield was unimpaired for at least 50 successive subcultures. In contrast, Hansen and Snyder (1944) had described the origin of a mycelial (M) type in single-spore cultures which showed reduced yield and an enhanced production of yellow pigment. The normal type was conidial (C). They attributed this dual phenomenon—the M and C types—to heterokaryosis but, as discussed earlier, it could be due to segregation in either a heterokaryon, or a hetero-plasmon (p. 120). At that time the causes of strain degeneration were not clearly understood and in subsequent work other phenomena arose; none have been accounted for in detail.

Selection of mutants

In the next phase mutation techniques were applied followed by selection and this has continued to the present time. *P. chrysogenum* has probably been more exploited by 'mutation breeding' than any other organism. The reason for this changed approach was that the screening and selection of

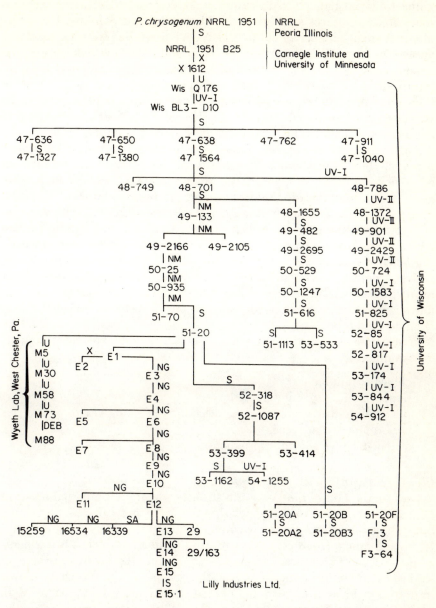

FIGURE 12.1 The phylogeny of the Wisconsin strains of *Penicillium chrysogenum* with some indication of their subsequent development. Key: Mutations brought about spontaneously, S; by X-irradiation, X; by u.v.-irradiation, wavelength unspecified, U; 275 nm, UI; 253 nm, UII; nitrogen mustard, NM; nitrosoguanidine, NG; diepoxybutane, DEB. (Based on Stauffer, 1961; Elander, 1967.)

natural variants had led to no further major increase in yield. For example, some 60,000 cultures were unsuccessfully assayed at Stanford University alone. A small number of cultures of NRRL 1951 B25 were exposed to a variety of mutagenic agencies; nitrogen mustards, ultraviolet and X-irradiation. One conidial isolate after X-irradiation, X1612, proved to show both a dramatic increase in yield, 300–500 units/ml, and to do better in fermenters than in laboratory culture. This strain was taken up by the Wisconsin group who, after u.v.-irradiation, obtained Q176 with a yield of nearly 1000 units/ml. This was generally released to industry later in 1945. A further irradiated selection from it, strain BL3-D10, lacked the yellow pigment, chrysogenin, which interfered with the recovery and purification of penicillin. From this strain all further mutant-selection lines were derived in the next decade (Figure 12.1, p. 239; Backus and Stauffer, 1955).

Three selection lines were established. One was largely based on selection of spontaneous variants, another on variants derived by u.v.-irradiation and the third, in 1948/9, largely from the use of a nitrogen mustard (methyl-bis(β-chloroethyl)amine). Wis 53-399 with a yield of 2500 units/ml was derived from this last lineage and was the highest yielding Wisconsin strain in all three lines. Since then other mutagens have been employed, e.g. nitrosoguanidine, nitrous acid, and diepoxybutane in Wyeth strains (Elander and others, 1973).

Programmes such as these raise a variety of problems of a general nature, for example:

(a) choice of mutagen,
(b) sampling and screening procedures,
(c) the scaling-up problem, i.e. selecting in one environment for performance in another,
(d) genetic basis of penicillin production.

Choice of mutagen

Because the best of the Wisconsin strains came from a nitrogen-mustard treated lineage, it has been claimed that this is the most effective technique, but for several reasons this view cannot be sustained. Firstly, not only did Reese and others (1949), starting with Q176, induce and select better penicillin producers from ultraviolet-irradiated lines in a test involving over 2000 strains, but subsequent selections from the best Wisconsin strain have shown improved yield with u.v.-irradiation, e.g. Wyeth strains M5, M30, M58, M70 (Elander and others, 1973) and Glaxo strains (Ball, 1971). Secondly, in the nitrogen-mustard lineage there were long periods where spontaneous variants were selected successively. Finally, it will be recalled that Q176 was derived from X1612, an X-irradiated derivative from NRRL 1951 B25, which out-yielded all other mutants derived from u.v.-irradiation and nitrogen-mustard treatment. This makes the point that the history of a strain, as well as its nature, can modify the apparent efficacy of a mutagen.

For example, an entirely separate strain from those already considered, derived by selection of spontaneous variants, was improved immediately by u.v.-irradiation which was more effective than nitrogen mustard (Stahmann and Stauffer, 1945).

In this work it was also noted that the highest mutation rate under u.v.-irradiation did not coincide with the highest rate for killing. It is generally accepted that lower doses, e.g. giving 25–30% survivors, are often more effective in inducing mutants with improved penicillin yields than a dose which gives maximum kill (Stauffer, 1961). The reasons for this are not entirely clear. One possibility is that after repeated cycles of mutation and selection with a particular mutagen, the strain becomes more sensitive to its lethal action (Brown and Elander, 1966).

Although observations such as these support Alikhanian's first three generalizations (p. 236), it is clear that little critical work has been done on the comparative efficacy of different mutagens. Mutant induction and selection is still a very empirical activity.

Sampling and screening

In a sense these are different aspects of the same problem, namely how to assess the efficacy of a selection programme. In general, the probability, p_m, of picking up a mutant in a sample size, N, when the mutation rate is α is:

$$p_m = 1 - \frac{1}{e^m} \quad \text{where } m = \alpha N.$$

The efficacy of sampling will be determined by p_m/n, i.e. the smallest effective sample size, and the most convenient size will be inversely proportional to the mutation frequency. The simple application of this relationship has, in practice, to be modified for a variety of reasons. Any increased yield shown by a mutant may be relatively small and its detection is therefore liable to a variety of errors. Two sources of error are of especial importance, namely, variation between batches of medium and the validity of comparisons between new and original strains. The latter will have been stored during the selection procedure for the new strains and so may have undergone spontaneous mutation. In addition, transitory variation may have been induced as a consequence of exposure to the mutagen. Ball (1973a) has pointed out that comparisons of original and new strains, both on medium from the same batch and from different batches, should be employed. Such controls of genetic variation are important but add to the overall logistic problems of space and time involved in testing. Davies (1964) carried out a computer simulation of different sampling procedures employing data from selection programmes and suggested that a cyclical two-stage testing procedure should be adopted. Initially, the most convenient size of sample for detecting a mutant, in accordance with the relationship already given, should be taken. For example, to detect a 20-fold increase in the average mutation rate with a 99% probability of detecting a mutant, some 500 samples need to be screened (i.e. $m = 0.1$, $p_m = 99\%$, $\alpha = 2 \times 10^{-5}$). If the strains to be carried on for further testing have to be reduced from N to n, then the second stage should be reduced to $\sqrt{N\,n}$, i.e. for $N = 500$ and $n = 5$, 50 strains should receive second-stage testing. This two-stage cycle can be repeated until no further yield is obtained.

A further useful, initial control is to judge the yield of a new selection and compare it with that of its conidial progeny. This is of especial value in guarding against the possibility of transitory variation induced by mutagenic treatment.

Screening procedures set particular problems for particular products but the penicillins illustrate some of the problems. It was not realized at first that several different penicillins exist (Figure 12.2).

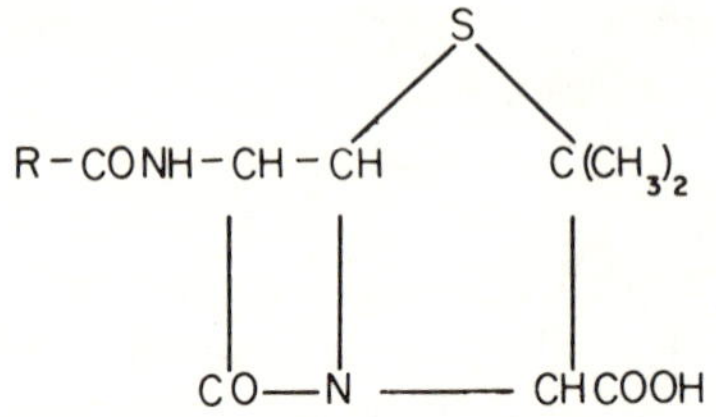

R can be

Δ^2 – pentenyl	– CH_2 $CH{=}CH$ CH_2 CH_3	Penicillin I or F
n – amyl	– CH_2 CH_2 CH_2 CH_2 CH_3	Dihydropenicillin F
benzyl–	– CH_2 C_6H_5	Penicillin II or G
p – hydroxybenzyl–	CH_2 C_6H_4 OH	Penicillin III or X
n – heptyl	– $CH_2(CH_2)_5 CH_3$	Penicillin IV or K

FIGURE 12.2 The basic penicillins.

P. chrysogenum, NRRL 1984 N22, developed through u.v.-irradiation, gave a yield by weight of Penicillin X of some 70%. Assay with *Bacillus subtilis* as test organism gave a similar estimate but assay by *Staphylococcus aureus* only 50%. This is because Penicillin X is less effective against *S. aureus* than against *B. subtilis*. Strain X1612 was preferred because it showed high activity against *S. aureus*, since it produced far more Penicillin G—that most effective against this bacterium—and only 20% by weight of Penicillin X. Thus the assay procedure must give the maximum amount of information to be effective.

Environment also influences microbial products. In most synthetic media Penicillin K predominates; even Q176 only produces a 50:50 mixture of Penicillins G and K. In corn-steep liquor, however, the balance is altered dramatically in favour of the desirable Penicillin G. If precursors of the side chain of this penicillin are added to a synthetic medium the same result can be achieved, e.g. phenylacetic acid or phenylacetamide. Consequently selection on Q176 and its successors was practised in the presence of such precursors by the Wisconsin Group. Therefore, the subsequent improvement of strains, e.g. 47-1564 (Figure 12.1) was readily achieved by selecting for their ability to incorporate side-chain precursors more effectively. Alikhanian (1962, 1965) suggested that once this capacity for incorporation was saturated no further progress would be achieved unless the pattern of selection was changed. Although he employed a strain with a high ability to incorporate a precursor, he selected *in its absence* and so obtained even higher yields by exploiting novel variation. This raises two issues. Firstly, it is possible to select successfully without realizing exactly what attribute is being selected for. Secondly, it is not clear for fungi whether it is better to (a) exploit selection for one trait to the limit and then to select for another desirable trait, or (b) to attempt selection for both traits simultaneously or in alternating sequence. Historically, in penicillin improvement, the first of the procedures has been followed.

Scaling-up

This is basically a problem of assessing genotype–environment interactions of a most complex kind, but potentially capable of study by methods such as those described for simple situations in Chapter 7 (pp. 140–143). The Wisconsin group assayed their strains in shake-flask cultures and found that the ranking achieved under these conditions was usually maintained in 30-litre fermenters although there were departures in commercial 40–50,000-litre fermenters. Some data concerning Lilly strains illustrates this (Elander, 1967; Table 12.2).

At that time attention was paid to searching for correlated characters. For example, vigorous and thickly growing cultures often do best in large-scale units but less well in shake flasks, e.g. E4 in Table 12.2 is a dense grower in

TABLE 12.2 Improvement in benzyl penicillin production by successive nitrogen-mustard derived variants of *Penicillium chrysogenum* as reflected in different conditions for fermentation. (From Elander, 1967.)

Strain	Percentage improvement over predecessor in		
	shake-flask culture	5 l stirred culture	30,000 l fermenter
51.20			
E1	48	42	10
E2	14	22	14
E3	18	−7	12
E4	17	8	25
E5	10	19	1
E6	2	15	2
51.20 to E6	109	99	64

deep fermentation. In general, the Wisconsin group appeared to have demonstrated a correlation between increased penicillin yield and reduced sporulation and growth. The greatest change was at the outset of selection. Between NRRL 1951 and Q176, sporulation and growth were reduced by 60% but yield increased sixfold. Later, a further threefold increase in yield was associated with only an additional 10–12% reduction in vegetative vigour. These changes could represent a correlated response due to linkage of loci determining growth characteristics with those determining yield, such as has been described in many selection experiments (see p. 192). If this were the case then it might be expected that high-yielding strains could be found showing vigorous vegetative growth. Lilly strain E15 grew and sporulated abundantly on minimal medium (Elander, 1967) and recently Ball (1971, 1973a) has described Glaxo strains with yields of several thousand units/ml where sporulation and faster growth could be achieved in a few cycles of selection. All these strains were derived ultimately from Wisconsin strains. It is difficult, however, from published data to judge how far such strains are comparable with earlier strains in the lineage. There does seem to be some, unavoidable, reduction in vegetative vigour associated with high penicillin yield. This could reflect conflicting, internal physiological or metabolic balances or pleiotropic effects of high-yield determining genes. In fact it is not possible to assess the genetic basis of genotype–environment interactions in connexion with penicillin yield because of lack of relevant information.

TABLE 12.3 Improvement in penicillin production in a family of mutant strains of *Penicillium chrysogenum* developed from strain NRRL 1951. (Data from Elander, 1967.)

Strain	Improvement in production over	
	Previous strain	Fleming strain
Fleming		
NRRL 1951	0·50	0·50
NRRL 1951 B25	2·70	4·05
X1612	0·06	4·90
Q176	0·52	9·00
47–1564	−0·04	8·55
48–701	−0·14	8·15
49–133	0·08	12·70
51–20	0·73	22·60
E15	1·05	50·64
E15–1	0·09	55·00
Cumulative total	5·55	176·09

Note:
1. The strains concerned can be identified in the family lineage in Figure 12.1, p. 239.
2. The measurements of production were made in shake-flask fermentation under modern conditions.

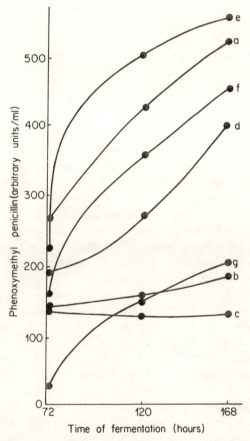

FIGURE 12.3 Penicillin production by strains involving *Penicillium chrysogenum* E15. Key: a, original haploid protroph (w^+); b, c, derived haploid auxotrophs, $(w\ ade)$ and $(y\ met)$; d, e, derived heterokaryon and diploid, $(w\ ade$ and $y\ met)$ and $(w\ ade\ ++/++\ y\ met)$; f, g, diploid mitotic segregants from original diploid, $(w\ ade\ ++/++\ y\ met)$ and $(++++/++\ y\ +)$; w, y, white and yellow spore colour, respectively; *ade*, *met*, auxotrophic for adenine and methionine respectively. (From Elander, 1967.)

test between a low-producing strain NRRL 1951-*pro* and a high-producing strain Wis 49-133-*nic* which had been derived from it in 9 successive steps. Homokaryotic segregants from this heterokaryon showed a clear association between the nuclear marker and yield, i.e. *pro* segregants were associated with low yield and *nic* segregants with high yield. Thus penicillin yield is evidently determined by nuclear genes and there is clear evidence from both heterokaryons and parasexual recombinants that several different genes are involved. For example, the 'New Hybrid' strain of Alikhanian and Borisova (1956) and the *y met* and *w ade* heterokaryon derived from the Lilly E15 strain (Figure 12.3. Elander, 1967) both show significantly higher yields than their parent homokaryons, suggesting complementation between non-allelic genes.

Caglioti and Sermonti (1956) attempted to map *pen-1*, a determinant for increased yield of penicillin, by mitotic recombination. They showed that its locus was probably distal to *pro-1* and *met-1*, two auxotrophic markers in one of the three linkage groups they tentatively identified. More recently, Ball (1971, 1973a, b) has confirmed, by haploidization analysis, the existence of three linkage groups in a Glaxo strain. He induced increased yield in five separate strains, each of which carried colour or auxotrophic markers. He designated the determinants for increased yield t_1 to t_5 and showed that t_2, t_3 and t_4 could be assigned to his linkage group I, t_1 to either his group II or III and t_5 to his group III. Neither the nature of the determinants nor specific locations were determined. However, since the three determinants in group I showed characteristic and different ranges of penicillin yields, they were presumably not identical (Figure 12.5, p. 251). Sermonti (1961) has also provided segregation data from a heterozygous diploid, which suggests that several genes are involved.

Two phenomena have been described which suggest that genes determining increased yield of penicillin are recessive and that independently induced mutations could be allelic. In most cases when a diploid is derived via the parasexual cycle between a low-yielding and a high-yielding strain, its yield is comparable with that of the former. Thus mutant genes which result in increased yield are largely recessive (Sermonti, 1957; Macdonald and others, 1963). Diploids obtained between strains derived from a common ancestor but carrying independently induced mutations for higher yield, gave yields not much lower than their parental strain but notably higher than the yield of the common ancestor (Macdonald and others, 1964). The simplest explanation of this observation is that the strains carried a number of allelic mutant sites in common, even though the sites had been independently mutated.

In summary, therefore, it looks as though there is a limited number of determinants, recessive in their expression and located on more than one chromosome, which determine increased penicillin yield in *P. chrysogenum*. Whether these determinants are oligogenes, polymeric genes, polygenes, or some combination of these, cannot yet be ascertained. Their expression is certainly modified by the genetic background, for most auxotrophic mutants reduce yield drastically. Macdonald and others (1963) found that gradually decreasing yields were obtained when single auxotrophs, deficient for vitamins, amino acids or adenine, were introduced into a strain. They speculated that a genetic block to adenine synthesis was likely to have more profound and widespread metabolic repercussions than blocks to either vitamin or amino acid synthesis. Hence adenine auxotrophs would exert a more profound effect on penicillin yield. Although possible, the argument is not convincing in the absence of any information on the biochemistry of the blocked syntheses. That reduced yield was a pleiotropic effect of the mutant allele was suggested by comparisons between auxotrophs and reverted protrophs.

Possibilities of genetic improvement

There is little evidence that any technique other than the induction and selection of mutants has been employed successfully with *P. chrysogenum*. This has not prevented suggestions concerning possible breeding procedures and some experiments being made. That they do not seem to have led to progress probably reflects the history of the existing industrial strains. Breeding involves the parasexual cycle and there are difficulties in achieving both this and unimpaired segregation. In addition, strain variability and instability frequently cause problems.

It has often proved difficult to initiate the parasexual cycle because heterokaryons are not produced easily. This is especially true of the Wisconsin strains and those others where selection for yield has resulted in poorly growing strains. In such circumstances auxotrophically marked strains can be grown on partially supplemented media, the supplement being either small quantities of complete medium or limiting quantities of the nutrients required by the auxotrophs (Sermonti, 1954). Another procedure has been to inoculate strains in saline solution spread over the surface of solid complete medium in plates. After 4 to 5 days growth, the mycelium is macerated, mixed with minimal medium and plated (Macdonald and others, 1963). Even if heterokaryons develop, they often sporulate poorly so that it may prove necessary to plate such conidia as develop, allow them to re-form heterokaryons and then to select the best sporulating areas on plates and repeat the process. Once sporulation has been increased, either diploid conidia can be selected in the usual way (see Chapter 2) or vigorous diploid sectors, which sometimes occur, can be used. Such problems need not arise if the initial strains selected are vigorous growers and good sporulators, e.g. Wyeth and Glaxo strains (Ball, 1971; Elander and others, 1973).

Even when a heterokaryon can be produced, difficulties have been experienced with the instability of diploids and their segregation. Instability may take a variety of forms. It will be recalled (Chapter 9, p. 190) that the phenomenon of 'population patterns' had been met in the early haploid selections of the Wisconsin series such as Q176. It is not clear whether haploids exhibiting high variability of this nature will give rise to equally variable diploids. The only data bearing upon this problem are those derived from a study of the Lilly diploid referred to earlier and derived from the production haploid E15 (Elander, 1967). The spontaneous percentage variation in the diploid compared with the haploid was 9·2% to 31·9%, as assessed by the proportion of colour types which segregated. This difference was even more striking after exposure for 2 minutes to u.v.-irradiation, when the diploid:haploid derived variants were 11·6% and 41·1% respectively. Here, therefore, the diploid is far more stable than its haploid progenitor.

In contrast, other diploids, such as that described by Ball (1971, 1973a), show greater instability than their haploid progenitors but within the same range. This variability is morphological and is expressed as differences in sporulation. Thus the diploid showed twice as many poor sporulating types as the haploid, while densely sporulating types were increased at least tenfold. Both Ball (1973a) and Roper (1973) have compared this phenomenon with that of 'mitotic non-conformity' described by Nga and Roper (1969) in *A. nidulans*. It will be recalled (p. 44) that this results from the existence of duplicate segments of quite small fractions of the genome and which have probably originated through translocations. It is not clear whether the type of instability described by Elander is comparable with that described by Ball.

Macdonald (1966) has described an even more surprising phenomenon in diploids. He synthesized 68 sister diploids from the same two parental haploids which had penicillin yields of about 3000 and 3500 units/ml, respectively. About one-third (24) of the diploids had yields comparable with those of the parents, ranging from 2400 to 3600 units/ml but the rest (44) ranged from 300 to 1200 units/ml, with a mode of about 600 units/ml. He suggested that this was due to the occurrence of spontaneous dominant mutations, resulting in reduced yield. These were either present in very low frequency in the parental haploids or arose during the synthesis of the diploids. In either case they must have had a high selective value in diploids. There is some evidence that such mutants do occur in haploid cultures. If this explanation were correct it would be expected that diploids homozygous for genes which increase penicillin yield would show similar behaviour to the heterozygous diploids studied. Evidence on this point is indecisive but some homozygous diploids do not show this phenomenon (Sermonti, 1957; Macdonald, 1966).

Finally, even if stable diploids are achieved with comparable yields to the best haploids, their segregation may be abnormal when compared with *A. nidulans* or *A. niger*, for example. Macdonald and others (1965) noted that there is a tendency for an unusually high proportion of the haploid segregants to possess the parental characters. They show so-called 'parental genome segregation'. A possible explanation for this phenomenon is the occurrence of different chromosomal rearrangements in the two haploid parents, so a reciprocal translocation in one parent would reduce or prevent recombination at haploidization. This phenomenon has been fully studied in *A. nidulans*, where many translocations have been detected after high doses of X- or u.v.-irradiation. Heterozygous translocations not only result in the deletion of certain genotypes but also can be associated with unstable strains and changes in sporulating ability (Käfer, 1962; Tector and Käfer, 1962). It is evident from other organisms that many mutagenic chemicals are radiomimetic and give rise to chromosome aberrations of various kinds (Kihlman, 1966).

When the history of the degree of exposure to mutagens which has characterized most industrially developed strains of *P. chrysogenum* is considered, it seems not improbable that they now contain various types of chromosomal aberration. This is supported by direct experimental investigation (Ball, 1973b; Table 12.4), by the small number of linkage groups detected—not more than three to date—and by the apparent restrictions on recombination which nearly all workers have encountered. It is consistent with the fact that Ball (1971) was successful in detecting recombination after inducing mutant markers with u.v.-irradiation at a level comparable with that which in *A. nidulans* had not been found to induce transloca-

TABLE 12.4 The production of recessive lethals and translocations in a diploid strain of *Penicillium chrysogenum* exposed to different mutagenic treatments. (Data of Ball, 1973a.)

Lesion and linkage group	Mutagens employed				
	UV	γ	EMS	NG	Totals
Single lethals					
Group I	3	1	1	2	7
Group II	1	1	0	0	2
Group III	3	4	1	2	10
Double lethals					
In homologues	2	6	8	11	27
In non-homologues	0	0	0	1	1
Translocations					
Groups I + II	0	1	0	0	1
Groups I + III	1	1	0	0	2
Groups II + III	0	0	0	0	0
Multiple	1	1	0	0	2
Lethal	0	0	2	2	4
No lesion detected	8	4	7	2	21
Totals	19	19	19	20	77

Note:

1. The diploid employed, $w:lys/br:nic:his$, had w and br in repulsion on Group I, lys and nic in repulsion on Group II and his marked Group III.

2. Lethal mutations were detected by the absence of a segregating marker; translocations were detected by the absence of expected recombinant genotypes.

3. Abbreviations indicate: w, white spore colour; br, brown spore colour; his, lys, nic, requirements for histidine, lysine and nicotinamide, respectively. EMS, ethylmethane sulphonate; NG, nitrosoguanidine; UV, u.v.-irradiation; γ, γ-irradiation.

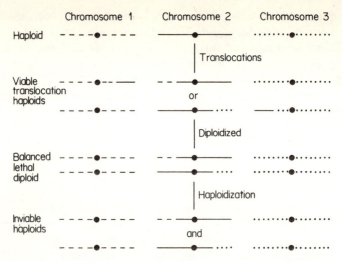

FIGURE 12.4 Diagram to illustrate the origin of a balanced lethal diploid such as could be developed for *Penicillium chrysogenum*. (Based on Ball, 1973b.)

tions in high frequency. No direct cytological observations of chromosomal aberrations have been made.

If the difficulties described in producing diploids can be overcome, they have many attractions for a breeder. Firstly, there is always the possibility of a heterotic effect on yield, to date only found sporadically (Sermonti, 1959; Figure 12.3, p. 245 and Elander, 1967).

Even if this does not occur, the possibility exists of selecting more productive segregants. If the latter is to be achieved, it would be necessary either to start with a strain lacking chromosomal rearrangements or to induce variants with increased penicillin yields from an existing strain, without inducing novel chromosomal aberrations. Thus the future improvement of existing industrial strains is likely to be dependent upon the use of mutagens causing minimal cytological disturbance. In so far as spontaneous de-diploidization can always occur, systems have been proposed to mitigate such effects. Macdonald (1964) suggested the use of parental haploids which grew poorly in complex fermentation media, so that parental segregates arising during fermentation would be selected against. Azevedo and Roper (1967) suggested an alternative system based on studies with *A. nidulans*. This was to induce recessive lethals in the diploid so that haploid segregants would be eliminated as they arose. Ball has discussed this with particular reference to *P. chrysogenum* (Figure 12.4). It has been suggested that the Lilly diploid referred to earlier (p. 248) is just such a strain, for it is not only extremely stable but it will be recalled that even after u.v.-irradiation the proportion of viable segregants it produced (*c.* 10>) was hardly increased. Provided that diploid stability and unrestricted parasexual recombination can be achieved, it should be possible to make further progress by selection. There is evidence that this can be achieved. Elander (1967) described Selection no. 2, a spontaneous segregant from the Lilly diploid, which produced nearly 25% more antibiotic than its parent, which itself yielded better than the parental production haploid E15. Ball (1973a, b) has also described recombinant segregants with improved yield (Figure 12.5).

In retrospect, it is clear that the first 25 years of scientific breeding of *P. chrysogenum* have demonstrated both the possibilities of 'mutation breed-

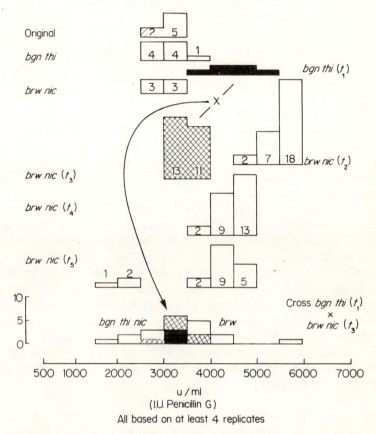

All based on at least 4 replicates

FIGURE 12.5 The penicillin production of some genetically marked strains of *Penicillium chrysogenum* and the yield of a cross between two of them (*bgn thi* t_1 × *brw nic* t_3). Key: Numbers in histograms indicate number of strains. Symbols: *bgn*, bright green and *brw*, brown colonies; *nic*, *thi*, auxotrophic for nicotinic acid and thiamin, respectively; t_1–t_5, determinants of penicillin yield. Solid histogram, *bgn thi* t_1 phenotype; cross-hatched histogram, *brw nic* t_3 phenotype, others indicated. (Based on data of Ball 1973a.)

ing' and the dangers inherent in it if mutagens are employed causing cyto-logical aberrations. The next 25 years should see a far more effective application of the parasexual cycle to breeding. Finally, it is worth recalling that such success as has been achieved can be traced back to the extra-ordinary narrow genetic basis of the single haploid nucleus of the conidium which gave rise to strain NRRL 1951 B25. It would seem prudent to start from a broader genetic base than this in future. It is probable also that there will be more exploitation and application of experimental studies on genotype–environmental interactions in order to deal with the perennial problem of all industrial fermentations, i.e. scaling up.

Yeast improvement

There is surprisingly little evidence for the application of genetics to yeast improvement, despite the fact that since Winge's (1935) classic paper on *Saccharomyces cerevisiae*, yeasts have played a role second only to *Neurospora* in the development of fungal genetics.

It is clear that there has been a long process of unconscious and conscious selection of locally maintained strains with particular attributes. This has proved reasonably efficacious, perhaps because the principal attribute, fermentative ability, is one presumably selected for in wild yeasts in nature. In this sense yeast improvement differs from selection for penicillin, for example, which is a secondary metabolic product that is probably not produced in great quantity by wild fungi. On the other hand, it is clear that yeast technologists still rely heavily for improvement on the manipulation of the fermentation environment and the great phenotypic plasticity of yeasts. The rise of continuous fermentation processes and the establishment of larger manufacturing groups by merger has led to a need for a more restricted range of uniform yeasts with precisely defined attributes. Here the breeding of yeasts is likely to be of greater importance. The production of new strains of baker's yeasts with specific properties is a case in point (Fowell, 1966).

Two features restrict such progress. The first is the problem of strain variability and degeneration, the second that of maintaining the recombina-tion processes effectively.

The maintenance of reproductive vigour

The properties of yeasts frequently alter with repeated serial transfers after several months or years. Thorne (1962) studied this in brewery yeasts (*S. cerevisiae*) and found that cultures showed increased flocculence and a loss of fermentation efficiency. Another commonly occurring condition is the loss of ability to sporulate after prolonged maintenance on artificial media. The causes of these 'degenerative' changes are not clear. Thorne (1962) invoked spontaneous double mutation in which he supposed an initial loss mutation was succeeded by one which enhanced the growth rate of the mutant so that it outgrew the normal cells. The evidence is not convincing nor, indeed, is that for the alternative hypothesis that the changes result from mitotic recombination (Emeis, 1965). The evidence from studies with growing yeasts is that the rate of production of variants is appreciably higher than 10^{-6} or 10^{-7} (Chester, 1963). Accordingly, the best way of reducing such degeneration seems to be to limit cell division and growth and for this, culture under mineral oil and storage at 4°C appears to be the best method (Hartsell, 1956). This method maintains viability far longer than other methods such as freeze drying, lyophilization or very low-temperature storage in liquid nitrogen. Low-temperature storage results in high cell mortality and may result in inadvertent selection for mutants (Wynants, 1962).

In those cases where mixtures of strains are employed, especially in brewing, continued culture can result in selection against particular components and so change the behaviour of the mixture (Hough, 1957). Another and quite common cause of strain degeneration is contamination with bacteria or wild yeasts. Re-selection is the only course then open.

Attention has already been drawn to the loss of sporulating ability. This is a widespread problem, particularly common in certain yeasts, e.g. bottom-fermenting yeasts (*S. carlsbergensis*), which sporulate poorly, if at all. Poor sporulation is often associated with low spore viability and both may be related to polyploidy. Emeis (1958), for example, only achieved an average of 4% sporulation and 0·07 to 2·00% viability in trials with 11 strains of *S. carlsbergensis*. The problem may be overcome in various ways and its resolution depends, in part, on its cause. For example, a triploid yeast clone gave $2n$ spores with 1% viability; selection from the few viable diploids gave rise to completely viable haploids capable of mating (Fowell, 1956). In other cases mitotic haploidization has been induced by p-FPA (Emeis, 1966). Another technique is essentially a method of selecting the spores in the culture and this depends on the slightly greater temperature tolerance of spores over vegetative cells (Wickerham and Burton, 1954). The culture is held at 50 to 60°C for a little longer than is necessary to kill the vegetative cells. When spread and plated out the majority of colonies are derived from spores. Successive treatments can result in re-selection for good sporulating ability.

Even when haploids have been obtained, their mating competence may be weak. The resolution of this problem depends upon finding the most suitable environmental conditions for successful mating. These will vary for different species and strains. For example, mating of *S. cerevisiae* is usually favoured by high aeration and glucose concentration, *c.* 10% (Jakob, 1962; Fowell, 1969), whereas in *Hansenula wingei* 1% glucose gives the best result (Brock, 1961).

Crossing technique

For industrial purposes it is essential to be able to recover crosses free from contamination and preferably in quantity. Thus the classical method of pairing spores in a droplet of wort (Winge and Laustsen, 1938) is not satisfactory and some modification of the mass-mating technique devised by the Lindegrens is desirable (Lindegren and Lindegren, 1943). Fowell (1969) has described such a method. This involves incubating for 24 hour at 30°C to obtain a vigorously growing culture and then mixing compatible mating types in a drop of water on a slide. A drop of this mixed suspension is transferred to an agar film of high-carbohydrate agar, e.g. 16% malt agar, on a cover slip. The slip is used as the roof of a moist chamber kept at 24°C for 3 to 4 hours, when the mixture is subcultured 2 or 3 times to encourage active growth and to enable diploid cells to outgrow haploids. The ovoid zygotes are usually larger than the spherical haploids and can be isolated with a micromanipulator. The method commonly employed by yeast geneticists of mating auxotrophs and selecting protrophic diploids on minimal medium is unsuitable. This is because the mutagenic treatment employed to induce the auxotrophs may impair the industrially desirable traits of the strains. Indeed, in contrast to the situation in *P. chrysogenum*, there is little evidence that induced mutants have been employed in yeast breeding.

Desirable and undesirable characters

The preferred methods in yeast breeding have been the selection of meiotic recombinants. The early studies of Winge at Carlsberg and the Lindegrens at Carbondale were concerned to investigate recombination of the different fermentation characteristics of hybrids between different species. They revealed a great wealth of genetic variability and breeding has largely exploited this natural reservoir. For example, most yeasts are grown in a complex medium containing a mixture of carbohydrates in which maltose is frequently an important constituent. The demonstration that maltose fermentation is due to at least four dominant, polymeric genes (Winge and Roberts, 1948, 1950, 1952; and pp. 125–126) was, therefore, of crucial importance in ensuring that industrial yeasts acquired maximum maltose-fermenting ability. Moreover, a species with several such genes is often found to have no more than one for fermenting sucrose and *vice versa*. This is because the *MAL* and *suc* loci are extremely tightly linked and are usually in repulsion (Winge and Roberts, 1952). Thus, for instance, baker's yeasts usually possess high invertase activity but the maltose activity is variable and is often the limiting factor. It is desirable, therefore, that such yeasts should possess a genotype which will confer the most efficient utilization of maltose on the cells, i.e. *MAL1* which, although adaptive, is most highly so and is also the most efficient (see Table 7.1, p. 126).

Fermentation efficiency, i.e. fermentation velocity/nitrogen content, is clearly a desirable character in a yeast, but like many industrially desirable characters, it is complex. Fermentation rates vary widely in different strains of yeasts. When the ability to produce CO_2 was used as a measure in a study of 12 diploid baker's yeasts and their progenies, it was found that the range was very great, suggesting that all the parental $2n$ strains were heterozygous (Lindegren and others, 1945). Later studies using heterozygous yeasts suggest that the rate of fermentation is associated with the ploidy of the culture, $3n$ and $4n$ types exceeding those of lower or higher ploidy (Emeis, 1964). Since, however, such types are liable to segregate during growth, they are probably not desirable industrially. It is clear that associated characters such as resistance to ethanol and sugar concentrations, both of which have been found to have a genetic basis, are of importance, especially in wine yeasts such as *S. cerevisiae* var. *ellipsoideus* and *S. oviformis* (Lescure, 1956; Martini, 1960).

One of the most important characteristics of a yeast is its flocculating ability. If a yeast flocculates too readily, the fermentation may not go to completion, whereas if it flocculates too little the fermentation goes too far and much yeast is left in suspension at the end. Flocculating yeasts are essential for modern continuous tower systems. The genetics of flocculence have received less attention than they deserve. Flocculence (F) appears to be dominant to inability to flocculate (f) but there is a tendency for F to mutate readily to f, especially in certain genetic backgrounds, such as in top yeasts. Thorne (1951) also suggested that three or more genes were involved. Recently, Lewis and Johnston (1973) have reinvestigated the problem and have found that there are two dominant linked genes, *Flo-1* and *Flo-2*, some 8 cM apart, as well as an unlinked recessive gene *flo-3*. The presence of any one of these genes confers flocculence on a haploid cell. Only *Flo-1* has been derived from a brewery strain. Since the flocculating properties of a yeast appear to depend upon the protein and carbohydrate, especially mannan, moieties of its wall (Rainbow, 1970), it may be sur- mized that investigations of cell-wall mutants might be of potential value to industrialists in this area.

Flavour is another important character which is determined, in part, by yeasts. It is of interest, therefore, that respiratory-deficient yeasts—*petites*— are apparently responsible for the excessive production of diacetyl which causes 'spoilage' of beer (Czarnecki and van Engel, 1959). It will be recalled that these arise spontaneously as either nuclear mutants or by the mutation of extranuclear factors (p. 101). Their frequency varies in different strains and this, presumably, is a reflexion of the genetic background. Clearly, that which confers the greatest stability should be used. Nevertheless, the effect of yeasts on beer flavour is still one of the least understood variables.

Hybridization of strains possessing desirable characters and selection against undesirable features are clearly the principal objectives followed in breeding yeasts. They are possible because of the great wealth of natural genetic variability in the yeasts and the opportunity for recombination and segregation via the sexual cycle. The relative ease with which different species of yeast can be hybridized experimentally provides even greater opportunities for developing novel recombinants.

Mushroom improvement

The cultivated mushroom of Europe, N. America and Australia, *Agaricus bisporus*, is probably the only completely domesticated fungus known. Although Singer (1961), in his monograph, suggests that wild races occur, the habitats are such that the possibility that these mushrooms are escapes or throw-outs from cultivation cannot be excluded. All that is known of the origins of the crop is that it was introduced by unknown horticulturalists in France about 1700. It was probably isolated from melon beds or garden areas with partially decomposed stable manure. Early selections were made by transferring spawn, especially 'virgin spawn', i.e. mycelium from the mushroom beds, before they had fruited for the first time. In 1894, however, Costantin and Matruchot took out a patent for the production of spawn from spores. This was a major advance since it provided relatively sterile spawn. The details were never fully published; presumably they were regarded as a commercial secret, although the Institut Pasteur acted as agent for spawn from 1894. In 1905 Duggar developed a technique, in the U.S.A., of regenerating mycelium from fragments of mushroom tissue and, about the same time, mycelium production from spores was also re-discovered in the U.S.A. by Ferguson (1902).

Many details of mushroom production continue to be treated as commercial secrets but the possibilities for improvement are now much clearer.

Spawn production

The starting material for spawn is nearly always a polysporous mycelium derived from basidiospores collected from a 'typical' basidiocarp as soon as possible after it commences spore production. Basically, therefore, phenotypic selection is practised. Genotypically, the product is likely to be heterogeneous. The mycelium of *A. bisporus* is composed of multinucleate cells (Colson, 1935). Two nuclei fuse in each basidium, which bears only two basidiospores, so that each of these receives two nuclei derived from meiosis (Evans, 1959). Raper and others (1972) demonstrated the occurrence of heterokaryosis in two commercial strains through the isolation of spontaneous auxotrophs, and they have claimed that a similar situation exists in at least two other commercial strains. The proportion of nuclei carrying auxotrophic mutants may be quite high. From a protrophic commercial strain with a high spore germinability, 5 out of 100 monosporous isolates were auxotrophs (Raper and others, 1972). It seems clear, therefore, that basidiospores may be homo- or heterokaryotic and that, depending upon the nuclear ratios in the hymenial cells of the basidiocarp, then one, two or several, genotypically different kinds of basidiospores may be produced. A further complication is that basidiospore germination is variable, ranging from as little as 1–2% to nearly 100% (Sarazin, 1952). This, therefore, will affect the composition of the nuclear components in the heterokaryotic polysporous mycelium. So far as can be ascertained, spawn producers have to exercise constant selection on the genotype of the spawn by testing—in pilot plants—its ability to produce mushrooms with the appropriate attributes. It is not surprising, therefore, that crises arise from time to time in spawn production when it 'runs out'. In addition to maintaining, in essence, a balanced heterokaryon, spawn producers have to contend with strain variability and instability from causes such as those discussed earlier in this chapter.

Selection

From time to time variants, presumably heterokaryotic, have arisen and been selected. A particularly striking form is the Snow-White strain which arose in a mushroom bed from the cream form, *A. bisporus* v. *albidus* of Singer (1961) (Lambert, 1938). It differs by remaining pure white, its pileus never turning cream-coloured at any stage. There is another colour variant, the brown form (*A. bisporus* v. *avellanus*). Attempts have been made to produce heterokaryons between the cream and brown forms but the results have been inconsistent and, since reliable markers were not employed, e.g. auxotrophs, it is difficult to assess the situation. The selection of spontaneous variants, however they may be constituted genetically, has been the principle means of mushroom improvement. Two other possibilities have arisen in the last decade. Fritsche (1967) in the Hamburg Mushroom Institute has selected single-spore cultures which give rise to abnormal fruit bodies. One of these, 59c, gave rise to irregular plectenchymatous masses of undifferentiated tissue weighing 350–500 g, compared with 5 g for a normal basidiocarp in the same conditions. This was propagated successfully by tissue culture and further selection gave rise to 59f. This culture yielded almost as much as a normal strain, 13 kg/m², compared with 14·15 kg/m², and with a high proportion of smooth, rounded tissue masses weighing, individually, 200 g or more. This type of production is of value for the production of mushroom material for flavouring and comparable uses.

The possibility of breeding

A more profound development is the elucidation by Miller (1971), Elliott (1972) and Raper and others (1972) of the mating system of *A. bisporus*. Miller isolated 12 basidiospores from rare 4-spored basidia in two strains. He showed that such cultures were incapable of developing basidiocarps but that when paired in all combinations, nine of the twelve isolates could form fertile heterokaryons; five were of one mating type, four of another. Elliott confirmed this behaviour and was able to isolate, propagate and mate the four spores of a tetrad; two were of one mating type, two of another. Thus *A. bisporus* has a unifactorial mating system with at least two alleles and is probably homodiaphoromictic. In the vast majority of the basidia, there must be some mechanism which ensures that two nuclei of complementary mating types enter each basidiospore (see p. 168). The work of Raper and others (1972) has shown that in this fungus yield is not apparently impaired by the inclusion of auxotrophic nuclei in the mycelium. Thus there is now a possibility, not only of analysing existing commercial strains, but of including known auxotrophic markers in them, so producing novel heterokaryons of known genotype, combining desirable characters. The ability to introduce recognizable auxotrophs also provides an opportunity of using such markers to monitor the mycelium utilized for spawn production. Here, therefore, the new understanding of the life cycle and mating system ensures that 'prospects for interstrain breeding are difficult but feasible' (Raper and Raper, 1972).

General conclusions

This chapter shows clearly how very far behind plant and animal breeders the fungal breeders lie. However, already several lessons for the future are apparent. Firstly, progress is most likely to come when the mating system is both fully understood and efficiently exploited. Secondly, if induced mutants are to be used then the mutagens employed should be such as cause the least structural chromosomal abnormalities. Thirdly, since yield is usually a complex quantitative character, it is desirable that more attention be given to the genetic basis for such characters in fungi. Finally, because most of the breeding procedures are inevitably at a laboratory level, a deeper understanding is necessary of genotype–environment interactions, to attempt to predict the outcome of scaling-up from laboratory to commercial production. Fungal breeding is likely to develop, not only by improved techniques, but also by the exploitation of many more species than have been utilized hitherto.

CHAPTER 13

Genetical aspects of fungal pathogenicity

The host ranges of different fungal pathogens differ greatly. For example, *Pythium debaryanum* infects over 250 species of flowering plants from several families; in nature *Phytophthora infestans* is restricted to several genera of Solanaceae, while *Puccinia buxi* is said to be confined to a single host species, the box (*Buxus sempervirens*). A similar pattern of behaviour is shown by fungi pathogenic on organisms other than flowering plants.

Pathogenic attack can be regarded as a particular form of interaction between two organisms and hence, in some way, it must represent an interaction between their respective genotypes. Attention has been concentrated, therefore, on those situations where differences between hosts and between pathogens, respectively, are minimal. Thus a beginning has been made to the genetical and physiological analysis of interactions between different populations of a species of fungus on different cultivars of a single species of host. This situation will be considered first in some detail and attention will then be given to less exacting situations. The whole topic has recently been reviewed by Day (1974).

PARASITISM ON A SINGLE HOST

Genetical aspects of virulence

Physiological races

Much attention has been paid to obligate pathogens such as rusts. Up to 1894 *Puccinia graminis* was regarded as a single species capable of infecting a wide range of cereals and grasses but, in that year, Eriksson demonstrated that it was best regarded as being composed of at least six *formae speciales*, differentiated by the principal hosts which each attacked (Table 13.1, p. 260).

In 1911, Barrus described an even finer degree of host specialization in the bean (*Phaseolus*) anthracnose organism, *Colletotrichum lindemuthianum*.

TABLE 13.1 The host plants and uredospore sizes of the principal *formae speciales* recognized in *Puccinia graminis*.

Formae speciales	Principal host	Other hosts	Uredospore size (μm)
1. *tritici*	*Triticum*	*Hordeum*, cultivated and wild; *Secale*, rare; *Agropyron*, *Bromus*, *Elymus* and many other wild grasses	32 × 20
2. *avenae*	*Avena*	Several wild grasses different from 1 and 3, e.g. *Dactylis*	28 × 20
3. *secalis*	*Secale*	*Hordeum*, and wild grasses comparable to 1	27 × 17
4. *phleipratensis*	*Phleum*	*Avena*, *Secale*, *Hordeum* weak parasite; *Festuca* spp.	24 × 17
5. *agrostidis*	*Agrostis*		22 × 16
6. *poae*	*Poa*		19 × 16

He showed that within the species two races, α and β, could be distinguished by their pathogenic response to different cultivars of *P. vulgaris*.

This approach was taken up and extended to the many available isolates of *Puccinia graminis* by Stakman and his co-workers in Minnesota (Stakman and Piemeisel, 1917; Stakman and others, 1919; Stakman and Levine, 1922). They showed that each *forma* of *P. graminis* comprised a number of 'physiological races' recognizable by the characteristic infective response elicited by the dikaryotic phase on a selected group of cultivars of wheat, rye or oats (Table 13.2), the so-called 'differential varieties'. Johnson and his co-workers in Manitoba carried the genetic analysis of such interactions further (Newton, Johnson and Brown, 1930a, b, Johnson and Newton, 1940). Firstly, they selfed several races of *P. graminis tritici* and showed that the majority were heterozygous (strictly heterokaryotic since it was the dikaryon that was being tested) for most pathogenic characters

TABLE 13.2 The standard differential varieties employed to identify physiological races of *Puccinia graminis tritici*. (From Stakman, Stewart and Loegering, 1962.)

Wheat species	Name of cultivar	Abbreviation
Triticum compactum	Little Club C.I. 4066 (or Hood)	LC
Triticum vulgare	Marquis C.I. 3641	Ma
	Reliance C.I. 7370 (or Kanred)	Rel (Kan)
	Kota C.I. 5878	Ko
Triticum durum	Arnautka C.I. 1493	Arn
	Mindum C.I. 5296	Mnd
	Spelmar C.I. 6236	Spm
	Kubanka C.I. 2094	Kub
	Acme C.I. 5284	Ac
Triticum monococcum	Einkorn C.I. 2433	Enk
Triticum dicoccum	Vernal C.I. 3686	Ver
	Khapli C.I. 4013	Kpl

Note: The numbers after the cultivars are U.S.D.A. accession numbers.

expressed on the differential varieties. Thus, in any one race, more than one pair of genetic factors were operative. Secondly, they demonstrated that some factors behaved as dominants, others as recessives. When sibs of Race 15 were selfed some only gave rise to the infection pattern of Race 15 while others gave patterns characteristic of Race 15 and Race 52. It may be inferred that the former were homokaryotic, the latter heterokaryotic, and the pattern of Race 15 dominant to that of Race 52. Thirdly, the number of genes involved was determined by normal Mendelian methods from the segregation data of the crosses and pathogenicity was probably recessive. Races 9 and 36 were shown to differ in their pathogenic responses to the cultivars Kanred, Mindum and Vernal. An F_1 was made between the races and from it an F_2 by selfing. The results of this experiment are set out in Table 13.3.

TABLE 13.3 Infection types derived from a cross between physiological Races 9 and 36 of *Puccinia graminis tritici* and the self-fertilization of the progeny of the cross. (Based on data from Johnson and Newton, 1940.)

Race	F_2 distribution % observed	F_2 distribution % expected	Kan	Arn	Mnd	Spm	Ver
Parents							
9			R	S	S	S	S
36			S	R	R	R	R
F_1 generation							
17			R	S	S	S	R
F_2 generation							
17	51	52·8	R	S	S	S	R
11	19·8	17·9	S	S	S	S	R
1	14·4	17·6	R	R	R	R	R
36	9·3	5·9	S	R	R	R	R
9	3·4	3·4	R	S	S	S	S
57	1·2	1·2	R	R	R	R	S
15	0·6	1·0	S	S	S	S	S
52	0·3	0·3	S	R	R	R	S

Note: R, resistant; S, susceptible. For key to cultivars see Table 13.2.

Comparison of parental behaviour with that of the F_1 suggests that reduced infection is dominant to strong infection on cv. Kanred and cv. Vernal but the opposite is true on cv. Mindum. However, this is not certain for neither Race 9 nor Race 36 were selfed to test whether they were heterokaryotic for pathogenicity. The F_2 data suggest that the first interpretation is correct, i.e. that strong infection is recessive to weak infection on both cv. Kanred and cv. Vernal, although in the former it is determined by one pair of alleles, in the latter probably by two. Pathogenicity on cv. Mindum is dominant and determined by a pair of alleles. It is clear that to obtain completely unequivocal results it is necessary to ascertain the genotypes of both the host and the pathogen in respect of resistance genes and genes for pathogenicity respectively.

The gene-for-gene concept

Flor attempted to determine both host and pathogen genotypes using flax (*Linum usitatissimum*) and flax rust (*Melampsora lini*) as host and pathogen. This combination had a number of technical advantages of which the most important were:

(i) The whole life cycle of the fungus occurs on flax and it is dimictic so that selfing or crossing can readily be achieved.

(ii) Flax is an inbreeding annual with a short life cycle so it can readily be propagated to provide large populations for scoring satisfactorily. It elongates terminally, so that successive leaf whorls can be inoculated with different races providing comparable data.

(iii) Rust reactions are reasonably clear cut (Figure 13.1) and give obvious differential reactions on different flax cultivars.

Cultivars employed to differentiate physiological races of *M. lini* were selected on a trial and error basis from over 200 tested. Eventually Flor believed he had obtained tester lines each possessing a single rust-resistant gene (Flor, 1935, 1940, 1954); such resistant hosts are the most efficient for classifying the genotypes of the races. In fact, this was not always the case but this situation will be described later. Flor investigated (a) the inheritance of rust resistance in crosses between pairs of cultivars showing differential responses to physiological races and (b) the inheritance of pathogenicity in crosses between the races. The crosses between *cv.* Ottawa 770B and Bombay and physiological Races 22 and 24 illustrate his approach (Flor, 1942, 1946, 1947 and Table 13.4).

TABLE 13.4 The inheritance of virulence towards two cultivars of flax by two races of *Melampsora lini* and of the resistance shown to the races by the two cultivars. (Data of Flor, 1956.)

Flax host	Flax rust races, generations, genotypes and frequencies						
	P_1		F_1	F_2			
	22 a_1A_2	24 A_1a_2	A_1A_2	A_1A_2	a_1A_2	A_1a_2	a_1a_2
Ottawa	S	R	R	R	S	R	S
Bombay	R	S	R	R	R	S	S
				78	27	23	5

Rust race	Flax cultivars, generations, genotypes and frequencies						
	P_1		F	F_2			
	Ottawa R_1r_1	Bombay r_1R_2	R_1R_2	R_1R_2	R_1r_2	r_1R_2	r_1r_2
22	S	R	R	R	S	R	S
24	R	S	R	R	R	S	S
				110	32	43	9

Note: R, resistant; S, susceptible. Gene symbols: *A/a* avirulence/virulence; *R/r* resistance/susceptibility.

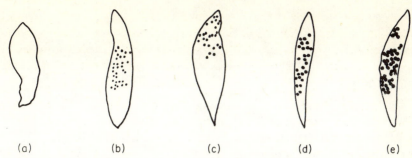

FIGURE 13.1 Types of rust infection produced by *Melampsora lini* on flax. (a) Type 0, no uredosori, sometimes some distortion of leaf. (b) Type 1, uredosori small, scattered, a little chlorosis or distortion. (c) Type 2, uredosori small to medium in heavily infected areas which become stunted or distorted. (d) Type 3, uredosori medium sized, vigorous not compound, little distortion. (e) Type 4, uredosori large, vigorous, may become compound, no necrosis or chlorosis. (After Flor, 1954.)

From these data it can be concluded that the F_1 and F_2 ratios obtained fit satisfactorily the hypothesis that:

(i) Resistance in the flax plant is dominant to susceptibility and is independently inherited.
(ii) Avirulence in the physiological races is dominant to pathogenicity (virulence) and is independently inherited.
(iii) The inheritance of the rust reaction in the host is paralleled by the inheritance of virulence in the pathogen and their interaction is conditioned by the two corresponding pairs of genes.

This last hypothesis is known as the gene-for-gene concept. Because it has subsequently been widely applied and in different ways, it has been redefined more precisely by Person and others (1962) as:

'A gene-for-gene relationship exists when the presence of a gene in one population is contingent upon the presence of a gene in another population, and where the interaction between the two genes leads to a single phenotypic expression by which the presence or absence of the relevant gene in either organism may be recognised.'

The relationship can be recognized most readily when the host cultivars and physiological races are isogenic at all loci other than those for resistance and pathogenicity. These are set out in detail in Figure 13.2a and can be summarized in a form which has been termed the 'quadratic check', Figure 13.2b (Loegering and Powers, 1962; Rowell and others, 1963; Loegering, 1971). In these figures resistance and avirulence have been shown as dominant, since this is the most common situation but, of course, similar diagrams can be prepared if different dominance relationships hold.

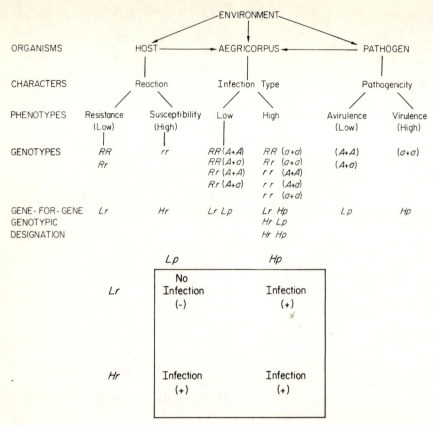

FIGURE 13.2 Diagrams to illustrate the genetical host–pathogen relationships where a gene-for-gene relationship obtains. The situation shown is based on a cereal rust where the pathogenic phase is dikaryotic $(n + n)$; the host, of course, is diploid $(2n)$. In cases where the pathogen is monokaryotic, such as barley mildew, the genotypes can only be A or a with a corresponding reduction of the numbers of possible genotypes. Only a pair of corresponding genes in each organism is considered, R/r representing host genes for resistance or susceptibility (low host reaction versus high host reaction) and A/a pathogen genes for avirulence or virulence (low pathogenicity versus high pathogenicity) Resistance and avirulence are shown as dominant but either or both could equally well be recessive. The symbols used to designate the gene-for-gene genotypes are Lr, Hr, low and high reaction, respectively; Lp, Hp, low and high pathogenicity, respectively. (After Loegering and Powers, 1962 and Loegering, 1971.)

In most circumstances the data will be imperfect or incomplete and it is not often easy to obtain isogenic lines either of host or pathogen. Person (1959), therefore, investigated the properties of the relationship theoretically in an ideal system based on five pairs of loci. He hoped to be able to deduce

TABLE 13.5 The theoretical patterns of resistance and susceptibility between tester hosts (T) and fungal races (F) when determined by gene-for-gene relationships involving three loci each with two alleles.

Tester hosts and Genotypes	Fungal Races								Number of races infecting tester
	F_1	F_2	F_3	F_4	F_5	F_6	F_7	$F8$	
	$A_1A_2A_3$	$a_1A_2A_3$	$A_1a_2A_3$	$A_1A_2a_3$	$a_1a_2A_3$	$a_1A_2a_3$	$A_1a_2a_3$	$a_1a_2a_3$	
T_1 $r_1r_2r_3$	S	S	S	S	S	S	S	S	8
T_2 $R_1r_2r_3$	R	S	R	R	S	S	R	S	4
T_3 $r_1R_2r_3$	R	R	S	R	S	R	S	S	4
T_4 $r_1r_2R_3$	R	R	R	S	R	S	S	S	4
T_5 $R_1R_2r_3$	R	R	R	R	S	R	R	S	2
T_6 $R_1r_2R_3$	R	R	R	R	R	S	R	S	2
T_7 $r_1R_2R_3$	R	R	R	R	R	R	S	S	2
T_8 $R_1R_2R_3$	R	R	R	R	R	R	R	S	1
Number of testers infected	1	2	2	2	4	4	4	8	

Key: Gene symbols: A, a—avirulence, virulence; R, r—resistance, susceptibility.

from this rules for recognizing the gene-for-gene relationship in less complete data. A system such as he considered is illustrated in Table 13.5 but for three pairs of loci only, since this is sufficient to illustrate the principles.

The data in Table 13.5 demonstrate the following features:

(i) A set of n host cultivars, each carrying a different single resistance gene, is sufficient to identify all races. There will be 2^n of these (i.e. T_2–T_4; $n = 3$; $2^n = 8$). Each host will be attacked by half the total number of races.

(ii) One host, the 'universal suscept', is susceptible to all races and one race is pathogenic to all hosts (T_1 and F_8, respectively).

(iii) If the host cultivars are arranged in a descending arithmetic series according to the number of resistance genes which they carry, i.e. 3, 2, 1, 0, the number of races pathogenic to each of the cultivars increases in the geometric series 1, 2, 4, 8 (e.g. T_8 and F_8; T_7 and F_7, F_8; T_3 and F_3, F_5, F_7, F_8; T_1 and F_1–F_8).

(iv) Physiological races possessing a single gene for pathogenicity can only attack two cultivars; the universal suscept and that host which carries a single resistance gene (e.g. F_2 and T_1 and T_2; F_3 and T_1 and T_3; F_4 and T_1 and T_4).

The consequences of any departure from this ideal situation can readily be deduced. For example, if one of the differential hosts carrying a single resistance gene, e.g. T_3, is lacking, then none of the races characterized by this tester will be identified and instead of 8 races being recognized only 4 will be distinguished, i.e. F_1, F_2, F_4 and F_6. Similarly, if a tester in fact carries two resistance genes instead of a single one, e.g. R_1, R_2 and R_1R_3, then only 6 races will be recognized, i.e. F_1, F_2, F_3, F_5, F_6 and F_8. A third possibility is that instead of a one-to-one relationship between host–pathogen genes, there may be a two-to-one relationship. Such would occur if two host loci showed duplicate or complementary gene action, i.e. either a_1 or a_2 overcome the effect of R_1, or both a_1 and a_2 must be present to overcome the resistance conferred by R_1. In such cases the number of recognizable phenotypes would be reduced to half the potential number of genotypes.

Person's analysis shows clearly how data accumulated routinely on the response of tester hosts to large numbers of physiological races can be analyzed, partially at least, and the probable numbers of resistance or pathogenic loci estimated. In this way Person was not only able to demonstrate the occurrence of more than one resistance gene in some of Flor's supposed single resistance-gene testers, but also to demonstrate that data available on *Solanum/Phytophthora infestans* interactions conformed to the gene-for-gene hypothesis. This latter combination has been further analyzed by Black (1952, 1957), who has been able to use available knowledge of resistance genes to predict the existence of physiological races, some of which were detected five years later (Table 13.6).

An even more striking example is the *Coffea/Hemileia vastatrix* situation, where normal genetic analysis is precluded through the absence of a sexual stage. Nevertheless, by applying Person's technique, Noronha-Wagner and Bettencourt (1967) were able to identify 12 physiological races, to assign them genotypes in respect of pathogenicity genes and to predict the occurrence of 4 other races.

The most effective application of the hypothesis is achieved where there is adequate knowledge of the resistance genes of the differential hosts; the lack of this has hindered the analysis of the genotypes of long established physiological races of *P. graminis tritici* in N. America, although a start has been made (e.g. Loegering and Powers, 1962). In Australia the policy over the last three decades of introducing specific resistance genes in differential cultivars has enabled the genotypes of new races to be predicted and assessed with considerable efficiency. Isogenic lines of wheat with known resistance genes are now being produced at various centres and this will facilitate the analysis of the genotypes of pathogens.

Some 18 situations have now been described where it is thought possible that a gene-for-gene relationship exists (Flor, 1971; Table 13.7, p. 268).

Some of these exhibit rather different features from those already described. For instance, in several cases the pathogen is haploid but not dikaryotic so that it carries a gene determining either pathogenicity or avirulence, e.g. *Venturia inaequalis*, *Erysiphe graminis*. Since both these fungi reproduce sexually, various recombinant genotypes can be produced and used for the prediction of resistance genes in the host, as in the *Malus/Venturia* combination (Boone and Keitt, 1957). In haploids lacking a dikaryon it is not possible to ascertain whether virulence is dominant or recessive to avirulence. However, an interesting observation made with *Malus/Venturia* is that if the pathogen possessed two loci which determined pathogenicity to the same cultivar, an avirulence gene at either locus was epistatic to a gene for virulence at the other.

TABLE 13.6 The interrelations of sixteen strains of *Phytophthora infestans* and four major genes (R_1–R_4) which determine resistance to potato blight in potatoes. (From Black, 1952.)

Genotypes and designations	A	B¹	H		D	G	E	B²		C	I			F		
	o	(1)	(2)	(3)	(4)	(1, 2)	(1, 3)	(1, 4)	(2, 3)	(2, 4)	(3, 4)	(1, 2, 3)	(1, 2, 4)	(1, 3, 4)	(2, 3, 4)	(1, 2, 3, 4)
r	+	+	+	+	+	+	+	+	+	+	+	+	+	+	+	+
R_1	−	+	−	−	−	+	+	+	−	−	−	+	+	+	−	+
R_2	−	−	+	−	−	+	−	−	+	+	−	+	+	−	+	+
R_3	−	−	−	+	−	−	+	−	+	−	+	+	−	+	+	+
R_4	−	−	−	−	+	−	−	+	−	+	+	−	+	+	+	+
R_1R_2	−	−	−	−	−	+	−	−	−	−	−	+	+	−	−	+
R_1R_3	−	−	−	−	−	−	+	−	−	−	−	+	−	+	−	+
R_1R_4	−	−	−	−	−	−	−	+	−	−	−	−	+	+	−	+
R_2R_3	−	−	−	−	−	−	−	−	+	−	−	+	−	−	+	+
R_2R_4	−	−	−	−	−	−	−	−	−	+	−	−	+	−	+	+
R_3R_4	−	−	−	−	−	−	−	−	−	−	+	−	−	+	+	+
$R_1R_2R_3$	−	−	−	−	−	−	−	−	−	−	−	+	−	−	−	+
$R_1R_2R_4$	−	−	−	−	−	−	−	−	−	−	−	−	+	−	−	+
$R_1R_3R_4$	−	−	−	−	−	−	−	−	−	−	−	−	−	+	−	+
$R_2R_3R_4$	−	−	−	−	−	−	−	−	−	−	−	−	−	−	+	+
$R_1R_2R_3R_4$	−	−	−	−	−	−	−	−	−	−	−	−	−	−	−	+

Note:
1. Note that + indicates susceptible, − resistant; bold face are observed data, light face are theoretical.
2. Within 5 years of the establishment of these interrelations all those shown here as 'theoretical' had been detected.
3. Some ten *R* genes are now known so that an up to date table would be far more complex.

TABLE 13.7 Fungi and host plant where a gene-for-gene relationship has been claimed to underly the parasitic relationship. (After Flor, 1971.)

Fungus	Host	Reference
Phycomycetes		
Phytophthora infestans	*Solanum*	Toxopeus, 1956
Synchytrium endobioticum	*Solanum*	Howard, 1968
Ascomycetes/Imperfecti		
Erysiphe graminis hordei	*Hordeum*	Moseman, 1957
E. graminis tritici	*Triticum*	Powers and Sando, 1957
Venturia inaequalis	*Malus*	Boone and Keitt, 1957
Cochliobolus	*Avena*	Wheeler, 1969
(*Helminthosporium*) *victoriae*		
Cladosporium fulvum	*Lycopersicon*	Day, 1956
Basidiomycetes/Ustilaginales		
Tilletia caries	*Triticum*	Metzger and Trione, 1962
T. controversa	*Triticum*	Holton and Hoffman, 1968
Ustilago avenae	*Avena*	Holton and Halisky, 1960
U. tritici	*Triticum*	Oort, 1963
Basidiomycetes/Uredinales		
Hemileia vastatrix	*Coffea*	Noronha-Wagner and Bettencourt, 1967
Melampsora lini	*Linum*	Flor, 1942
Puccinia graminis avenae	*Avena*	Martens, McKenzie and Green, 1970
P. graminis tritici	*Triticum*	Loegering and Powers, 1962
P. helianthi	*Helianthus*	Sackston, 1962
P. recondita	*Triticum*	Samborski and Dyck, 1968
P. striiformis	*Triticum*	Zadoks, 1961

While all these cases share the basic properties of the gene-for-gene relationship, a majority of them also share a number of other features. In most cases, where it has proved possible to test it, virulence has behaved as recessive to avirulence. Exceptions are uncommon but are known in *P. graminis tritici* and *P. graminis avenae* (Green, 1964, 1965). Loci for pathogenicity are not often closely linked and even the exceptions are doubtful, e.g. *Venturia inaequalis* (Keitt and others, 1943; Boone, 1971). In contrast, there is considerable evidence that resistance genes in host cultivars often show multiple allelism, e.g. barley (Moseman, 1966), flax (Flor, 1960). It is also remarkable that in only 3 out of the 18 examples listed in Table 13.7 is the host an outbreeder and, in fact, these three: *Coffea*, *Malus* and *Solanum*, are frequently propagated clonally when cropped. Biological significance has been attributed to this since it has been claimed that a highly specific interaction, such as that shown in the gene-for-gene relationship, is greatly favoured by the indefinite perpetuation of particular genotypes (Griffiths, 1958). Since most of the pathogens involved reproduce sexually and their pathogenic genotypes are maintained by selection (see pp. 274–275), it seems more probable that the particular examples listed have been studied because of certain technical advantages which they possess and that the apparent correlation with inbreeding is largely fortuitous.

Gene action and interaction

The high specificity of the gene-for-gene relationship has suggested some highly specific form of gene action and interaction and this has been looked for. The most remarkable developments of this nature have been made with the *Lycopersicon/Cladosporium* system (van Dijkman, 1972; van Dijkman and Kaars Sijpesteijn, 1971, 1973; van Dijkman and others, 1973). These workers used tomatoes carrying known resistance genes and isolates of *C. fulvum* of known virulence. The latter were grown in shake culture and the culture filtrates subjected to gel filtration through 'Sephadex' G-25 columns. The material so isolated possessed a molecular weight greater than 1000; 27–60% was protein and the rest possibly carbohydrate. Tomato leaves were labelled with ^{32}P and then 8 mm discs were cut from their leaves. The discs were infiltrated into buffered extracts of culture filtrates and the subsequent leakage of ^{32}P from the discs was measured and compared with an appropriate control (Table 13.8).

TABLE 13.8 The correlation between resistance *in vivo* of tomatoes carrying different resistance genes and leakage of materials from leaf disks exposed *in vitro* to selective toxins from races of *Cladosporium fulvum*. (Data from van Dijkman and Kaars Sijpesteijn, 1973.)

Resistance genes in tomato	Resistance (R) Susceptibility (S) or Leakage (L) or none (N) Virulence genes in pathogen						
		a_1	a_2	a_4	a_1a_2	$a_1a_2a_3$	$a_2a_3a_4$	$a_1a_2a_4$
	SN	SN	S	SN	SN	S	S	SN
Cf_1	R	SN	RL	R	S	S	R	S
Cf_2	RL	RL	SN	RL	S	SN	S	S
Cf_3	RL	R	R	RL	R	S	SN	R
Cf_4	RL	R	RL	SN	R	RL	S	SN
$Cf_1 Cf_2$	RL	RL	R	R	SN	S	R	SN
$Cf_2 Cf_4$	RL	R	R	RL	R	RL	S	SN

Note:
1. Leakage occurs if plant is resistant to a particular fungal race.
2. No leakage occurs if plant is susceptible.
3. Combinations not tested shown only as resistant or susceptible.

It can be seen that leakage is greatest in those combinations where the tomato resists infection. *In vivo* such infections are manifested as strongly chlorotic areas which necrose, the size of the area being largest with Cf_1 and Cf_3 but mere pinpoint areas with Cf_2, Cf_4 or combinations of resistance genes including them, e.g. $Cf_2 Cf_3$. Fungal growth is then stopped but in what way is not known. It was hypothesized that each resistance gene determines a specific receptor in the plasmalemma which reacts with specific toxins, each determined by an avirulence gene of the fungus to give, in cases of resistance, the hypersensitive reaction. On this hypothesis virulence genes have lost the ability to produce the specific toxin, so that the receptor site in the host is unaffected and no leakage results; fungal growth can then continue.

There is some support for this kind of interaction of the host and pathogen genes from the *Malus/Venturia* combination. Raa and Sijpesteijn (1968) placed the petioles of leaves of resistant and susceptible plants in the culture filtrate from a pathogenic isolate of *V. inaequalis*; the former wilted. Necrosis could also be induced in the leaves of resistant plants which had either been infiltrated with culture filtrate or on whose surfaces filtrate had been placed. Pellizzari and others (1970) demonstrated a twofold increase in leakage of electrolytes in leaves which showed a hypersensitive response *in vivo*, compared with susceptible leaves. Hadwiger and Schwochau (1969) proposed an 'induction' hypothesis in general terms which they applied to the *Linum/Melampsora* system, namely that avirulence genes determine compounds which induce activity in host genes, resulting in the localized alteration or disintegration of resistant host cells. No experimental data were given but the results of Varns and Kuć (1972) with the *Solanum/Phytophthora* system, although briefly reported, are in keeping with the general hypothesis. For 11 cultivars and 6 races they showed that hypersensitive response was associated with rapid cellular necrosis and increased accumulation of the terpenoids rishitin and phytuberin within 12 hours of inoculation. These responses were suppressed in susceptible situations. Once a hypersensitive response had been elicited it could not be suppressed by inoculation with a virulent race and vice versa. They postulated a 'suppression' hypothesis, namely that susceptibility depends upon the suppression of a resistance response but, of course, this could equally well result from inability of the virulent pathogen to trigger off such responses. Protection against infection by prior inoculation with avirulent strains is not an uncommon phenomenon, e.g. *C. fulvum* (Curren, 1967), *M. lini* (Littlefield, 1969), *Fusarium oxysporum* f. sp. *melonis* (Mas and others, 1969). Such situations are amenable to either the 'induction' or the 'suppression' hypothesis.

Other evidence indicates that infection is due to the production of toxic materials by virulence genes and an inability to produce them by genes for avirulence. Even in *V. inaequalis* melanoproteins have been shown to be produced by pathogenic strains which result in necrosis and interfere with solute transport in susceptible plants (Kirkham and Hignett, 1966; Hignett and Kirkham, 1967). No information is available concerning melanoprotein production by avirulent strains. Similarly, it is clear that pathogenicity to oats by *Helminthosporium victoriae* is associated with the production of phytotoxin and that there is a single gene difference between pathogenic and non-pathogenic strains. Wheeler (1969) has claimed that this combination conforms with a gene-for-gene relationship. Resistant oats tolerate a concentration of the toxin more than 400,000 times greater than susceptible seedlings (Kuo and others, 1970).

Another type of response is shown in the *Erysiphe/Triticum* combination. Here it has been shown that different resistance genes in wheat operate at different times and at different sites in response to infection (Ellingboe, 1972). This makes it less probable that there is a common underlying mechanism of gene interactions to the gene-for-gene relationship. Indeed, there is always a danger that a unifying concept at one level, such as the gene-for-gene concept at the population level, will be applied, illogically and illegitimately, at another level (here that of the biochemical description of gene action and interaction). It seems probable that different cellular mechanisms will, indeed, underlie the uniformity of the gene-for-gene relationship.

The origin of virulent races

Four mechanisms have been suggested for the origin of novel, virulent races, namely mutation, heterokaryosis, parasexual recombination and sexual recombination (Day, 1960, 1974).

Mutation

The origin of new virulence genes through mutation, both spontaneous and induced, is fairly well authenticated.

Some of the least ambiguous data is available for *P. graminis tritici* in Australia. Field data have shown successions of single-step changes from avirulence to virulence which regularly parallel and follow the successive introductions of single resistance genes into wheat (Waterhouse, 1952). That such mutations can indeed occur spontaneously has been shown by Watson (1957), employing four different races marked by spore colour, virulence or both, under conditions carefully controlled to prevent con-tamination. Three spontaneous virulent mutants were obtained from uredospore cultures. Finally, EMS has been employed to induce mutations possessing virulence to the resistance gene *Sr5*. Such mutants occur relatively frequently in the field and could be induced to a frequency of a little over 1 in 2000. Zimmer and others (1963) have also provided unambiguous evidence for high spontaneous mutation to virulence in *Puccinia coronata* and were also able to demonstrate that mutation rates vary at different loci, e.g. for Race 202, mutation to virulence against *cv.* Ascencao was detected once in every 200 infections but to *cv.* Ukraine 1 the frequency of detection was 1 in 6450 infections.

Data on induced mutations are often more difficult to interpret. For example, Day (1957) employed an avirulent strain of *C. fulvum* marked by induced red pigmentation, of Race O ($A_1 A_2 A_3 A_4$), which he exposed to X-irradiation. Conidia so treated inoculated on to *cv.* Vetomold (*cf*$_1$ *cf*$_2$ *cf*$_3$ *cf*$_4$) did not infect it but, after a post-irradiation period of growth, one mutant capable of infecting this cultivar was obtained. Thus mutation had presumably occurred from $A_2 \rightarrow a_2$. By making reconstruction experiments with mixtures of known proportions of conidia carrying genes for virulence and avirulence, Day estimated the mutation rate at about $1 \cdot 7 \times 10^{-5}$. This particular experiment indicates clearly how essential it is to ensure that conditions are such that a lesion due to a mutant condition in the conidia can be detected.

Studies with *M. lini* are of special interest. Schwinghamer (1959a, b) employed X-rays, γ-rays, fast neutrons and u.v.-irradiation on uredospores of Race 1 of *M. lini*. This race infects *cv.* Bison but is avirulent on *cv.* Dakota which could, therefore, be used to detect induced mutants. He found that the kinetics of mutation induction by fast neutrons and X-rays were comparable with those associated with chromosomal aberrations in higher plants, whereas u.v.-irradiation suggested a 'point-mutation' effect. The relative effectiveness in inducing mutations for the three types of radiation was as 7:5:1, corresponding to average frequencies of mutation of 2·0, 1·5 and 0·3%. Nitrogen mustard gave results comparable with those after u.v.-irradiation. Schwinghamer suggested that the mutational change $A \rightarrow a$ had probably been brought about by a deletion of the chromosome including the A locus, i.e. by a loss mutation. Later Flor (1960) selfed two of the mutants obtained in this way which showed virulence to *cv.* Koto. The original Race 1 of *M. lini* is heterokaryotic for virulence alleles so that in these mutants $(Ap + ap) \rightarrow (ap + ap)$, i.e. they were homokaryotic for *ap*. As expected all the 198 selfed cultures were, indeed, virulent. In addition, however, and quite unexpectedly, 16 were found to be pathogenic to Abyssinian and Leona, cultivars to which Race 1 was originally homokaryotic for virulence. Thus in these mutants the full extent of the detected change was:

$$\begin{bmatrix} Ap + ap \\ Ap_2 + Ap_2 \\ Ap_3 + Ap_3 \end{bmatrix} \longrightarrow \begin{bmatrix} ap + ap \\ ap_2 + ap_2 \\ ap_3 + ap_3 \end{bmatrix}$$

That a change of this magnitude could be due to gene mutation seems improbable. Taken in conjunction with Schwinghamer's suggestion that the mutational events were probably deletions, it seems most plausible to suppose that the loci Ap, Ap_2 and Ap_3 were linked and had been lost together in similar chrosomal aberrations in both nuclei of the uredospore dikaryon, e.g.

$$
\begin{array}{ccc} Ap & Ap_2 & Ap_3 \end{array}
$$
$(\underline{}) + (\underline{})$

with positions ap, Ap_2, Ap_3 above the second line.

$$\text{Mutational} \downarrow \text{events}$$

$(\underline{}) + (\underline{})$ with ap above the second line.

Two interesting consequences stem from this hypothesis. Firstly, it provides evidence suggesting that some loci determining pathogenicity may be quite closely linked (a view contrary to most of the other available evidence, but see p. 210). Secondly, it suggests that avirulence compared with virulence is, indeed, the difference between a functional and a non-functional state, respectively. This, of course, is the kind of difference postulated by van Dijkman and Kaars Sijpesteijn in their theory of gene action in *C. fulvum* (pp. 269–270). On the other hand, it is difficult to believe that genes for virulence all represent loss of function as pointed out by Day (1966). For instance, in *V. inaequalis* this would suggest that the most virulent race would have suffered a loss of function at some 19 loci! Until the general architecture of the pathogenicity genes of *M. lini* on the one hand and the gene action and interaction of such genes on the other is better understood, it is not possible to make rational suggestions as to the nature of the change from avirulence to virulence.

Heterokaryosis and parasexuality

Heterokaryosis is certainly a way in which avirulent strains may acquire virulence. One of the best documented cases is that of *Thanetophorus cucumeris*. Heterokaryons between avirulent strains, both spontaneous and induced, or between virulent and avirulent strains, were all highly pathogenic, in the latter case more so than either component (Garza-Chapa and Anderson, 1966; Flentje and others, 1967; McKenzie and others, 1969). This is not always the case. Reference has already been made to the fact that *Fusarium fujikuroi* is a natural heterokaryon whose three components differ in their gibberellin production (a character closely paralleling virulence). The heterokaryon of all three is, in fact, less virulent than its most virulent component.

Heterokaryosis may give rise to races or strains with novel characteristics. One of the best documented examples is *Puccinia striiformis* (Little and Manners, 1969a, b). Uredospores of Races 2B and 8B were mixed and inoculated on *cv.* Strubes Dickkof, susceptible to both. Single-spore cultures were made from uredospores derived from the infection on a universal suscept and then tested on a range of differential wheat cultivars. Thirty spores out of about 150 were established, 17 were Race 2B or very similar, 10 were Race 8B and 3 showed new reactions, two of them virtually the same. Thus the combination of two dikaryons had given rise to four genotypes, two parental and two novel. Watson (1957) had pointed out that this represented the maximum reassortment of 4 nuclear types derived from two dikaryons, i.e.

$$(A + B) + (C + D) \longrightarrow (A + B) + (C + D) + (A + C) + (B + D)$$

The relatively high frequency of the recombinant *P. striiformis* races, *c.* 10%, and the fact that germ tube fusions were shown to be possible and to result in hyphal networks, suggested that heterokaryosis was likely. More variation in nuclear number was noted in the uredospores of the new races than in 2B or 8B and this could reflect the existence of mixed cytoplasm in the new heterokaryons. That all the nuclei were of a similar size is in keeping with the notion that all were haploid, and hence that parasexuality had not occurred. But it will be recalled that diploid nuclei need only have a transitory existence to give rise to mitotic recombinants (p. 80).

Heterokaryosis has been thought to be of more significance as the first step in the parasexual process (Nelson and others, 1955). Nevertheless, most of the evidence for the origin of new pathogenic races by parasexual recombination is circumstantial at best, and at worst, invoked because no other explanation seems possible. For example, Watson and Luig (Watson, 1970) obtained more than 16 different and easily recognizable strains over a ten year period by mixing uredospores of Races 111 and NR2 of *P. graminis tritici*. An even more striking case was that where 33 different physiological races, 17 of them not previously known, were derived from mixing uredospores of a red-pigmented Race 2 with a yellow-pigmented Race 122 of *P. recondita tritici* (Vakili and Caldwell, 1957). In neither of these cases was normal reproduction possible. Some of the races derived from the 2 + 122 mixture could be accounted for by nuclear re-assortment (Vakili, 1959). However, in both cases, the numbers of types obtained exceeded the possible maximum due to reassortment of two, so that somatic recombination was invoked. There are some puzzling features about these results. Firstly, two investigators have failed to repeat the results obtained with *P. recondita* (Barr and others, 1964; Bartos and others, 1969). Bartos and others took great care to exclude contamination and failed with *P. recondita* yet, at the same time, were able to show apparent parasexual recombination in *P. coronta avenae*. Secondly, there is no *a priori* reason why somatic recombination should not take place between the nuclei of the dikaryon of a single race. If the nuclei were heterozygous for virulence genes, and this is often the case, then new races should arise from the progeny of a single uredospore. Yet, as Ellingboe (1961) has pointed out, this has never been detected. Somatic recombination in the rusts, if that is indeed the process involved, appears only to be manifested after different races have been mixed. This situation deserves further study.

Although parasexuality has been demonstrated in several other pathogenic fungi, e.g. *Fusarium oxysporum*, *Cochliobolus sativus*, *Pyricularia oryzeae*, *Verticillium* spp., *U. maydis*, no unequivocal demonstration has been provided with any of them that new races can arise through somatic recombination. Although a real possibility exists, the probability must await further study.

Sexual recombination

The origin of new physiological races through meiotic recombination is well attested in many pathogenic fungi, rusts, smuts, powdery mildews and the potato-blight fungus. Two particularly striking examples illustrate the point clearly. In 1949 Stakman and Loegering (in Stakman and Harrar, 1957) identified 43 different races in the immediate vicinity of three barberry bushes in Pennsylvania but only 5 from non-barberry areas of the state. They also noted that in 1940 they identified 16 races from or near barberry bushes which they could not detect elsewhere in the U.S.A. or Mexico. It will be recalled that the sexual reproductive stage of *P. graminis* is located on the barberry. In an analogous manner, it has been noted by

Gallegly and Niederhauser (1959) that only in central Mexico are both mating types, *A1* and *A2*, of *Phytophthora infestans* found and here there occurs a relatively large number of specialized races. Indeed, central Mexico has been described as the gene centre of the potato-blight organism (Niederhauser and others, 1954).

Data from rusts also illustrate the enormous genetic potential for novel pathogenic types which sexual recombination releases. For example, selfing Race 49 of *P. graminis tritici* gave rise to 7 other recognizably distinct races (Johnson, 1954). Loegering and Powers (1962) obtained 108 F_2 progeny from a cross between Races 111 and 36 of the same species. Tested on 20 differential wheat cultivars, they were able to demonstrate segregation for virulence at eight independent loci. Similarly, a cross of the N. American Race 6 and the S. American Race 22 of *M. lini* gave rise to 54 different races amongst an F_2 of 67. Since the parent races differed in pathogenicity on 12 of the supposed monogenic testers, potentially 2^{12} races, at least, could have been generated from this cross (Flor, 1955). This example also illustrates the enormous potential wealth of virulence genes which may be made available when pathogens which have evolved in different geographical areas are brought together.

In pathogens, such as *Venturia inaequalis*, which are both haploid and possess sexual reproduction yet do not readily form heterokaryons, sexual recombination must be the source of new pathogenic races. Some 19 genes are known for pathogenicity in this fungus but only *p-8*, *p-9* and *p-12* are known to be loosely linked. Although $n = 7$, the frequencies of 2nd division segregations for the pathogenicity genes are all high, suggesting that they are some distance from the centromere. It seems probable, therefore, that although each chromosome must carry 2 or 3 pathogenicity genes, they are likely to show fairly extensive recombination and hence permit a maximum production of new races. The fact that some of them, e.g. *p-1*, *p-2*, *p-3*, are apparently more common in the U.S.A. than others must presumably reflect selection for certain genotypes of the pathogen (Boone, 1971).

Other mechanisms originating races

There are several anomalous situations which have resulted in the formation of new physiological races. Some cases are known where extrachromosomal factors are apparently involved. Johnson and his co-workers showed both for *P. graminis tritici* and *P. graminis avenae* that in reciprocal crosses between certain races the pathogenic behaviour reflected that of the aecidial (i.e. 'maternal') parent (Newton and Johnson, 1932; Johnson and others, 1934; Johnson and Newton, 1940). A more recent study has revealed another, more complex situation. Pathogenicity in *P. graminis avenae* is determined by two genes and an extrachromosomal factor. Normally *Ve* is a dominant gene for virulence, *ve* being a recessive allele for avirulence. If, however, *Ve* enters the cross on the maternal side, i.e. the aecidial parent, in a homokaryon also possessing *D*, the dikaryotic progeny are avirulent. It is supposed that D modifies the cytoplasm in such a way that Ve's expression is reversed. The gene D has another effect. In any dikaryon homokaryotic for D but heterokaryotic for Ve, i.e. $[(D + Ve) + (D + ve)]$, the range of expression of Ve is variable, the pustules varying in size and development from those typical of a susceptible reaction to those typical of a hypersensitive response (Green, 1965). A not dissimilar variability has been detected in *P. recondita* Race 229. On the *cv.* Saia, single uredospore subcultures gave approximately equal numbers of single-spore cultures avirulent (Race 228) and virulent on the cultivar. The former, when subcultured, remained avirulent but the latter again segregated to give approximately equal numbers of virulent and avirulent progeny. This kind of instability is reminiscent of 'mitotic non-conformity' described earlier (pp. 44 and 248) or mitotic non-disjunction associated with aneuploidy (p. 74). This case has not been adequately explained, however (Bartos and others, 1969).

The adaptation of *P. infestans* after serial passaging of zoospores through hosts initially resistant to the races employed has already been described but it will be recalled that other workers have not been able to effect these changes. The origin of these new pathogenic races must remain in doubt, therefore, and this also applies to the experiments of Malcolmson (1970). She obtained new races after inoculating tubers with mixtures of zoospores of various races but, since neither heterokaryosis nor somatic recombination have been detected in *P. infestans* and since only mating type $A1$ is present in Europe, the results are not yet explicable.

Genetical aspects of aggressiveness

Although this phenomenon is well known to plant pathologists, in the field as well as in the laboratory, little attention has been paid to its genetic aspects. Most pathogenic fungi, whether or not they show a differential interaction with different host cultivars, also show a more or less continuous range in pathogenicity to any one or to several hosts. van der Plank (1968) has utilized published data on *Fusarium oxysporum lycopersici*, the cause of tomato wilt, to illustrate this point (Table 13.9a, b).

In Table 13.9a it can be seen that, although there is a correlation between disease severity and certain morphological features, the ranking of the isolates is identical on both *cv.* Bonny Best and Marglobe: both are susceptible. However, if cultivars derived from *Lycopersicon pimpinellifolium* are included, *F. oxysporum lycopersici* can be divided into 2 races on the basis of their differential response (Table 13.9b). The gene-for-gene relationship relates only to the resistance gene in *L. pimpinellifolium* and its absence in *L. esculentum*. Aggressiveness is essentially a quantitative character and this can be seen from the individual disease ratings of the 28 isolates of *F. oxysporum lycopersici*, cf. Table 13.9a on *cv.* Marglobe; they are:

1·8:2·1:2·5:3·1:3·2(2):3·3:3·7(2):4·0:4·1:4·2:4·7(2); 4·9:5·0(2):5·1:5·3: 5·5:6·1:6·4:6·6:6·9:8·0:8·8:9·1.

TABLE 13.9 Aggressiveness and virulence shown by *Fusarium oxysporum* f. sp. *lycopersici* to different tomato cultivars. (Data of Wellman and Blaisdell, 1940; Henderson and Winstead, 1961.)

(a) *Aggressiveness*

Morphology of strains	Average disease severity	
	cv. Bonny Best	*cv*. Marglobe
Fluffy	10·4	7·5
Fluffy-sclerotial	8·7	6·3
Intermediate fluffy	8·3	4·5
Intermediate appressed	6·3	4·1
Appressed and slimy	4·7	3·0

Note: Disease severity measured on an arbitrary scale from 0, no disease, to 15, severe disease.

(b) *Virulence*

Cultivar and resistance genes		Physiological race and virulence genes	
		Race 1 A_1A_2	Race 2 a_1A_2
Bonny Best, Marglobe + 41 cultivars		S	S
Manalucie + 54 cultivars	R_1	R	S

Note: Key: R, resistant; S, susceptible.

The way in which genes for aggressiveness may interact with those for virulence has recently been illuminated by studies with the covered smut of barley, *Ustilago hordei*. Pathogenicity of this fungus to barley was shown to be variable but, in fact, it proved possible to demonstrate the existence of a recessive virulence gene *Uh v-1* and a corresponding resistance gene in the *cv*. Vantage (Thomas and Person, 1965; Sidhu and Person, 1971, 1972). Emara (1972), was also able to show that the natural variability in aggressiveness of the 13 physiological races recognized in *U. hordei* to the common suscept host *cv*. Odessa could be accounted for if it was supposed that aggressiveness was determined polygenically (Table 13.10, p. 278). Thus, there is evidence in this fungus for pathogenicity being determined by major genes and polygenes.

Emara and Sidhu (1974) then selected two teleutospores, E and F, from different collections, both being homozygous for *Uh v-1*, from each of which they derived the four meiotic products, the basidiospores. These were crossed in all combinations within the limits imposed by the dimictic mating system and the 16 dikaryons obtained were used to infect *cv*. Vantage. Thus all dikaryons were virulent to *cv*. Vantage but the degree of infection they showed, i.e. their aggressiveness, ranged from 20·1% to 68·9%. An analysis of variance revealed that:

(i) One teleutospore contributed more aggressiveness than the other.

(ii) Dikaryons derived from crosses between basidiospores from different teleutospores are more aggressive than those from crosses between

TABLE 13.10 The natural variability in aggressiveness of the thirteen physiological races of *Ustilago hordei* to the universal suscept barley *cv*. Odessa. (Data of Emara, 1972.)

Mating type and mean value	Physiological races and % infection												
+	27·3	25·3	23·6	18·8	5·6	16·8	19·0	17·6	19·0	21·6	21·6	19·3	22·9
−	25·0	25·1	20·6	15·0	13·5	16·2	18·4	13·1	14·9	25·2	25·6	20·0	25·1
± (Mean)	26·1	25·2	22·1	16·9	9·6	16·5	18·9	15·4	16·9	23·4	23·6	19·6	24·0

Partitioning of variance

	Phenotypic variance	Genotypic variance	Components of V_G		Environmental variance	Heritability h^2	
			Additive	Interactive			
	V_P	V_G	V_A	V_I	V_E	$\dfrac{V_A}{V_P}$	$\dfrac{V_G}{V_P}$
Actual	72·14	36·96	30·35	6·61	35·18	0·42	0·51
As %	100	51·23	42·07	9·16	48·77		

Note: A monosporidial line of each mating type of each race was sampled. Infections as % of rows of 100 plants.

basidiospores from the same teleutospore. (Mean values: E × E 32·2; F × F 48·9; E × F and reciprocal 52·9.) This suggests that genes determining aggressiveness are dominant and that there may be some heterosis.

(iii) Some 65% of the total variance was genotypic and two-thirds of this was contributed by additive genetical action, one third by interaction and dominance effects ($V_A = 43·9\%$; $V_I = 21·3\%$; $V_E = 34·8\%$).

Thus the polygenes determining aggressiveness act as modifiers of the virulence gene *Uh v-1* on a susceptible host. It remains to be determined how far selection for increased or decreased aggressiveness could result in a modification of the phenotypic expression of the major virulence gene on other types of host.

It has been claimed by van der Plank (1968) that:

'There is no evidence for a positive correlation between aggressiveness and virulence; as far as is known, greater aggressiveness is not linked with greater virulence. But there is a negative correlation: increased virulence may go with decreased aggressiveness . . .'

Were this always the case it would be of considerable importance but there is no evidence that this is so. Naturally, in cases where virulence genes are lacking, such as in many wood-destroying fungi, more aggressive races will be selected although monospore isolates often show a wide range of pathogenic ability. This was demonstrated for *Polyporus betulinus*, restricted to *Betula* spp. by Bell and Burnett (1966). Nevertheless, this fungus is capable, not only of forming dikaryons, but also of genetical 'mosaics' as a result of fusions of dikaryons with other dikaryons or with monokaryons (Burnett and Partington, 1957). The wood-destroying ability of such 'mosaics' appears to be at least as high as the best of its component monokaryons, and frequently better (Burnett and Bell, unpublished). If this is the general rule for such fungi, then aggressiveness will be determined both by the individual genes in each nucleus and by their interaction, as in heterokaryons of the type usually studied.

There is evidence, however, that in fungi with new virulence genes selection operates to promote both virulence and increased aggressiveness of the isolate. Some of this evidence has already been discussed in the chapters on selection (pp. 196–199 and 206–207). For example, Race 21-1,2 of *P. graminis tritici* arose in Australia from Race 21-2. In Western Australia it became the dominant race between 1961 and 1969, but in eastern Australia it has never competed successfully with Race 21-1 and is unimportant. The difference between these two races has been attributed to differences in aggressiveness (Watson, 1970). Here therefore, is a clear and highly significant correlation between virulence and aggressiveness which must be attributed to the action of selection.

One other suggestion has been made which challenges the view that these two attributes are not correlated. McIntosh and others (1967) and Knott (1968) have studied field resistance shown by wheat cultivars and thought to be non-specific. They have shown that resistance can be accounted for by several genes which confer specific resistance, either at the seedling stage or in the adult or in both. This implies that the rusts will have, on the gene-for-gene hypothesis, matching virulence genes for the specific resistance genes. However, if there are several of these, and if they operate at different times and in different ways, the phenotypic response of different genotypes will appear to be quantitative rather than qualitative. This will be the case especially if precise control of the environment is not maintained, and also if isogenic lines, save for genes at particular loci, cannot be obtained. Both of these conditions are probable and, until they are fully controlled, it will not be possible to ascertain whether virulence and aggressiveness always have a dissimilar genetic basis.

Parasitism on Several Hosts

Information on this topic comes from three sources: (a) the analysis of crosses between related pathogens naturally confined to different hosts, (b) the adaptation of a specialized pathogen to another host and (c) the results of crossing isolates of a pathogen from different hosts and testing their pathogenicity on all, or a wide range of, its hosts. Three examples will be considered, the *formae specialis* of cereal rusts, the adaptation of *Phytophtora infestans* from potato to tomato and the results of crosses within the genus *Helminthosporium*.

Crosses between formae of P. graminis

It will be recalled that these were established on the basis of the principal host genus which they parasitized (see p. 260, Table 13.1). Many attempts have been made to cross them on their common alternate host, barberry and the results have been summarized by Johnson and Newton (1946). The crosses have been:

	Cross	Host A	Hosts B
(i)	*tritici* × *agrostidis*	Wheat	*Agrostis* spp.
(ii)	*tritici* × *avenae*	Barley	Oats
(iii)	*tritici* × *secalis*	Many grasses	Rye, barley, grasses as *tritici*

(i) The first of these can be achieved with difficulty and only if *tritici* is the aecidial parent; in the reciprocal cross, hybrid aecidia never develop. The progeny in one case would only attack *Triticum* and, since *tritici* was the 'female' parent, it was debatable whether they were genuine hybrids (Stakman and others, 1930). Johnson and others (1932) obtained one aecidium from 14 attempted crosses and its progeny could infect both *Triticum* and *Agrostis* to a slight extent, and also *Hordeum*.

(ii) Here too there was evidence of considerable intersterility and hybrids were only obtained on the *tritici* side. However, their host range was wider than that of either parent, including *Triticum*, *Hordeum*, *Avena* and some grasses. But the width of host range was at the expense of pathogenic vigour and no host was attacked as heavily as it had been by the parental races (Johnson and Newton, 1933).

(iii) This cross has given the most successful and varied results. It will be noted that there are several hosts common to both *formae*. Race 36 of *tritici* was crossed with Race 11 of *secalis*. In one case 6 of the progeny resembled *tritici*, 2 resembled *secalis* and 2 were intermediate (Stakman and others, 1930). In another, 4 isolates were said to resemble *tritici* but they showed weak pathogenicity to both wheat and rye: this resembled the two intermediates obtained in the other cross (Johnson and others, 1932). These intermediate types could also infect barley. In contrast to these results, Levine and Cotter (1931) obtained a vigorous rust capable of attacking wheat, rye and barley quite severely.

The conclusions to be drawn from these limited experiments can only be tentative, for in no cases were the crosses taken beyond F_1, nor were the resistance alleles of the hosts known. It does seem that there are separate virulence genes specific for some, at least, of the different host genera. Thus infection to a wider host range than either parent is possible. The frequent reduction in pathogenicity is presumably due to an unbalanced genotype and it may be suggested that this is particularly associated with genes determining aggressiveness. Thus, although the hybrids acquire wider virulence, it is associated with reduced aggressiveness.

A further case, of some practical significance, has been reported from Australia (McIntosh and others, 1973a). It was known for many years that *P. graminis tritici* regularly attacked *Agropyron scabrum* and *P. graminis secalis* was believed not to occur. About 1950 a rust resembling this latter species was detected in southern New South Wales and is now widespread. It can be isolated from rye, barley, *A. repens* and *A. scabrum*. In areas where *A. scabrum* is common, it can be infected with *P. graminis tritici*, *P. graminis secalis* or both, as well as rusts specifically adapted to it. Evidently the variability of the rusts on *A. scabrum* has increased greatly since 1950 and it is believed that this is due to somatic hybridization (Figure 13.3). Tests in the laboratory have shown that the putative hybrid progeny are virulent to resistance genes in *Triticum* which normally prevent infection by *P. graminis secalis*.

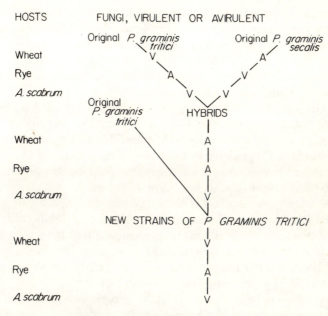

FIGURE 13.3 The possible origin of new physiological races of *Puccinia graminis tritici* involving the grass *Agropyron scaber* in Australia. V, indicates virulent; A, avirulent.
(After McIntosh and others, 1973a.)

Apart from demonstrating the practical importance of hybridization between *formae*, this study demonstrates that different sets of genes are involved in determining virulence to resistance genes in different host species. Indeed, supporting evidence for this had come from Watson and Luig's (1962) studies on inbreeding *P. graminis secalis* on barberry. Derived inbred lines had shown the ability to attack wheat as well as rye.

Adaptation of P. infestans to tomato

Attention has principally been given to the virulence of *Phytophthora infestans* to potatoes and 11 virulence loci have now been identified (Malcolmson and Black, 1966). In 1955, Gallegly and Marvel demonstrated the existence of a further locus for virulence to the Ph_1 (then called *TR1*) resistance gene in *Lycopersicon pimpinellifolium*. Thus two physiological races pathogenic to tomato were recognized; these were 0 and 1, the latter being capable of overcoming Ph_1 resistance which was lacking in the commercial varieties. Wilson and Gallegly (1955) showed that the locus controlling virulence towards tomatoes was independent of virulence to potatoes. They also showed that tomato races showed differences in aggressiveness to tomatoes. Moreover, as had been known since 1919, potato races were highly infectious to tomatoes but not *vice versa*. It was supposed, therefore, that races pathogenic to tomatoes arose from races pathogenic to potatoes.

A number of investigators started with potato races and showed that after 6 to 7 passages through tomato leaves of massive zoospore inocula, isolations were identical with those derived from races naturally strongly pathogenic to tomatoes. Moreover, this character was maintained in the passaged strains after being continued for a further 6 months on potato-tuber tissue (Mills, 1940). Comparable results were obtained by de Bruyn (1951), Graham and others (1961) and Kishy (1962). The last named, however, noted that initially isolates against the tomato *cv.* Ponderosa showed slight, moderate or severe infection. He was unable to adapt those showing slight pathogenicity to tomato and they died out after 4 passages, whereas those showing moderate infection could be 'trained' to equal the severe, typical tomato-pathogenic type of infection.

Between 1965 and 1968 late blight of tomatoes developed to epidemic proportions in some glasshouse areas in the Netherlands, although prior to 1965 it was hardly a problem. It was believed that the rise in importance of the pathogen reflected the reduced use of fungicides in the same period adopted because other fungal diseases of the tomato, notably *C. fulvum*, had been controlled by breeding resistant varieties.

Turkensteen (1973) carried out field experiments to see if *P. infestans* from potatoes could be the source of the epidemic on tomatoes. Using known races of *P. infestans*, namely (1, 2, 3); (1, 3, 4) and (1, 4, 10), she inoculated susceptible potatoes amongst which tomato plots were growing, including varieties both lacking and possessing the Ph_1 resistance gene. After 75 days *cv.* Moneymaker, lacking Ph_1, was infected by Race (2,4) and by 112 days tomatoes carrying Ph_1 had succumbed. A repetition of the experiment in the following season gave much the same result, without any evidence of change in the *P. infestans* in its pathogenicity to potatoes. In both years it was evident that contamination as a source of the tomato-pathogenic strain was virtually impossible. In the second experiment the race had become weakly pathogenic to tomatoes carrying Ph_1 after 37 days; about a week before all the potatoes had been totally destroyed. In a further 64 days the strain had become strongly pathogenic to tomatoes. This period in the field corresponds with a maximum of 6 to 9 serial passages. Turkensteen demonstrated, by frequent sampling, that the changed pathogenicity could be accounted for if it were supposed that a mutation had occurred by day 37 and thereafter had been selected in the population.

No breeding data are available from any of these experiments. The most plausible hypothesis is, however, that *P. infestans* carries at least one gene avirulent towards *Lycopersicon*. Provided that a mutation to virulence occurs at this locus, it can become associated with genes which increase its aggressiveness in selectively favourable situations, such as when propagated on tomato. Thus the three classes of isolate found by Kishy would correspond with (i) unmutated, hence avirulent, *P. infestans*: the slightly pathogenic types, (ii) mutated, hence virulent but non-aggressive types, showing intermediate pathogenicity and (iii) types both virulent and aggressive and hence fully capable of exploiting the tomato environment. The three types ($+/-$ indicating polygenes for aggressiveness) can be represented as:

$$A_{ph_1} + - + - - \qquad a_{ph_1} + - + - - \qquad a_{ph_1} + + + + +$$

Here, therefore, the extension of a host range to a new host is a comparable process to the extension of a physiological race from one cultivar to another cultivar of the same host. This is perhaps not surprising, since *Solanum* and *Lycopersicon* are clearly genera as closely related as is *Triticum* to *Secale* or *Agropyron*. It will be recalled that data from crosses between *formae* of cereal rusts suggested that the pathogens possessed appropriate virulence genes but were usually lacking in aggressiveness. The question therefore arises whether a similar situation occurs in a pathogen with a wide host range. This will now be discussed.

Pathogenicity in Helminthosporium

Helminthosporium is a large genus of species ranging from saprophytes through weak parasites to strong parasites, mostly on grasses or cereals. It has been extensivly studied by Nelson and his co-workers (Nelson, 1961; Nelson and Kline, 1962, 1963, 1969a, b). They have studied only a single qualitative character, namely whether or not an isolate was able to develop a leaf lesion in specified conditions some 10 days after being sprayed with a spore inoculum. Neither the number, size, speed of production of lesions nor the damage achieved was reported. In effect, therefore, they studied the effective virulence of their strains. Their data are very extensive and only the most significant can be considered.

Firstly, employing representatives of five species, *H. carbonum*, *maydis*, *victoriae*, *sorokianum* and *oryzeae*, they tested their ability to develop infections on 25 spp. of grasses. They found a great range of intraspecific variation. For example, two of the 10 isolates of *H. carbonum* could only infect *Sorghum sudanense* which was a universal suscept to all isolates, whereas two others could infect 9 grass species (Table 13.11).

TABLE 13.11 The infectivity of a sample of 44 strains of 5 *Helminthosporium* species to 25 different species of grass or cereals. (Data of Nelson and Kline, 1962.)

Grass or cereals	H. carbonum										H. maydis										H. victoriae										H. sorokianum								H. oryzae					
	1	2	3	4	5	6	7	8	9	10	1	2	3	4	5	6	7	8	9	10	1	2	3	4	5	6	7	8	9	10	1	2	3	4	5	6	7	8	1	2	3	4	5	6
Agrostis palustris																																									+			+
Alopecurus arundinaceus			+																												+	+	+	+	+				+	+	+	+	+	+
Avena fatua																															+	+	+	+					+	+	+	+	+	+
Avena sativa				+	+																										+			+				+	+			+	+	+
Axonopus affinis													+	+	+																		+		+									
Bromus willdenovii				+	+			+			+	+	+	+	+	+	+	+	+	+											+	+	+	+	+	+	+	+	+	+	+	+	+	+
Cynodon dactylon								+			+	+	+	+	+	+	+	+	+												+	+	+	+	+	+				+		+		+
Dactylis glomerata			+					+		+	+	+	+	+	+	+	+	+	+	+											+	+	+	+	+	+	+	+	+	+	+	+	+	+
Digitaria sanguinalis																																												
Eleusine indica																			+	+																								
Festuca elatior																				+											+	+	+	+	+	+	+	+	+	+	+	+	+	+
Hordeum vulgare			+	+	+		+				+	+	+	+	+	+	+	+	+	+	+	+									+	+	+	+	+	+	+	+	+	+	+	+	+	+
Lolium multiflorum	+	+			+					+	+	+	+	+	+	+	+	+	+	+							+			+	+	+	+	+	+	+	+	+	+	+	+	+	+	+
Oryza sativa	+			+						+	+	+	+	+	+	+	+	+	+	+											+	+	+	+		+			+	+	+	+	+	+
Panicum virgatum															+	+	+	+		+																								
Paspalum dilatatum																		+																										
Paspalum notatum																																												
Pennisetum spicatum	+			+	+	+	+			+	+	+	+	+	+	+	+	+	+	+					+						+	+	+	+	+	+	+	+	+	+	+	+	+	+
Phalaris arundinacea																															+	+	+	+	+	+	+	+					+	+
Phalaris tuberosa					+			+		+	+	+	+	+	+	+	+	+	+	+							+	+			+	+	+	+	+	+	+	+						
Poa pratensis						+				+	+	+	+	+	+	+	+	+	+	+											+	+	+	+	+	+	+	+	+	+	+	+	+	+
Setaria viridis	+	+		+						+	+	+	+	+	+	+	+	+	+	+					+						+	+	+	+	+	+	+	+		+		+		+
Sorghum sudanese	+	+			+			+		+	+	+	+	+	+	+	+	+	+	+					+						+	+	+	+	+	+	+	+	+	+	+	+	+	+
Triticum aestivum																		+		+											+	+	+	+	+	+	+	+	+	+	+	+	+	+
Zea mays					+	+	+	+	+	+	+	+	+	+	+	+	+	+	+	+											+	+	+	+	+	+	+	+	+	+	+	+	+	+

The most significant feature in all the species was that very few isolates shared a common host range. Even closely related species, as judged on morphological criteria, showed little pathogenic similarity, e.g. *H. victoriae* and *H. sorokianum* (Table 13.11). It seemed, therefore, that different species possessed different pools of pathogenicity genes and that even intra-specific crosses should give rise to isolates with different host ranges. This possibility was explored further with *H. carbonum* (=*Cochliobolus carbonum*). The data obtained are summarized in Table 13.12.

TABLE 13.12 Grasses for which 19 strains of *Cochliobolus* (*Helminthosporium*) *carbonum* possesses virulence genes with an indication of their numbers and linkage. (Data of Nelson and Kline, 1963, 1969a; Nelson 1970.)

Virulence determined by 1 gene to each grass
Linked genes
    4 *Cynodon dactylon—Festuca elatior—Poa pratensis—Lolium multiflorum* (These are linked to the mating-type, *A/a*, locus.)
    3 *Oryza sativa—Eleusine indica—Axonopus affinis.* (The virulence gene for *A. affinis* is loosely linked to the other two.)

Genes segregating independently
    6 *Buchloe dactyloides*; *Dactylis glomerata*; *Lolium perenne*; *Panicum virgatum*; *Phalaris arundinacea*; *Ph. tuberosa stenoptera.*

Virulence determined by 2 genes to each grass
10 × 2   *Agrostis palustris*; *Avena sativa*; *Eragrostis curvula*; *Paspalum notatum*; *Pennisetum glaucum*;
= 20   *Phleum pratense*; *Secale cereale*; *Sorghastrum nutans*; *Triticum aestivum*; *Zea mays*
Total number of virulence genes—33

Some 19 isolates were crossed in various combinations, the ascospore progeny isolated and their pathogenicity to 25 graminaceous hosts assessed as already described. The hosts usually included differential hosts, i.e. those to which one parental isolate was pathogenic, the other non-pathogenic, and in some cases to hosts resistant to both parental isolates.

In all some 33 distinct genes for pathogenicity were identified. The haploid number of *C. carbonum* is 8, so that not surprisingly some genes were found to be linked, although data are not available to show how closely. In almost half the cases pathogenicity to a host species is apparently deter-mined by two genes, as indicated by 3:1 segregations for pathogenic:non-pathogenic progeny, or vice versa.

In a more limited study of 5 isolates of *H. maydis* a minimum of 13 genes for pathogenicity to nine graminaceous hosts was detected. Here pathogen-icity to 5 species was controlled by 5 different genes and that to the other 4 species by 4 different sets of 2 genes each. As was to be expected, the host range of a number of the progeny was wider than that of the parental isolates and in some cases progeny were capable of causing lesions on hosts to which neither of the parents were pathogenic. For example, neither of the parental isolates of *H. carbonum* or *H. victoriae* employed were capable of infecting either *Cynodon dactylon* or *Triticum aestivum*, but their progeny could infect both these species. This implies epistasis between avirulence and virulence genes, a phenomenon found in other species, e.g. *Venturia inaequalis.*

The striking feature of these studies has been the demonstration of how many genes determining pathogenicity to different hosts are distributed through the gene pool of a population. In view of the small size of the samples investigated, the actual numbers of genes determining pathogenicity to different hosts must be very large indeed. In the samples studied no isolate had more than two genes determining pathogenicity to one host, although in some physiological races of rusts studied the number of loci involved approached ten. If this were applicable to *H. carbonum*, there would be about 250 loci in the genotype associated with the determination of pathogenicity alone! When it is recalled that Nelson's work deals with only one aspect of pathogenicity, effective virulence, it can be seen that a large proportion of the genotype of a wide-ranging pathogenic fungus must be devoted to genes determining virulence and aggressiveness.

GENERAL CONCLUSIONS

The genetical analyses of pathogenicity in fungi have been carried out at a range of different levels on both highly specialized obligate forms like rust fungi and on general, weakish to strong, facultative forms such as species of *Phytophthora* or *Helminthosporium*. It seems as though the basic ingredients are the same in all these cases, although their quantitative contributions differ. These ingredients are (a) virulence genes which determine the gene-for-gene relationship between genes in the pathogen and the corresponding resistance genes in the host, and (b) many others, both oligogenes and polygenes, which determine the effectiveness of expression of the virulence genes by conditioning aggressiveness. Selection operates on one or on the other type or both, and determines the balance between them. In highly effective, specialized pathogens it results in the reinforcement of virulence genes for a particular host species by selection for high aggressiveness. On the other hand, it may be surmised that the opposite extreme is, perhaps, represented by *Helminthosporium*, where selection has favoured a large number of virulence genes for different hosts, no one of which is greatly reinforced by the supporting action of genes for aggressiveness. Provided that there is an effective recombination mechanism such species are always potentially dangerous. For, at any time selection may favour virulence genes in respect of a particular host and, through the build-up of supporting genes for aggressiveness, so develop the pathogen that it emerges as a highly active form, showing specialized pathogenicity to a particular host. Thus for example, the additional susceptibility to Race T of *Helminthosporium maydis* imparted to maize in the U.S.A. by the widespread use by breeders of Texas male-sterile cytoplasm provided just such a selective situation. A major epidemic of southern leaf blight rapidly developed. That it has not been maintained is due to various factors but perhaps virulence was not sufficiently associated with aggressiveness for the high pathogenicity to be sustained. Ultimately, disruptive or directional selection could result in a virulent, aggressive isolate acquiring the status of a *forma speciale*, when isolating mechanisms would operate to channel it along a path distinct from the rest of the species. The converse is probably true. It is, therefore, quite plausible, as Green (1971) has pointed out, that races derived from crossing *P. graminis tritici* and *P. graminis secalis* (pp. 280–281) which show reduced pathogenicity but a wider host range than their parents, may resemble primitive rusts from which the modern *formae* have been derived by selection. *Helminthosporium* on this view would be a genetically primitive parasite, *P. graminis tritici* a highly advanced one.

Although these conclusions are speculative, based as they are upon a very limited amount of data, they do at least enable the pathologist to begin to see how the known facts of fungal population genetics can be applied to an understanding of how the different kinds of pathogens have arisen. Of more importance, further studies should provide an insight whereby the pathologist can confidently predict the probable course of pathogen evolution and take the necessary steps to prevent it from becoming economically disastrous.

CHAPTER 14

Fungi and general genetics
I. Recombination

Fungi have made notable contributions to an understanding of general genetical problems such as the mechanism of recombination, the nature of gene action and its regulation in eukaryotes. Other areas where it seems that fungi are likely to contribute in the future are, for example, processes involved in differentiation and development, the study of genotype–environment interactions (Chapter 7), selection experiments (Chapter 9) and ideas on speciation and evolution (Chapter 8, 10, 11).

In these final chapters attention will be concentrated principally on the contribution of fungal genetics to general genetics at the cellular level; specifically, in this chapter, on the mechanism of recombination.

Contributions from fungi to the understanding of recombination fall roughly into three phases: (a) confirmation and extension of early hypotheses of crossing-over, (b) the study of non-reciprocal recombination and (c) hypotheses of repair models for recombination. The basic reason why fungi have proved so rewarding in these studies is that they provide readily accessible tetrads and, in particular, in the Pyrenomycetes, ordered tetrads. Fungal tetrads offer a unique possibility of examining all the products of a single meiosis between complete, haploid genomes as directly as possible. In diploid organisms a further generation is necessary in order to detect recessive alleles which have segregated because their phenotype has been masked through dominance. Such data as they provide are statistical and indirect. This is true also for bacteria and viruses but they possess the additional disadvantages that there is no certainty that all the genome has been involved in recombination, that all the products of recombination have been recovered, and that one or several recombinational events may not have occurred.

The mechanics of crossing-over

Intergenic recombination

The earliest contributions to this topic came from the study of the linear, ordered ascus of *Neurospora* spp. This arose from Wilcox's (1928) demonstration that the disposition of ascospores in the ascus of *N. sitophila* did, indeed, reflect the planes of division of the two successive meiotic divisions. He showed that the mating-type factor could segregate at either division and, a year later, Lindegren showed, for *N. crassa*, that the ratio of first- to second-division segregation was constant at 8:15 under his experimental conditions. Lindegren extended these studies to other characters, both linked and unlinked (1932, 1933). This provided the first unequivocal evidence that the normal condition is that crossing-over occurs at the four-strand stage of meiosis and involves only two of the four homologous chromatids, of different origin, at any one place. He suggested that the frequency of second-division segregation for a character pair reflected the frequency of cross-overs between the gene locus and the centromere, and so was related to the distance between them. This required the first-division segregation of the centromeres of homologous chromosomes. His views were supported by linkage data from *Neurospora*, derived in the usual way.

The ability to determine with precision the positions and types of exchanges, suggested that fungi would be of particular value for studies on interference. In its general form, namely the situation in which a cross-over in the region either prevents or promotes the occurrence of another cross-over in an adjacent region, it had already been studied in diploids such as *Drosophila*. However, the linear, ordered tetrad enabled the question to be investigated with greater precision at the chromatid level. Thus the effect of a cross-over between two chromatids at one point on the occurrence of cross-overs either between the same, or different chromatids in an adjacent region, could be examined, i.e. chromatid interference. Such studies required at least three-linked loci and had, therefore, to await adequately marked fungal stocks. Nevertheless, the earliest data on chromatid interference were provided by the Lindegrens, using *N. crassa* (Lindegren and Lindegren, 1937, 1942) and fungi still provide the greater part of the data on this topic.

Chiasma interference occurs in the fungi, as in other organisms, but data from different species have demonstrated that it can occur with different intensities in different species. For instance, in *Neurospora* and *Saccharomyces*, interference is positive within chromosome arms but it probably does not occur across the centromere (Lindegren and Lindegren, 1939; Shult and Lindegren, 1959; Bole-Gowda and others, 1962). Interference is usually greatest between closely linked loci but decreases with distance (Perkins, 1962). On the other hand, some data for closely linked loci in *Aspergillus nidulans* give no evidence of interference (Strickland, 1958).

The position with regard to chromatid interference is less clear, as has already been mentioned (p. 69). Most available data suggest little chroma- tid interference but, particularly in the region of the centromere, there often seems to be a departure from random occurrence. This is usually manifested, as in the original data for *N. crassa* (Lindegren and Lindegren, 1942), by an excess of 2-strand exchanges from the expected ratio of $1:2:1$ for two: three:four-strand exchanges (p. 68) and a deficit in 4-strand exchanges. Thus, two-strand plus four-strand exchanges just about equal the three- strand exchanges.

Compilations of data relating to both types of exchange have been pro- vided by Esser and Kuenen (1965) and Fincham and Day (1971).

The use of fungal tetrads, therefore, has played an important role in supporting and confirming the chiasmatype hypothesis which had been put forward to account for crossing-over. On this hypothesis chiasmata located the sites of crossing-over. Two mechanisms had been proposed to account for the exchange of chromosomal material. One, proposed by Darlington (1934, 1935), suggested that chiasmata located identical sites where chromatids had broken and rejoined, not necessarily to the same broken end. The other, due to Belling (1933), proposed that gene replica- tion occurred at pachytene and was followed by the formation of connexions between them. If, therefore, homologous chromosomes were coiled round each other at one point, the new genes formed adjacent to the old ones would copy some from one, up to the twist, and then some from the other. When the new genes joined together they would produce a 'cross-over chromatid'. Both these hypotheses were based upon cytological observa- tions and the fungi, with their small chromosomes, technically difficult to see or stain, contributed little to such ideas.

However, recently they have provided some direct evidence bearing on these issues. Firstly, Rossen and Westergaard (1966) showed that in the Discomycete *Neotiella* (*Humaria*) *rutilans* the DNA doubled in the haploid nuclei in the ascus *prior to fusion followed by meiosis*. This effectively precludes any general mechanism such as that proposed by Belling to account for both chromosome replication and crossing-over. Further studies with this fungus revealed the presence of well-developed synaptinemal complexes. First discovered by Moses (1956) in a crayfish, these are structures, visible only in the electron microscope, which assume their maximum complexity at the pachytene stage of meiosis. They consist of two lateral components, some 50 nm in diameter, their outer sides embedded in chromatin. Between them is an electron-translucent region, in the centre of which is an amor- phous central component some 20 nm in diameter. Only a very small portion of the chromatin of each homologous pair of chromosomes is in contact with the synaptinemal complex; data from *Neurospora* suggest about 0·3% of the total genome (Gillies, 1972). Its development in leptotene and its subsequent history after pachytene have been documented in several organisms but data from *Neotiella* and *Neurospora* have been especially informative (reviewed, Westergaard and von Wettstein, 1972). Westergaard and von Wettstein (1968, 1970) were the first to claim that at early diplotene chiasmata comprised short lengths of synaptinemal complex, which gradually changed their appearance. The central region became fibrillar, until finally only continuous chromatin bridges could be seen (Figure 14.1).

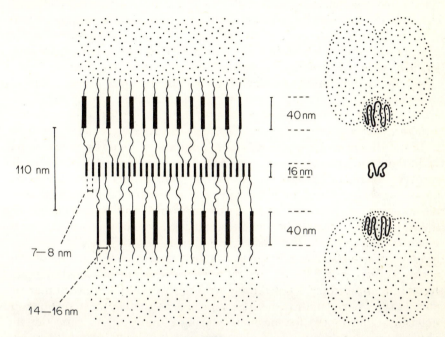

FIGURE 14.1 Diagrammatic L.S. and T.S. through a synaptinemal complex of *Neotiella rutilans*. Chromatid material is stippled. The complex consists of regularly spaced, lateral units which alternate in thickness, being joined by fine fibrils through an electron-translucent region to the central rod-like elements. In the T.S. the lateral and central units are represented imaginatively. (From Westergaard and von Wettstein, 1968.)

They believe that the complex mediates chromosome pairing and the physical exchanges involved in crossing-over. As Westergaard and von Wettstein expressed it:

'The synaptinemal complex in eukaryotes is a structure of great evolutionary stability, that must be intimately related to the universality of four-strand crossing-over at meiosis. The synaptinemal complex in meiotic cells is the vector for chromosome pairing and crossing-over, as is shown by the genesis of the chiasmata from the synaptinemal complex.'*

The presence of the complexes is not a sufficient condition for chiasmata formation, but it is a necessary one. For many years it has been known that for several organisms changes in temperature at appropriate stages in meiosis can alter the frequency of chiasmata. Recently, Lu (1970) has been able to show for *Coprinus* that this effect can only be brought about when the synaptinemal complex is present.

Fungi are contributing notably, therefore, to an understanding of both the genetics and cytology of crossing-over, as envisaged in the generally accepted manner. Moreover, fungi have provided novel data which have led to new hypotheses. One of these, to be treated in the next section, is gene conversion, the other is intragenic recombination and negative interference.

Intragenic recombination

The ability of fungi to produce large numbers of recombinants, coupled with the technique of selective plating, enabling rare protrophs to be recovered from crosses between closely linked auxotrophic mutants, opened up the study of closely linked or allelic loci. Such studies enabled truly allelic genes, or homoalleles, to be distinguished from extremely closely linked alleles of similar function, heteroalleles. Some of the earliest experiments employed *Aspergillus nidulans*, e.g. the bi_1–bi_3 'alleles', where bi_1 and bi_2 gave protrophs with a frequency of 1 in 2000 and bi_1 and bi_3 of 1 in 5000 (Roper, 1950). Roper (cited in Pontecorvo, 1959) noted that the recombinant protrophs showed a higher frequency of recombinant markers outside the *bi* locus than would be expected. A similar phenomenon was noted by Giles (1952) for inositol-requiring mutants of *N. crassa* but the phenomenon was most clearly demonstrated by Pritchard (1955) at the adenine-requiring *ade-8* locus in *A. nidulans*.

Pritchard noted that the closer two heteroalleles were together, the higher was the frequency of cross-overs between them, and that the effect was highly localized. In other words, *over very short regions of the chromosome*, about 0·3–0·4 map units, *the formation of one cross-over appears to promote others*. This phenomenon was termed 'localized negative interference'. Pritchard suggested that 'effective pairing', i.e. the condition which permits the occurrence of crossing-over, only occurs over a few short segments in a small fraction of the cells in meiosis but, in such regions, the probability of recombination is very high. The phenomenon is only detected when very closely linked heteroalleles are studied and has been amply confirmed in other fungi, e.g. *Neurospora* (Mitchell, 1956; St. Lawrence, 1956; Case and Giles, 1958) and in prokaryotes, such as phage T4 (Chase and Doermann, 1958). In normal mapping, when the marker genes are several units apart, the incidence of crossing over in effective pairing segments is underestimated, provided that single exchanges are more frequent within them than multiple exchanges, which seems to be the case (Pontecorvo, 1959). This failure to detect exchanges is the same well-known phenomenon seen on a larger scale, when linkage data between two distant loci are compared with summed linkage data derived from intervening loci.

Thus Pritchard demonstrated the presence of highly localized regions in which crossing-over is extremely frequent but, before the interpretation of this phenomenon can be usefully discussed, another important, novel feature of crossing-over in fungi needs to be examined.

Non-reciprocal recombination

Tetrad analysis of fungi in the vast majority of cases has demonstrated that crossing-over results in the formation of reciprocal recombinant types, i.e.

$$Ab \times aB \longrightarrow AB + ab,$$

if there is a cross-over between the A and the B loci.

However, in 1930 Winkler described cases of non-reciprocal recombination in moss and Basidiomycete tetrads. He explained them as being due to the conversion of genes from the dominant to the recessive state, or *vice versa*, in heterozygotes. The term and notion of 'gene conversion' (p. 72) was applied by Lindergren (1949) to those tetrads of *Saccharomyces* where alleles segregated in ratios other than $1:1$. For example, in the cross

$$ade^+ \ pan^+ \ inos^+ \ pyr^+ \ thi^+ \times ade \ pan \ inos \ pyr \ thi,$$

the asci were recovered showing abnormal ratios, such as

Spore 1:	ade^+	pan^+	$inos^+$	pyr^+	thi^+
Spore 2:	ade	pan	$inos$	pyr	thi^+
Spore 3:	ade	pan	$inos$	pyr	thi
Spore 4:	ade	pan^+	$inos^+$	pyr^+	thi

i.e. $3 \ ade : 1 \ ade^+$, all other loci $2:2$.

Since adenine-deficient auxotrophs develop a pink pigment, the meiotic products could be readily scored as 3 pink:1 white colony. Lindegren claimed that this was because an ade^+ gene had been converted to its ade allele in the heterozygous diploid cell.

Since $3:1$ ratios can arise from a variety of causes (p. 72) which must be eliminated before conversion can be accepted, there was disbelief and disagreement at first. However, in 1955, M. Mitchell described three unambiguous examples at the *pdx* (pyridoxin) locus in *N. crassa*. This locus lies between pyrimidine–1 (*pyr-1*) and colonial-4 (*col-4*). From 585 asci analysed from the heterozygote *pyr-1*$^+$ *pdxp col-4*/*pyr-1 pdx col-4*$^+$, three were abnormal in that they included protrophic recombinants (Table 14.1, p. 294).

In all cases the double mutant *pdx pdxp* was lacking and in only one was there recombination of the outside markers. Here, therefore, conversion and crossing-over appeared to have occurred together.

Non-reciprocal recombination showing these and other features has now been found and studied in several fungi, including yeast, *Neurospora*,

TABLE 14.1 Aberrant segregation of heteroalleles in *Neurospora crassa*.
(Data from M. B. Mitchell, 1955.)

| Cross | pyr^+ $pdxp$ col × pyr pdx col^+ | | |
| Spore | Ascus segregations | | |
pairs	1	2	3
1 and 2	pyr^+ $pdxp$ col	pyr^+ $pdxp$ col	pyr^+ $+$ col^+
3 and 4	pyr^+ $+$ col	pyr^+ $+$ col	pyr^+ $pdxp$ col
5 and 6	pyr $pdxp$ col^+	pyr pdx col^+	pyr pdx col^+
7 and 8	pyr pdx col^+	pyr pdx col^+	pyr $+$ col

Note: Key to symbols: *pyr*, pyrimidine requring; *pdx* and *pdxp*, pyridoxin requiring; *col*, colonial growth.

Sordaria and *Ascobolus*. These last two have a particular advantage since they possess several genes which determine, autonomously, the colour of the mature ascospores and thus segregation can be determined by inspection. As a result of these studies, certain general features associated with non-reciprocal recombination have emerged. These are:

(i) The frequency of conversion of heteroalleles is far higher than the spontaneous mutation rate in somatic nuclei or in crosses between strains carrying either of the same genes as homoalleles.

For example, Mitchell (1955) has shown that no protrophs were recovered from 21,825 ascospores from the cross *pdx* × *pdx*, nor from 22,747 spores from the cross *pdxp* × *pdxp*, whereas in heteroallelic crosses 44 pdx^+ spores were recovered from a sample of 21,577 ascospores.

(ii) Intergenic recombination is almost exclusively reciprocal, whereas most intragenic recombination is non-reciprocal, although reciprocal cross-overs also seem to occur. The most striking demonstration of the reciprocity of intergenic events and the non-reciprocity of intragenic recombination comes from work of Rizet and others (1960) with *Ascobolus immersus*. Here they were able to detect several loosely linked mutant sites, between which reciprocal recombination occurred. Within the sites, reciprocal recombination was either almost entirely lacking, e.g. 'Series 46' (Lissouba and Rizet, 1960), or it declined as the heteroalleles became more closely situated and was increasingly replaced by non-reciprocal events, e.g. 'Series 75' (Rossignol, 1967, cited by Fincham, 1970). A similar situation has been reported in yeast for the *arg-4* locus (Fogel and Mortimer, 1969 and p. 295).

(iii) The frequency of intragenic conversion is characteristic of the site at which it occurs, of its origin and of the organism.

In yeast there is good evidence from many loci that conversion from wild type to mutant and *vice versa* is about equal (Mortimer, cited in Roman, 1963). In *Sordaria fimicola* different heteroalleles at the *g* (grey) locus differ greatly in their conversion frequencies, e.g. $g_1 \rightarrow g_1^+ : g_1^+ \rightarrow g$ is approximately 5:1 (Kitani and Olive, 1967). Leblon (1972) followed up similar observations in *Ascobolus immersus* and showed that the conversion rates of mutants appeared to be correlated with their mutagenic origin. For example, NG-induced mutants showed an excess of conversion to wild type in 12 cases and equal frequencies of conversion in the two directions in 6 cases. The mutagen ICR-170, an acridine, which is believed to cause base additions and/or deletions, gave rise to mutants which showed the opposite behaviour. Non-reciprocal recombination has been detected with different frequencies in different fungi. It appears to be rare in *A. nidulans*, uncommon in *N. crassa*, more frequent in *S. cerevisiae* and relatively common in *A. immersus* and *S. fimicola*.

(iv) Gene conversion may include adjacent heteroallelic sites, i.e. so-called coincident conversion. This phenomenon has been detected in *Sordaria*, *Ascobolus* and *Saccharomyces* and there is suggestive evidence from *Neurospora*. It can be detected when doubly or triply heteroallelic mutant strains are crossed with wild-type strains. For example, Kitani and Olive (1969) have studied crosses involving grey (g) and hyaline (h) ascospore colour mutants. In several cases there were coincident conversions of $h \to h^+$ and $g^+ \to g$ but this was confined to the h sites most closely linked to the g locus. Fogel and Mortimer's (1969) study of the coincident conversion of heteroalleles at the *arg-4* locus has been related to the number of DNA base pairs between sites. This is possible because this locus has been mapped by the X-ray method of Manney and Mortimer (1964) in which the map units have been correlated with physical distance. These sites, with the numbers of base pairs between them, have been mapped as:

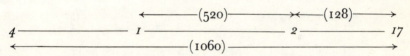

Double-site conversions were most frequent between sites *2* and *17* and single-site conversions rare (27 double:9 single). The opposite was true for crosses between sites *4* and *17* (3 double:46 single). Crosses of the *1* × *2* type showed approximately equal numbers of both types (23 double: 27 single). Incidentally, the number of reciprocal recombinations increased from 0 between *2* and *17* to 9 between *4* and *17*. Thus conversion can apparently involve sites even of the order of 1000 DNA base pairs apart but it becomes highly probable when the sites are separated by one-tenth of this distance.

(v) When heteroalleles within a locus are considered, there is often a gradient in the frequency of conversion from one end of the locus to the other: the so-called 'polaron effect'.

The first and most striking example of this phenomenon was described by Lissouba and Rizet (1960) in *A. immersus* at the '*Series 46*' locus. This is a group of 5 spontaneous heteroalleles all giving rise to pale-coloured ascospores rather than the normal, wild-type dark-brown ones. The tetrads can be scored easily, since each ascus discharges its spores as a group of 8, which can be collected on a sheet of glass above the apothecium.

The ten crosses between each pair of heteroalleles were made. In every case the majority of asci showed an 8:0 pale-coloured:dark-brown ratio, but some 6:2, pale:brown asci were also found. The brown ascospores proved to be wild type but the pale ascospores arose either from reciprocal, or from non-reciprocal recombination, as below:

$$m_1 + \times + m_2$$

┌────── reciprocal ──┴── non-reciprocal ──────┐

$$\text{recombinant} \begin{cases} \text{pale} & m_1 + & & m_1 + \text{pale} \\ \text{brown} & + + & & + + \text{brown} \\ \text{pale} & m_1 m_2 & & m_1 + \text{pale} \\ \text{pale} & + m_2 & & + m_2 \text{pale} \end{cases} \text{conversion}$$

The genetic make-up of pale ascospores could be determined by back-crossing the progeny with the parental types. In the case of a double mutant $(m_1 m_2)$ the result of such a test cross is an 8:0, pale:brown segregation. No such reciprocal recombinant asci were found but it was possible to map the different heteroalleles by the frequency of 6:2 asci between each pair. It was also found that in these non-reciprocal segregations the parental mutant type which occurred twice, the so-called 'majority parent', (m_1 in the example above) was (a) always the same and (b) always that to the left of the map as drawn in Figure 14.2.

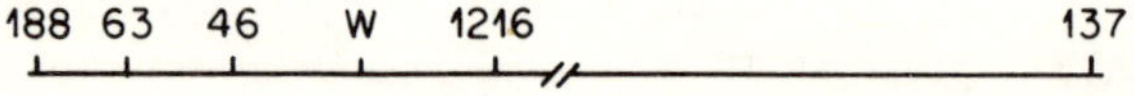

```
188  63   46    W    1216                       137
 ├────┼───┼─────┼─────┼──//──────────────────────┤
```

FIGURE 14.2 The linkage relationships of the hetero-alleles of the *Series 46* locus of *Ascobolus immersus*.

Data such as this suggest that whatever process causes conversion, it is polarized, i.e. it proceeds in one direction only.

Polarized conversion has also been detected in *Aspergillus* (Siddiqi, 1961) and in *Neurospora* at the *me-2* locus (Murray, 1963) and others. At the *me-2* locus conversion is most frequent in the heteroalleles at the ends of the locus and declines towards its centre. In 1968 Murray studied the same phenomenon at this locus, when it was inverted relative to the centromere of its chromosome and translocated from chromosome I to V. She found that this had no effect on the absolute configuration of the gradient of conversion frequency, which remained unchanged. Thus the direction of polarized conversion is a property of the chromatid segment involved.

(vi) Conversion can apparently occur in half chromatids only, resulting in post-meiotic segregations.

This effect is readily detected in 8-spored asci segregating for spore colour. Kitani and others (1962) described asci in *S. fimicola* where the gene order, from apex to base for $g^+ \times g$ crosses, includes:

$$
5g:3g^+ \quad
\begin{matrix} g \\ g \\ g^+ \\ g \\ g \\ g \\ g^+ \\ g^+ \end{matrix}
\qquad
5g^+:3g \quad
\begin{matrix} g \\ g \\ g^+ \\ g^+ \\ g^+ \\ g^+ \\ g^+ \\ g^+ \end{matrix}
\qquad
\begin{matrix} 4g^+:4g \\ \text{abnormal} \\ \text{arrangement} \end{matrix}
\begin{matrix} g \\ g \\ g^+ \\ g \\ g^+ \\ g \\ g^+ \\ g^+ \end{matrix}
\qquad
\begin{matrix} 4g^+:4g \\ \text{abnormal} \\ \text{arrangement} \end{matrix}
\begin{matrix} g^+ \\ g^+ \\ g^+ \\ g^+ \\ g^+ \\ g \\ g \\ g \end{matrix}
$$

Similar 5:3 segregations or *vice versa* have also been detected in *Ascobolus* and *Neurospora* (Case and Giles, 1964; Gajewski and others, 1968). Their detection in 4-spored tetrads such as yeast is improbable. The frequency of such segregations, as in the case of whole chromatid conversion, varies with the site and with the fungus. For example, two strains, the French and the Californian, of *A. immersus* differ in their frequencies of half-chromatid conversion, it being rather rare in the former but not uncommon in the latter (Emerson and Yu-Sun, 1967; Gajewski and others, 1968). Half-chromatid conversions show a polaron effect in the French strain of *A. immersus* (Gajewski and others, 1968).

(vii) Crossing-over in the same chromatids associated with regions of intragenic recombination is correlated, in up to 50% of the cases, with recombination of outside markers.

The evidence for this correlation comes from the simultaneous study of recombination between markers outside the intragenic region and of recombination within it. It was noticed in Mitchell's (1955) first account of non-reciprocal recombination at the *pdx*-locus in *Neurospora* and has been found in more extensive data from the same fungus, e.g. the *cys*-locus (Stadler and Towe, 1963). Much the largest body of data, however, is available from yeast. Here Hurst and others (1972) investigated 11,023 unselected yeast tetrads from 12 diploids and analysed the conversion of six genes in three separate linkage groups in 907 asci. The marker genes flanking the conversion sites were not more than 20 map units apart, and in 445 cases (49·1%), conversion was associated with reciprocal recombination of the flanking marker genes. Earlier studies by Fogel and Hurst (1967) at the histidine locus, *his-1*, had provided further evidence of a correlation between the two types of recombinational event. The segment of chromosome involved was marked on either side of the *his-1* locus as below:

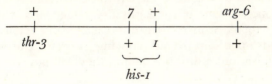

They were able to show that there was a strong tendency for conversion of the *his-1* (*1*) site to be associated with crossing-over to the right of the *his-1* (*7*) site and *vice versa*. Thus, conversion of *1* → + was associated with 49 cross-overs between *7* and *arg-6*, out of a total of 52 single cross-overs and, for *7* → +, 141 between *1* and *thr-3*, out of a total of 225 single cross-overs. Thus gene conversion seems also to be highly correlated with crossing-over in a region adjacent to the converted site. Supporting evidence for this view has come from studies at the *g* locus of *S. fimicola* (Kitani and Olive, 1969).

Clearly some of the conversion data are amenable to an explanation similar to that originally described by Belling and already mentioned. Indeed, Lederberg (1955) suggested, from work with bacteria, that recombination occurred by 'copy-choice', in which genetic material was replicated by copying parental template material, and that synthesis could switch from one parental strand to the other, where they were closely paired. Such a mechanism could account for non-reciprocal conversion, involving adjacent heteroallelic sites or for high negative interference, if rapid switching occurred. Polarized recombination would not be unexpected if it were supposed that the DNA strand replicated in a specific direction, a highly plausible consideration. Moreover, if in eukaryotes the chromatid was multistranded, then an explanation for half-chromatid conversion becomes available.

Despite these powerful arguments, there were some equally telling defects in the hypothesis. Firstly, it should be recalled that the vast majority of recombinational events studied are reciprocal and that it is rare to find gene conversion not associated with some reciprocal recombination. This suggested that both types of events could occur. Secondly, simple copy-choice could not account for three- and four-strand double cross-overs without further assumptions such as some breaking and rejoining of chromatids. Thirdly, since pre-meiotic DNA synthesis occurs, simple copy-choice is highly improbable. Lastly, since the two kinds of events were not mutually exclusive and were frequently correlated, it seemed likely that they had a similar basis.

The general recognition that:

(a) copy-choice by itself was not sufficient,
(b) strand-exchange probably occurred but was also insufficient alone,
(c) copy-choice and strand-exchange could both be involved and, indeed, were manifestations of the same process,

has led to attempts to reconcile the two processes and several hypotheses have been proposed. One of the earliest was suggested, with few details, by Pritchard (1960), who envisaged breakage and rejoining of strands being imposed upon an overall copying process. Current hypotheses represent modifications of this notion and they will now be considered.

Repair models for recombination

Perhaps the only area concerning recombination to which fungi have not yet made major contributions is that of the biochemistry and molecular genetics of DNA. In this area the contributions from prokaryotes, especially *Escherichia coli* K12 and the bacteriophages, are outstanding. As a consequence, enzyme systems are now known to be capable of opening, shortening, extending and joining DNA molecules *in vivo* and in several cases the detailed biochemical mechanisms are understood in outline. Much of this information has come from the study of DNA replication but another important contributory source has been the study of the repair of damage to DNA after irradiation with u.v. light and ionizing radiations. Such damage and its repair is comparable, in a sense, to the kind which has to be postulated to account for the breakages and rejoining of DNA strands or copy-choice mechanisms in spontaneous recombination processes. Current hypotheses of recombination, therefore, seek to relate such knowledge from prokaryotic organisms to genetic information derived from eukaryotes, especially fungi, and to cytological and cytochemical studies, mostly with higher eukaryotes. The fundamental assumption is that the mechanism of recombination is common to all organisms, chromosomal and non-chromosomal. The principal justification for this assumption is that DNA is the universal genetic material, save for the RNA viruses.

Some hypotheses are concerned almost exclusively with possible molecular mechanisms involving DNA and are biased towards explaining recombination in prokaryotes (e.g. reviews by Hotchkiss, 1971; Radding, 1973). Others are based on the molecular biology of DNA but attempt to account for phenomena described from eukaryotes. Two important similar, contemporary hypotheses have been put forward by Whitehouse (1963, 1965, 1966, 1967, 1970; Whitehouse and Hastings, 1965) and Holliday (1964a, 1968); both are geneticists who have worked principally with fungi.

They start from the same assumption, that each chromatid includes only a single double-stranded (duplex) DNA molecule, or molecules, joined longitudinally. There is no unequivocal evidence for this. The cytological evidence suggests that some chromatids may be multistranded structures (Wolff, 1969). Nevertheless, if the DNA were multistranded, appropriate modifications of the hypotheses could accommodate this feature (Whitehouse, 1967; Holliday, 1968).

The details of the hypotheses are illustrated in Figure 14.3a and b (p. 300).

It can be seen that these hypotheses account for much of the data described in the two previous sections.

Both hypotheses require opposite breaks to occur in the pair of chromatids which recombine. Both hypotheses also involve the unwinding of one nucleotide chain from each chromatid which then re-winds with a complementary chain from the other chromatid—either an existing chain (Holliday) or a newly synthesized one (Whitehouse). The result will be a stretch of 'hybrid DNA'. Any mis-matched base pairs in the 'hybrid DNA' region are then corrected by being removed and replaced by processes similar to those believed to occur in repair systems after u.v.-irradiation

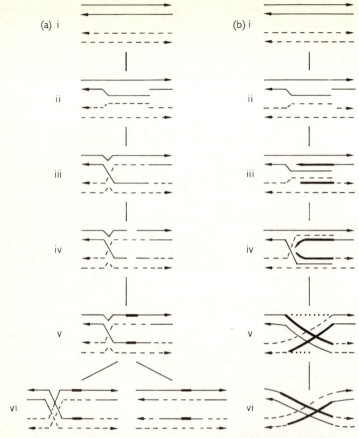

300

FIGURE 14.3 Diagrams to illustrate the hypothesis (a) of Holliday and (b) of Whitehouse to account for recombination in eukaryotes. Key: Lines indicate nucleotide chains, arrows the direction of the chain in the double helix; solid lines, new DNA. (a) *Holliday model*: i. Paired chromatids. ii. First break in equivalent chains and unwinding of chains. iii. Annealing and rejoining of break, twisting of chains. iv, v. Repair excision followed by repair synthesis of new DNA. vi. Alternative resolutions of twisted chains by breaking and joining of chains; 1 and 4 (left) or 2 and 3 (right). This affects the apparent exchange of flanking markers adjacent to the break regions. (b) *Whitehouse model*: i. Paired chromatids. ii. Break and unwinding of chains; the broken chains need not be equivalent. iii. Production of new DNA from broken end in polarized manner. iv. Pairing of old broken chains with new, homologous DNA. v. Excision of old mis-matching strands. vi. New strands fully associated involving exchange of flanking markers adjacent to break region.

etc. The mutational origin of the sites undergoing conversion could be affected by DNA repair processes in specific ways. Hence site specificity could be accounted for in principle.

There is a genuine conversion, on these hypotheses, in the sense that a region is actually removed and replaced, rather than being replicated by some sort of copy-choice mechanism. These mechanisms, therefore, place no restrictions on the chromatid strands involved in recombination, so that three- and four-strand doubles are possible. This, it will be recalled, was an unavoidable difficulty in the simple copy-choice hypothesis. Regions of 'hybrid DNA' are likely to be relatively short, but there is no reason why they should not extend over more than one site within a locus and so account for coincident conversion. It is perhaps a little easier to envisage this for the Whitehouse model than for the Holliday model.

Direction is given to the conversion provided that it is assumed that 'hybrid DNA' is initiated at fixed points at fairly short intervals along the nucleotide chain. If this is so, the nearer a site is to the initiation point, the higher is the probability that it will be converted.

Post-meiotic segregation is also explicable by these hypotheses if it be supposed that conversion can occur in a single nucleotide chain rather than in a pair. Thus, if a correction occurred in the mutant chain, viz.:

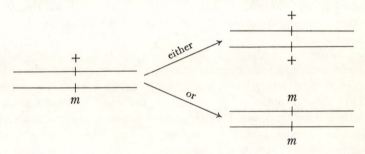

then 6:2 or 2:6 asci would ultimately result. If no correction occurred, segregation of the chains at the mitotic division subsequent to meiosis would lead to 5:3 and 3:5 ratios—truly half-chromatid segregations. Whitehouse (1965) has in fact shown that the *Sordaria* data of Kitani and others (1962) for the frequency of different kinds of segregating asci fit an assumption that the frequencies of correction of $+ \rightarrow m$ and $m \rightarrow +$ are equal. An explanation for these *Sordaria* data can be given on the Holliday model but it is more complex and involves additional assumptions (Emerson, 1966).

Holliday's model provides a simpler explanation of the observed fact that intragenic recombination is only sometimes accompanied by the recombination of the flanking markers. On his model, if the two chains broken at the second breakage are identical with those broken previously, then outside markers show no recombination. If the second breakage is in the other two chains, then a reciprocal exchange is detected in the flanking markers. If it be supposed that either of these events is equally likely, then the 50% reciprocal exchange associated with conversion, found in *Saccharomyces* for example, would be explicable. Any bias could increase or decrease this frequency. On Whitehouse's original hypothesis, reciprocal recombination would always be expected, so that it has had to be modified. If chains dissociate on both sides of a break point, rather than on one as originally supposed, there would be simultaneous cross-overs between the same chromatids in adjacent regions and so there would be no recombination of the flanking markers (Whitehouse and Hastings, 1965). In both cases, therefore, a double cross-over is, in effect, involved but in Whitehouse's model additional assumptions are required.

These hypotheses suffer from common defects. The whole process is initiated by approximately coincidental breaks in two nucleotide chains in different chromatids. It is not clear how or why this should occur. There is no direct evidence in fungi for the formation of 'hybrid DNA', nor for the subsequent processes. On the other hand, the synthesis of a little DNA at the appropriate stage of meiosis has been detected in a newt and in some flowering plants; also there are data from transformation in bacteria which can be most plausibly explained if 'hybrid DNA' had been formed (Wimber and Prensky, 1963; Fox and Allen, 1964; Bodmer and Ganeson, 1964; Hotta and others, 1966). A further problem is the high frequency of localized exchanges, in particular those of reciprocal intragenic recombination. No clear explanation of this has yet been given.

It can be seen that the study of recombination in fungi has generated much important information concerning the process and has also led to the development of hypotheses which attempt to account for all recombination data. That these hypotheses are not wholly successful is not surprising and, in any case, they are limited by the present lack of understanding of the possible biochemical events accompanying recombination. This is still largely based on prokaryote-derived data.

There is reason to suppose that the fungi will contribute further to an understanding of recombination. It will be recalled (pp. 156–157) that in several fungi genes have now been discovered which regulate crossing-over and conversion. Some cause loss of recombination, often associated with increased sensitivity to radiation damage, e.g. *uvs-2* in *U. maydis* (Holliday, 1965, 1967). In others there are changes in the frequency of recombination at sites distant from the *rec* loci, which in some cases are on different chromosomes, e.g. *rec-3(1)* affects the *am* locus in linkage group V in *Neurospora* (Catcheside, 1966). Since *rec* genes were not initially the subject of a particular search, it seems probable that many more exist. Clearly they could be used to dissect still further the mechanism of recombination. Analysis of their biochemical effects might be even more rewarding. Success obtained with fungi in unravelling gene action in other contexts has been notable, as illustrated in the next chapter; this augers well for the future analysis at the molecular level of the mechanism of recombination.

CHAPTER 15

Fungi and general genetics
II. Gene action and regulation

Fungi have contributed notably to recent ideas on the primary action of genes in determining proteins, and prokaryotes to ideas on the regulation of gene action. As a result of the detailed knowledge available in fungi concerning genetically controlled biochemical syntheses, it was natural that they should be examined to see whether prokaryotic regulating mechanisms also occurred in eukaryotes. These two aspects, gene action and its regulation, will be considered briefly, the former because it is now well established, the latter because so little is known.

The one gene–one enzyme concept

Beadle and Tatum (1941) formulated their famous 'one gene–one enzyme' hypothesis (p. 29) as a consequence of studying nutritional mutants of fungi. Beadle (1973) has written:

> '*Neurospora* returned once again to my scientific life when in 1940 Edward Tatum and I were finding the going rough in our attempts to work out the genetic-biochemical relations of eye-pigment relations in *Drosophila*. As I sat one day in Tatum's Comparative Biochemistry lecture, it suddenly came to me that it would be much more efficient to produce mutants concerned with known biochemical reactions than to fathom the chemistry of genetically known characters such as eye-pigment synthesis in *Drosophila*. It was immediately obvious that *Neurospora* was a logical organism for such an approach, for Nils Fries had by then shown that many other filamentous fungi were capable of synthesising such building blocks as amino acids and vitamins in culture media of defined chemical composition.'

The technique they developed has been widely applied to every kind of organism for the elucidation of biochemical pathways and their genetic

Arg V

control. As a result of the coincidence between specific mutants blocking specific steps in biochemical synthesis, Beadle (1945) assumed that every biochemical reaction was directed by a specific gene which determined the appropriate enzyme. An early example was the synthesis of arginine, studied by Srb and Horowitz (1944). Knowledge has now been greatly extended but Table 15.1, based on the original data, illustrates the principle employed.

TABLE 15.1 The behaviour of 15 mutants of *Neurospora crassa* which required arginine for normal growth. (Data of Srb and Horowitz, 1944.)

Mutant	Substrate enabling normal growth to occur			
	Proline	Ornithine	Citrulline	Arginine
arg-8, -9	+	+	+	+
arg-4, -5, -6, -7	−	+	+	+
arg-2, -3	−	−	+	+
arg-1	−	−	−	+

Biosynthetic sequence determined from data above

arg-8, -9   arg-4, -5, -6, -7   arg-4, -3   arg-1
——— Proline ——————— Ornithine ——— Citrulline —— Arginine

Since the early days of the 'one gene–one enzyme' hypothesis, exceptions have been discovered. For example, it appeared as though some single-gene mutants of *N. crassa* required both valine and isoleucine for growth, while others required both methionine and threonine. The explanations of these double requirements have proved to be quite different and yet both are in keeping with the 'one gene–one enzyme' concept. In one case the double requirement arises because the step blocked by the mutant is the production of a common precursor—homoserine—giving rise to threonine and, indirectly, to methionine (Teas and others, 1948). In the other case it seems that the same enzymes are responsible for the last three steps in the synthesis of valine and isoleucine, although they have no common precursor. If, therefore, any one of these enzymes is deficient, both pathways are blocked. Thus, any of the mutants *iv-1, iv-2* or *iv-3* results in a double growth requirement (Wagner and others, 1958; Radhakrishnan and others, 1960; Figure 15.1).

Studies of this type are greatly facilitated in fungi by their ability to form heterokaryons permitting the performance of complementation tests. This has provided greater insight into the way genes specify their protein products.

Complementation and its interpretation

An admirable short monograph on this topic has been prepared by Fincham (1966). It will be recalled (pp. 96–99) that complementation can be used to distinguish allelic from non-allelic mutants, and this principle has been extended to groups of closely linked 'allelic' mutant sites. In this

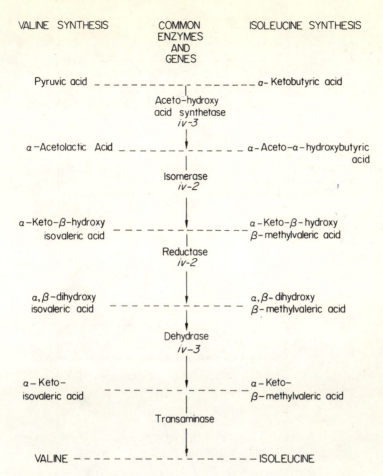

FIGURE 15.1 The biosynthetic sequences involved in the valine and isoleucine pathways. Note that certain enzymes are common to both pathways.

way complementation maps can be prepared of such gene clusters and it often appears that different functions can be ascribed to each distinct region. Fincham and his colleagues were largely responsible for developing a functional hypothesis to account for such data at the *am* locus (Fincham and Pateman, 1957; Pateman and Fincham, 1958; Fincham, 1959; Fincham and Coddington, 1963; Coddington and Fincham, 1965). These mutants lack glutamic dehydrogenase and are, therefore, incapable of amination (hence the symbol *am*). Complementation studies showed that they could be represented by a linear map although the level of enzyme activity in complementing heterokaryons was low. Indeed, in certain cases, e.g. (*am-1* + *am-2*), (*am-1* + *am-3*), the enzyme produced differed from the normal in exhibiting either low thermostability or reduced affinity for

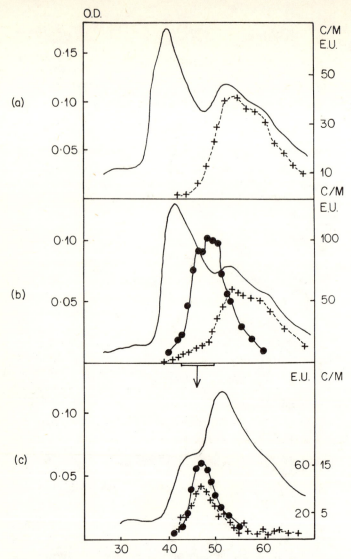

FIGURE 15.2 *In vitro* complementation of proteins produced by *am-1* and *am-3* mutants of *Neurospora crassa*. (a) Mixture of *am-3* and ^{35}S-labelled *am-1* proteins fractionated on DEAE–cellulose column. (b) The same mixture after complementation has occurred, note enzyme activity. (c) Selected fraction from (b) (see arrow and box) refractionated with excess of non-radioactive *am-1* protein. Key: ———, absorbancy (O.D.) at 280 nm, indicates protein; --+--, radioactive counts per minute (C/M); —●—, enzyme units (E.U.) formed by complementation. Note, the size of the samples counted differed for the three sets of data. In a further experiment ^{35}S-labelled *am-3* protein and non-radioactive *am-1* protein showed complementation, the *am-3* protein being incorporated into the active product. (From Fincham, 1966; after Coddington and Fincham, 1965.)

glutamate. Two of the mutant strains, *am-1* and *am-3*, were radioactively labelled in turn with ^{35}S, their enzymatic proteins purified and mixed *in vitro*. Not only was *in vitro* complementation comparable with that demonstrated *in vivo*, but the interaction product contained approximately equal amounts of the ^{35}S-labelled moiety (Figure 15.2).

This led to the hybrid protein hypothesis. It supposes that in complementing strains defective protein subunits can associate in such a way as to produce something resembling the normal protein enzyme which is multimeric, i.e. composed of several, normally identical, polypeptide chains.

Thus complementation can be seen to arise whenever the protein specified by the gene is multimeric, provided that mutant alleles can make defective proteins possessing properties which in the 'hybrid protein' resemble, or can be corrected to resemble, the normal enzyme protein. A linear complementation map indicates that the mutant sites are arranged in a similar linear order and so give rise to a sequence of non-overlapping defects in the protein. This would be expected if the nucleotide sequence in the gene determined the amino acid sequence in the protein. In many cases the order of mutants in a linear complementation map is virtually the same as the heteroallelic order derived by recombination analysis. Such comparisons (e.g. Figure 15.3) suggested that the sequence of the nucleotides was co-linear with that of the amino acids in a protein. In fact, further studies revealed that complementation maps were more often non-linear and it is now supposed that they reflect in some way the tertiary structure and conformational properties of multimeric proteins (Fincham, 1966; Gillie, 1966).

Nevertheless, initially in *E. coli* and T4 phage, the principle of co-linearity of gene and polypeptide was established by the direct comparison of the mapped sites of mutants at a locus and the order of amino acid substitutions which they caused in a protein. Similar data now exist for fungi, notably the *cyt-1* locus in *Saccharomyces cerevisiae* which determines the protein *iso-1* cytochrome *c* (Figure 15.3, p. 308).

The study of the genetic determination of this cytochrome has been particularly rewarding in other ways, notably in respect of the genetic coding for this protein.

Genetic coding for protein structure

The complete sequence of the 108 amino acids of *iso-1* cytochrome *c* from *S. cerevisiae* is known as well as the positions where the iron pophyrin, haem is attached (Narita and Titani, 1969). It will be recalled that the *petite* mutants of yeast, both extrachromosomal and chromosomal, are all deficient in some way for this protein (p. 101). Those chromosomal mutants which are deficient for cytochrome *c* are designated *cyt*; almost all can be grown on a yeast-extract medium containing non-fermentable substrates.

A number of *cyt-1* mutants were found to be situated at the extreme left end of the linkage map and none of them produced any protein whatsoever.

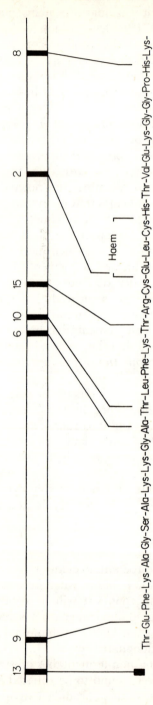

FIGURE 15.3 The location of heteroalleles in the *cyt-1* gene of *Saccharomyces cerevisiae* in relation to the amino acid residues of *iso-1* cytochrome *c* which they affect. The initiation region, which may be a methionyl residue is shown as ■ (From Sherman and others, 1970.)

This suggested that the protein chain could not be initiated. A study was made of the 45 intragenic revertants from this group. Some of these produced cytochromes identical with the normal ones save that either they were longer by two amino acids to the left of threonine at position 1, or shorter by four commencing at alanine at position 5. The *cyt-1–13* mutant and its revertants *13A* and *13Q* illustrate this, viz.:

Normal cytochrome c	*Cyt-1*	Thr–Glu–Phe–Lys–Ala–etc.
No protein	*cyt-1–3*	————
Longer protein	*cyt-1–13A*	Met-Ile-Thr-Glu-Phe-Lys-Ala-etc.
Shorter protein	*cyt-1–13Q*	Ala-etc.

(Met = methionine; Ile = isoleucine; Thr = threonine; Glu = glutamic acid; Phe = phenylalanine; Lys = lysine; Ala = alanine.)

When the genetic code established for *E. coli*, i.e. the codons of RNA nucleotide triplets which code for different amino acids (Morgan and others, 1966), is applied to these situations, it becomes reasonable to suppose that the triplet which indicates that a protein is to be initiated is AUG (Adenine-Uracil-Guanine). Thus the data given above can be rewritten employing codons as follows (bold codons changed through mutations):

$$Cyt\text{-}1 \longrightarrow cyt\text{-}1 \quad \text{XXX–AUG:} \quad AC^{U}_{C\,A\,G}\text{-}CA^{A}_{G}\text{-}UU^{U}_{C}\text{-}AA^{A}_{G}\text{-}GC^{U}_{C\,A\,G}\text{-etc.}$$

(mutation) ↓ **C**

XXX–AUU: **A** ————

(mutation) ↓ C / A

$$cyt\text{-}1 \longrightarrow cyt\text{-}1\text{-}13A \quad \textbf{AUG}\text{–AUU:} \quad AC^{U}_{C\,A\,G}\text{-}CA^{A}_{G}\text{-}UU^{U}_{C}\text{-}AA^{A}_{G}\text{-}GC^{U}_{C\,A\,G}\text{-etc.}$$

(mutation) ↓

$$cyt\text{-}1 \longrightarrow cyt\text{-}1\text{-}13Q \quad \left[\text{XXX-AUU}^{C}_{A}\right] \qquad \text{AUG:}GC^{U}_{C\,A\,G}\text{-etc.}$$

It also becomes clear from data such as this (and revertants of other *cyt-1* mutants supported this interpretation) that the codon which results in the initiation of a protein chain is:

(a) The nucleotide triplet AUG.
(b) This is preceded by another codon XXX which can be readily mutated to AUG but in the normal form is non-functional.

This was the first unambiguous demonstration of the codon for protein chain initiation.

From other *cyt-1* mutants it proved possible to demonstrate that if certain substitutions occurred then the protein chain would be terminated. If this occurred in position 2, normally glutamic acid as in *cyt-1–9*, no cytochrome *c* was produced. Five different kinds of revertants were found after mutagenic treatment of these mutants; in these the amino acids substituted for glutamic acid at position 2 and their codons were:

Cyt-1 ⟶ cyt-1–9		⟶ Revertants of *cyt-1–9*	
		Lysine	AAA (or AAG)
Glutamic		Glutamine	CAA (or CAG)
Acid	(UAA?)	Tyrosine	UAU or UAC
GAA or		Leucine	UUA (or UUG)
GAG		Serine	UCA (or UCG)

The simplest solution to account for all these changes is that the glutamic acid codon, GAA, has been changed to UAA in mutant *cyt-1–9*. Thereafter, in the revertants, substitution of A or C for uracil would encode for lysine and glutamine respectively; substitution of the middle adenine by U or C would encode for leucine or serine, and substitution of the final adenine by either U or C would encode for tyrosine. Hence UAA encodes for chain termination and, from similar reasoning with revertants of another mutant, *cyt-1–179*, it became clear that UAG has the same function. Both these codons had been shown to be chain terminating in *E. coli*. These important studies have demonstrated that the genetic code based on work with prokaryotes is applicable to a eukaryote in respect of the coding for specific amino acids and for the initiation and termination of protein molecules (Sherman and others, 1970).

Thus fungi have contributed to ideas on protein structure and the genetic control of their structure. In this respect studies on prokaryotes and fungi have demonstrated a quite remarkable underlying similarity. This is not true, so far, of comparisons between the genetic regulation of protein synthesis in prokaryotes and eukaryotes, as exemplified by fungi.

The operon concept

A variety of studies on the regulation of protein synthesis, especially in bacteria, was provided with a coherent, unifying basis in the theory of the operon, first enunciated by Jacob, Monod and their co-workers (Jacob and others, 1960; Jacob and Monod, 1961, 1962). They suggested that in *E. coli* there were *structural genes*, each responsible for determining the amino acid sequence in a protein, occurring in closely linked groups and associated with a region of DNA termed an *operator*. An operator controls the action of its associated structural genes in a co-ordinated manner and the whole functional unit is termed an *operon*. They also recognized *regulatory genes*, which produce proteins that interact specifically and allosterically with the operator, preventing its operation and hence regulating the action of the operon as a whole. Subsequently, a region adjacent to the operator was recognized which was responsible for initiating operon expression; this was termed a *promoter* (Jacob and others, 1964). It was also discovered that regulatory genes could act in ways other than the repression of operator activity. In some cases the operon is normally inactivated for other reasons and the regulatory protein results in the removal of this inhibition or acts as an inducer of gene action.

An operon has a number of properties by which it can be recognized, namely:

(a) close linkage of structural genes, often concerned with a single bio-synthetic sequence;

(b) co-ordinated regulation of action of the structural genes;

(c) the existence of mutants with pleiotropic effects such that several genes in an operon may be affected, often in a polar manner, i.e. the effect declines in those genes furthest from the mutant site;

(d) regulatory genes occur which affect the operon as a whole.

Figure 15.4 illustrates, diagrammatically, the structure and function of an operon.

There is good evidence for the existence of several operons in both bacteria and viruses and they have been looked for in fungi and other

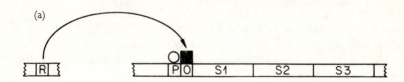

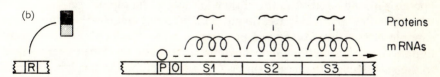

FIGURE 15.4 Diagram to illustrate a typical operon. (a) The regulator protein has produced a repressor protein (■) which combines allosterically with the operator so preventing the RNA polymerase (○) from transcribing the DNA of the structural genes. (b) The repressor protein has combined with a metabolite (□) so that the RNA polymerase can transcribe the genes in the operon. Alternative modes of regulation also occur. Key: P, promoter, O, operator; S1–S3, structural genes; R, regulator gene.

tes. A number of regulatory systems have been examined but so far
proved to possess all the properties of an operon. Most studies
gulation of protein synthesis in fungi have been made with *Asper-*
...eurospora or yeast. Several comprehensive reviews are available
which can be consulted for details (Gross, 1969; Fincham and Day, 1971;
Calvo and Fink, 1971; de Robichon-Szulmajster and SurdinKerjan, 1971;
Metzenberg, 1972). Here three regulatory systems in fungi will be described
to illustrate the features involved.

The gal *system in* Saccharomyces

This system involves several genes (Douglas and Pelroy, 1963; Douglas
and Hawthorne, 1964, 1966) but three of them, *Gal-1*, *Gal-7* and *Gal-10*,
occur as a tightly linked group in linkage group II. They are responsible for
three enzymes catalysing successive steps in the conversion of galactose
to UDP-glucose, viz.:

$$\text{Galactose + phosphate} \xrightleftharpoons{\text{galactokinase}} \text{galactose-1-P}$$

$$\text{Galactose-1-P + UTP} \xrightleftharpoons[\text{uridylyl transferase}]{\text{galactose-1-P}} \text{UDP-galactose + PP}_i$$

$$\text{UDP-galactose} \xrightleftharpoons[\text{epimerase}]{\text{UDP-galactose}} \text{UDP-glucose}$$

All three enzymes are induced by galactose. However, there are also two
genes, *gal-3* and *gal-4*, which block the adaptation system in some way,
i.e. the initiation of the enzymes coded for by the linked group. Two loci
have been detected which affect *gal-3* and *gal-4* respectively. One, i^-, was
discovered by selecting for cells capable of fermenting galactose in haploid
yeast carrying *gal-3*. Cells carrying i^- *gal-3* produce the three enzymes
constitutively so it was concluded that in i^+ cells, the *i* locus was coding for a
repressor of the gene cluster. A search was therefore made for dominant
operator mutants linked to the cluster, by subjecting diploid *gal-3* strains to
selection after mutagenic treatment. This was not achieved although cells
capable of constitutive enzyme production were obtained. The dominant
mutation responsible for this behaviour, *C*, turned out to be closely linked
to the *gal-4* locus. It emerged that *C Gal-4/c gal-4* diploids develop the three
enzymes but *C gal-4/c Gal-4* diploids need to be induced. It was also ascertained, by means of complementation tests, that *gal-4* probably produced
a protein. Thus *C* is behaving like an operator for the structural gene *Gal-4*,
which, in turn, is somehow regulating the action of the *Gal-1*, *Gal-7*, *Gal-10*
cluster. But, of course, the *i* locus is also involved and presumably i^- produces a repressor which either acts directly on the gene cluster in addition
to repressing *C* action, or acts on *Gal-4* itself. The latter is more probable

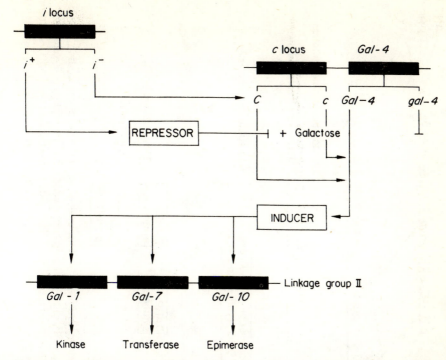

FIGURE 15.5 Diagram to illustrate the interaction of genes in *Saccharomyces cerevisiae* which regulate galactose utilization. The linkage groups of *Gal-4*, *C* and *i* have not been ascertained.

since i^- alleles are not expressed save in the presence of *Gal-4* (Figure 15.5).

Thus, although there are some similarities with a bacterial operon, the regulating system appears to be more complex. There is no evidence of co-ordinated regulation or of mutants with pleiotropic effects.

The arom *system in* Neurospora

In 1960 Gross and Fein demonstrated the basic features of the genetic control exercised on the synthesis of the important aromatic amino acids, tryptophan, tyrosine, phenylalanine, and also *para*-aminobenzoic acid in *N. crassa*. It is now known that 9 genes, *arom-1* to *arom-9*, control the sequence from the combination of phosphoenolpyruvate and erythrose-4-phosphate to chorisimic acid, the common precursor of the aromatic amino acids already mentioned (Giles and others, 1967). These steps, the genetic control exercised on them and other genetic data are set out in Figure 15.6 (p. 314).

It can be seen that *arom-1, -2, -4, -5 and -9* are contiguous in linkage group II, a few map units from *arom-3*; *arom-7* and *arom-8* are on opposite sides of the centromere in I and *arom-6* is in VI. The gene cluster is concerned with

314

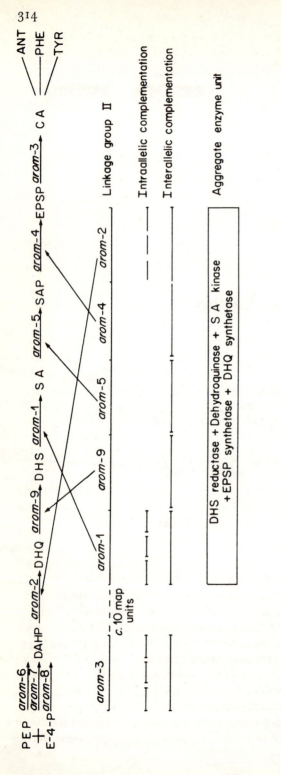
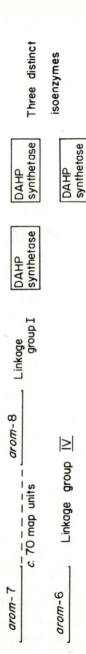

Figure 15.6 The biosynthetic sequences involved in aromatic acid synthesis in fungi with an indication of the genes controlling them in *Neurospora crassa* and their relationships. Key: Reactants involved; PEP, phosphoenolpyruvic acid; E-4-P, erythrose-4-phosphate; DAHP, 3-deoxy-D-arabinoheptulosonic acid-7-phosphate; DHQ, 5-dehydroquinic acid; DHS, 5-dehydroshikimic acid; SA, shikimic acid; SAP, shikimic acid-5-phosphate; EPSP, 3-enolpyruvyl shikimic acid-5-phosphate; CA, chorismic acid; ANT, anthranilic acid; PHE, phenylalanine; TYR, tyrosine.

five sequential biosynthetic steps but the genes concerned are not arranged in the same order in the linkage group. Five mutually complementing groups of mutants each lack one enzyme and map contiguously. However, pleiotropic mutants are known which lack all five of these enzymes; they are found to map in *arom-2*. It is to be noted that *arom-2* is at one end of the linked cluster and determines the first of the five sequential ateps. Moreover, other mutants are known which lack two or more of the possible enzyme activities and show reduced levels of the others. When complementation tests are made between these, or between them and mutants lacking a single enzyme, the complementation map does not correspond with the known gene order. Many pleiotropic mutants complement *arom-4* mutants but this is the second site in the cluster, having *arom-2* beyond it.

The pathway prior to this sequence is controlled by three DAHP iso-enzymes determined by *arom-6*, *arom-7* and *arom-8* and allosterically in-hibited by tyrosine, phenylalanine and tryptophan respectively.

In some respects the *arom* cluster is reminiscent of an operon but there are distinct differences. The five enzyme activities are all associated with a single protein dimer of M.Wt. *c*. 230,000 (Burgoyne and others, 1969; Case and Giles, 1971). Mutants which lack two or more of the enzyme functions contain a protein usually of lower molecular weight than the normal aggregate enzyme. The correlation between these molecular weights and the position of the gene is not very strong. Such a correlation might be expected if precise co-linearity occurred between the components of the gene cluster and the complex protein.

In *Neurospora*, therefore, there are in effect three separate pathways for synthesizing the aromatic amino acids and a part of the pathway common to all three is controlled by an enzyme complex. This complex, although determined by a linked cluster of genes, is very different from the sequen-tially acting series of enzymes found in bacterial operons. The clustered *arom* mutants interfere with the action of the aggregate enzyme as a whole; there is no evidence that individual components respond to co-ordinated regulation. Moreover, in this system there is neither evidence for an opera-tor site nor for a regulator gene.

It has been suggested that a group of tightly linked genes is in some way necessary for the production of an enzyme complex. There is support for this from *S. cerevisiae* where there is a similar enzymatic unit controlled by a gene cluster (de Leeuw, 1967; Ahmed and Giles, 1969). But this is not always so. In *N. crassa*, *trp-1* is located in linkage group III and *trp-2* in group VI. These genes are responsible for an aggregate enzyme possessing three activities, anthranilic acid synthetase and isomerase, and indoleglycerol phosphate synthetase. The last two are apparently coded for by *trp-1* but the synthetase activity is coded for by both genes. The functional diversity of these genes is illustrated diagrammatically in Figure 15.7 (p. 316).

This example differs from the *arom* cluster, however, in that the reactions determined by the aggregate are not sequential.

316

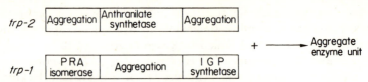

FIGURE 15.7 Comparison of the functional relationships determined
by two genes, *trp-1* and *trp-2* in the synthetic pathway to tryptophan in
Neurospora crassa. Note that *trp-1* is in linkage group III, *trp-2* in group
I but that their products are combined in a single functional aggregate
enzyme. Key: CA. chorismic acid; AA, anthranilic acid; PRA,
phosphoribosyl-anthranilate; DRP, 1-(*o*-carboxyphenylamino)-1-de-
oxyribulose 5-phosphate; IGP, Indoleglycerol phosphate.

Nitrate reduction in Aspergillus

The biochemistry of nitrate reduction was pioneered by Steinberg's
(1937) discovery that molybdenum was required when nitrate was supplied
as the sole nitrogen source for the growth of *A. niger*. Work with other fungi
resulted in a reasonably clear understanding of the sequence of biochemical
events (reviewed Nicholas, 1965). Pateman, Cove and their co-workers
have now greatly increased understanding of the genetic regulation of the
process in *A. nidulans* (Pateman and others, 1964, 1967; Pateman and Cove,
1967; Cove, 1966, 1969; Cove and Coddington, 1965; Cove and Pateman,
1963, 1969; Arst and Cove, 1969; Arst and others, 1970).

In *Aspergillus* two enzymes appear to be sufficient to complete the reactions,

$$NO_3 \xrightarrow{\text{Nitrate reductase (NAR)}} NO_2 \qquad\qquad \text{(i)}$$

$$NO_2 \longrightarrow [NH_2OH] \longrightarrow NH_4 \qquad\qquad \text{(ii)}$$
$$\text{Nitrite reductase (NIR)}$$

Molybdenum is required for the activity of nitrate reductase (NAR) and
its activity is closely paralleled by cytochrome *c* reductase (CR). Nitrate
normally induces both NAR and CR, the latter in constant proportion to
the amount of the former. Ammonium ions repress both activities.

Mutants incapable of carrying out this reduction were sought and four
major types were recovered. One group lacked NAR activity; another
lacked NIR activity; a third lacked NAR, CR and NIR activity; and a
fourth group lacked NAR and xanthine dehydrogenase activity but possessed
NIR and CR activity. The first two groups were about 10 map units apart
in linkage group VIII at loci designated *nia* and *nii*; they were presumed

to be the structural genes for NAR and NIR, respectively. The third group mapped some forty units away from the first two and the locus was designated *nir* to indicate that it was a regulatory gene locus for the *nia* and *nii* loci.

Other mutants at this locus, such as *nir*^c were later discovered; these enabled nitrate reductase to be produced constitutively in the absence of nitrate and even to some extent in the presence of NH_4^+ which normally represses such activity. In order to explain the facts that *nir*⁺ resulted in the inducible formation of NAR and NIR, *nir*^c resulted in their constitutive formation and *nir*[−] in their total repression, it was supposed that the normal regulator protein could exist in two forms. In one form it acted as an inducer of NAR and NIR but in the other form, determined by *nir*[−], it was inactive. Three heterozygous diploids, *nir*^c/*nir*⁺, *nir*^c/*nir*[−] and *nir*⁺/ *nir*[−], and the homozygote *nir*⁺/*nir*⁺, were made. The first two produced less NAR and NIR constitutively than the homozygote. After induction *nir*⁺/*nir*[−] also produced less NAR and NIR. These changes, reflected in the gene dosage, suggested that the permanent inducer, determined by *nir*^c, or the inducible inducer, determined by *nir*⁺, produced their product (presumably a protein) in limiting amounts. The system proposed is illustrated in Figure 15.8.

The figure also depicts the action of the fourth group of mutants, the

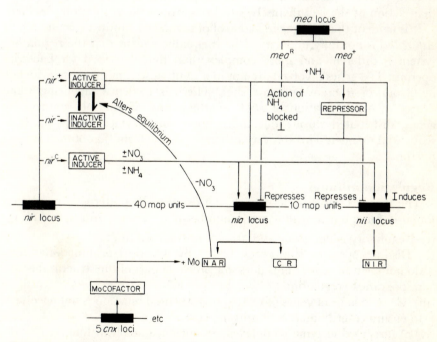

FIGURE 15.8 Genes involved in the production and regulation of enzymes concerned with nitrate reduction in *Aspergillus nidulans*.

cnx mutants which mapped at five distinct loci. They are believed to control the synthesis of a molybdenum-containing co-factor necessary for both NAR and xanthine dehydrogenase. Such mutants, therefore, lack NAR activity. It must be supposed that the associated CR activity is capable of expression even in the absence of the molybdenum-containing co-factor, since *cnx* mutants possess normal CR activity.

A further property of mutants at the *nia* or *cnx* loci was that they produced NIR even in the absence of nitrate. Thus it was supposed that in some way NAR is involved in the regulation of its own synthesis in the normal situation. A possibility is that NAR interacts with the inducer protein produced by *nir*$^+$ in the absence of NO_3^- ions, causing the complex to repress NAR and NIR production. In the presence of NO_3^- ions, however, the ions interact with the NAR-inducer protein complex so that it is no longer effective as a repressor.

Finally, a further, fifth, group of mutants, mapping at yet another locus— *mea*, are known. The mutant *mea*^{R} is not only resistant to methylammonium but also removes the susceptibility of both NAR and NIR to repression by NH_4^+ ions. It is proposed, therefore, that the normal allele *mea*$^+$ gives a product which, in the presence of ammonium, acts as a repressor of the *nia* and *nii* loci. However, *mea*$^+$:*nir*^{c}-carrying strains are less repressed by ammonium than are *mea*$^+$:*nir*$^+$-carrying strains. It is possible, therefore, that there is an interaction between the products of these two loci, rather than action at the loci themselves to bring about the regulation observed.

It is evident that the genetic control of regulation of nitrate reduction in *Aspergillus* is not fully understood. It is equally evident that the regulatory system is different and more complex than that described for bacterial operons. The genes involved are not in a tight cluster, nor is there evidence in favour of discrete operator sites. There is evidence favouring some degree of autoregulation of NAR synthesis and of multiple controls by genes, distributed throughout the genome, of enzymes which act in a sequential manner. This is a very different picture from that of the classical operon.

Regulation in fungi

The three examples described do not exhaust the patterns of regulation found in fungi but they illustrate a number of common features, including:

(i) Regulatory genes occur and are not uncommon in fungi.

(ii) Distinct operator sites recognizable by genetical or functional tests do not seem to occur. This does not preclude part of the structural genes responding to regulation.

(iii) Close linkage of genes of related or sequential function is not necessary to ensure co-ordinated action in the cell.

(iv) Aggregated enzyme complexes are not uncommon in fungi. They do not necessarily require to be coded for by a closely linked sequence of genes.

(v) Regulation may occur as a result of interactions elsewhere than at the site of the chromosomes.

In general, therefore, regulation in eukaryotic cells looks as though it will prove to be somewhat more complex than the operon model. This is not surprising since not only are eukaryotes more complex but even their individual cells are larger than those of most prokaryotes. This implies an important necessity for some kind of functional compartmentation within the cell. Hence the occurrence of isoenzymes involved in different but related biosynthetic sequences which provide separate channels, and the occurrence of aggregate enzyme complexes which provide convenient vehicles for common pathways.

The complexity of the eukaryotic cell is likely to result not only in increased compartmentation but in more complex regulation of transport both into and out of the cell. There is good evidence that such complexity exists in fungi. In yeast, for example, Grenson (1973) has demonstrated that at least six genes code for permeases which regulate the inward transport of amino acids. One of these, *gap*, regulates a transport system of broad specificity, another regulates transport of groups of amino acids, e.g. *dcp-1* for all the L-dicarboxylic amino acids, and yet others are highly specific for single amino acids, e.g. *agp-1* and *htp-1* for arginine and histidine, respectively. Mutants capable of the excretion of specific amino acids under various conditions were also isolated.

Regulation in eukaryotes is, therefore, likely to be complicated because of increased cell size and complexity and it seems probable that different organisms could have evolved different regulatory mechanisms. For example, Figure 15.7 illustrated the functional diversity of the genes *trp-1* and *trp-2* which determine an aggregate enzyme complex in *N. crassa*. There is a similar complex in yeast but it is not associated with anthranilic acid synthetase which is specified by a separate gene, *trp-2*. In yeast, *trp-1* only codes for anthranilic acid isomerase and *trp-3* for the indole-phosphate synthetase and the aggregation of these two enzymes in a single complex. Again, it will be recalled that *nia* codes for NAR and CR in *A. nidulans* but in *N. crassa* two genes are known, *nit-1* and *nit-2*. Mutant alleles of the former show CR activity and those of the latter are deficient in CR. This suggests that the two enzymes, NAR and CR, are specified by separate polypeptides in *N. crassa* (Sorger, 1963, 1965, 1966).

It seems probable that a considerable amount of the genetic material in eukaryotes may be responsible for determining genetic regulation. In the threonine–methionine pathway in yeast it has proved possible to demonstrate regulation of six enzymes, each coded by an unlinked structural gene, through the multiple effects of exogenous methionine. Then there are two unlinked genes, *eth-2* and *FS-296*, whose products act pleiotropically as repressors on three further unlinked genes concerned in the synthesis of methionine itself. This kind of study has led to the suggestion that there are not less than 15 regulatory genes concerned with the seven structural genes coding for enzymes in the complete biosynthetic sequence (de Robichon-Szulmajster and others, 1973; Cherest and de Robichon-Szulmajster, 1973). Since Hartwell (1967) has claimed that only 10% of yeast DNA is concerned with the specification of all essential enzymes and structural molecules, it may be suggested, by analogy with the yeast example just quoted, that not less than 20% could be concerned with genetic regulation.

EPILOGUE

These observations make it clear that the regulation of gene action in eukaryotes is not yet understood. It is likely to be complex and it will involve not only more precise genetical and biochemical knowledge, especially of protein structure, but also knowledge of the spatial organization of the cell and of the means whereby compartmentation is achieved. It is probable that fungi will continue to be favoured organisms for such studies for, apart from the last two requirements, so much is already known about them which can be further built upon. Structurally, fungi are relatively simple eukaryotes, yet they exhibit sufficient differentiation in time and space to ensure that such problems can be studied usefully. In this respect they have many advantages over both protozoa and protophyta. It is, therefore, a reasonable prediction that fungi will continue to be used for the study of gene action and regulation at the cellular level which will, in turn, throw light on cellular differentiation and, ultimately, on differentiation in the organism as a whole.

APPENDIX

Principal fungi studied genetically

Phytophthora

Species studied: *P. infestans*, *P. cactorum*, *P. dreschleri* and others; parasites but can be grown as saprophytes.

M.M. Sucrose, salts—asparagine + thiamin; 18–22–25°C used, dark preferred.

C.M. Potato–dextrose, oat extract plus supplements.

Hyphae coenocytic, no septa or hyphal fusion, growth rate fairly rapid.

Heterokaryons not demonstrated but presumptive mating-type heterokaryons occur.

Asexual reproduction by sporangium or motile uninucleate zoospores, promoted by medium supplemented with extract of appropriate host plant.

Parasexuality suspected but not demonstrated.

Sexual reproduction stimulated by natural extracts or addition of sterols, especially β-sistosterol, but some species specificity.

Mating system basically homomictic in many spp. but complex control (see Figure 8.2); in *P. infestans* dimictic with *A1* and *A2*; always involves oogonia and antheridia. Oospore difficult to germinate, may be given temperature shocks or digested by live snails and collected in excreta, germination by germ tube.

Chromosome number probably high; meiosis probably in gametangia.

Mutants, few auxotrophs, some drug-resistant; best mutagens NG or spontaneous selection to resistance. Natural series of virulence genes.

Genetic analysis by analysis of oospore germ tubes or zoospores derived from germ sporangium.

References: Elliott and others, 1966; Shaw and Elliott, 1968; Shaw and Khaki, 1971; Elliott and MacIntyre, 1973; Khaki and Shaw, 1974.

Phycomyces

Species studied *P. blakesleeanus*; saprophyte.

M.M. Glucose, salts—asparagine and thiamin or components; 15–25°C.

C.M. Various, potato dextrose agar, Sabouraud's medium.

Hyphae coenocytic, no septa or hyphal fusion. Growth rate rapid, can be made colonial with sorbose.

Heterokaryosis by mutation or hyphal grafting, especially of young sporangiophores.

Asexual reproduction by sporangiospores, mostly 3/4 nucleate (range 1–7), may be heterokaryotic, no uninucleate stage.

Parasexuality not known.

Sexual reproduction stimulated by hypoxanthine but not essential. Optimum 17°C.

Mating system dimictic with homogamous gametangia. Zygote difficult to germinate, gives germ sporangia with homokaryotic sporangiospores, uninucleate initially.

Chromosome number unknown, cytological, details of meiosis and reproduction obscure.

Mutants, spontaneous morphological; induced by NG, include auxotrophs, carotene-less, physiological lacking tropisms.

Genetic analysis by random-spore isolation.

Reference: Bergman and others, 1969.

Saccharomyces

Species studied: *Saccharomyces cerevisiae, Schizosaccharomyces pombe* and other spp.; saprophytes.

M.M. Salts, nitrogen, biotin and carbohydrate; last may vary with species. 25–27°C, optimum 30°C.

C.M. Dextrose, peptone–yeast extract.

Growth form typically toruloid, haploid or diploid cells; growth colonial. Cells not coloured, except *Rhodotorula*

Heterokaryosis of limited duration possible in special conditions.

Asexual reproduction by budding in *Saccharomyces*, fusion only in *Schizosaccharomyces*. Buds uninucleate before division.

Parasexuality possible and occurs spontaneously; variable with stocks.

Sexual reproduction enhanced by reduced nutrients and increased by acetate; special sporulation media employed.

Mating system dimictic, with irregularities (homomictic in some species). Initiated by cell fusion. Ascospores in tetrads usually treated as unordered; ordered in some cases; require a micromanipulator to separate. Meiotic cytology still obscure. Haploid, diploid and tetraploids available.

Mutants of all types, especially fermentation of carbohydrates, auxotrophs; few morphological. Ultraviolet-irradiation effective.

Genetic analysis by mitotic recombination, unordered tetrad analysis and random spore analysis.

Extrachromosomal mutants well studied, induced acriflavine or ethidium bromide. Mitochondrial DNA mapping possible.

References: Ephrussi, 1953; Mortimer and Hawthorne, 1969; Fowell, 1969; Coen and others, 1971.

Aspergillus

Species studied: *A. nidulans* (*A. niger*, *A. heterothallicus* and others); all saprophytes.

M.M. Glucose, salts inc. inorganic N, 20–37°C.

C.M. Peptone–yeast extract, casein hydrolysate.

Hyphae septate with central pore, cells multinucleate, hyphal fusions. Growth naturally colonial. Hyphae pigmented.

Heterokaryosis by mutation or hyphal fusion, nuclear migration possible. Heterokaryon incompatibility groups occur.

Asexual reproduction by uninucleate conidia (multinucleate in some spp.). Conidia pigmented.

Parasexual cycle possible and effective.

Sexual reproduction favoured by increased glucose and reduced nitrate.

Mating system homomictic (*A. heterothallicus*, dimictic; *A. niger* and others asexual), initiated by hyphal fusion or fusion at sites of cleistothecia. Ascospores in unordered tetrads, not discharged.

Chromosome number 8. Meiotic cytology known.

Mutants, all types, including colour mutants induced by u.v.-irradiation, various chemical mutagens.

Genetic analysis by heterokaryosis, mitotic recombination, random meiotic products, unordered tetrads.

Extrachromosomal mutants, spontaneous and induced esp. u.v.-irradiation.

References: Pontecorvo, 1953; Aspergillus News Letter—annual; Barratt and others, 1965.

Neurospora

Species studied: *N. crassa*, *N. sitophila*, *N. tetrasperma* (other species); all saprophytes.

M.M. Sucrose, salts, inorganic N + biotin. 20–35°C.

C.M. various, Fries' medium etc.

Hyphae septate with central pore, cells multinucleate, hyphal fusions. Growth rapid, spreading can be made colonial with sorbose. Hyphae not pigmented.

Heterokaryosis by mutation or hyphal fusion, nuclear migration frequent and effective. Heterokaryon incompatibility occurs.

Asexual reproduction by uninucleate microconidia or multinucleate macroconidia, 4/5 nucleate (2–8), homo- or heterokaryotic. Conidia contain pigments.

Parasexuality not significant, disomics known and can be produced.

Sexual reproduction favoured by limiting nitrogen, 25°C.

Mating system dimictic (*N. tetrasperma*—homodimictic; homomictic species known). Initiated by fusion of microconidium with trichogyne of protoperithecia. Ascospores in ordered tetrads (8/ascus), discharged violently.

Chromosome number 7, details of meiotic cytology well known.

Mutants, all types, induced by various mutagens.

Genetic analysis by heterokaryosis, random ascospore analysis and ordered tetrads—dissectable by hand using binocular microscope.

Extrachromosomal mutants, mostly induced. Maternal inheritance.

References: Singleton, 1953; Barratt and others, 1954; Emerson, 1955; Bachmann and Strickland, 1965; Neurospora News Letter—annual.

Other ascomycetes

Sordaria. Olive, 1956.

Podospora. Esser, 1969.

Ascobolus. Lissouba and others, 1962.

Ustilago

Species studied: *U. maydis, U. hordei, U. violacea*. Parasites, but last grows saprophytically.

M.M. Glucose, salts–inorganic N. Optimum 30°C.

C.M. Various; *U. maydis* on *Zea mays*.

Hyphae fine, septate with central pore. Cytological condition and vegetative fusions not well known or studied. Growth in culture often colonial and toruloid.

No asexual reproduction.

Parasexuality possible, unreduced 2n teleutospores give solopathogenic lines in *U. maydis*.

Sexual reproduction best in host. *Z. mays* for *U. maydis*, *Hordeum* spp. for *U. hordei* and *Lychnis* or *Melandrium* spp. for *U. violacea*.

Mating system diaphoromictic, 2A, several B factors in *U, maydis*; dimictic in other species. Initiated by hyphal fusion in host, therefore, inoculate compatible mating types.

Chromosome numbers not known, certainly >2 despite early reports.

Mutants auxotrophs known, induced by u.v.-irradiation and some mutagens, EMS, NG. Principal characters in *U. hordei*, virulence/avirulence genes.

Extrachromosomal situation known only in *U. maydis*, basis uncertain.

Genetic analysis by mitotic recombination of unordered tetrads. Tetrads dissectable by micromanipulator.

References: Holliday, 1961; Day and Jones, 1968, 1969.

Coprinus

Species studied: *C. cinereus* (= *C. lagopus* of many workers and other species). Saprophytes.

M.M. Glucose, salts—asparagine thiamin. Optimum 37°C.

C.M. M.M. + yeast extract, hydrolysed casein, hydrolysed nucleic acids, malt extract.

Hyphae septate with dolipore, monokaryotic or dikaryotic, latter with clamp connexions. Hyphal fusions regulated by mating-type factors. Growth spreading.

Heterokaryosis of common *A* or common *B* or common *A–B* possible, nuclear migration, may be unilateral.

Asexual reproduction on monokaryons only by uninucleate oidia.

Parasexuality possible in common *A*, *B* or *A–B* heterokaryons.

Sexual reproduction requiring light phase, reduced temperature, 25–27°C, and best done on sterile dung. (Organic Fe compound required otherwise.)

Mating system bifactorial diaphoromictic, several *A* and *B* factors with evidence of α and β loci in each. Initiated by dikaryon formation (different *A* and *B* mating-type factors). Basidiospores in unordered tetrads, can be picked off by dry needle or relatively coarse micromanipulator.

Mutants, mostly auxotrophic, some morphological; mating-type factors form natural allelic series.

Some extra-chromosomal mutants, not exploited.

References: Day, 1960; Day and Anderson, 1961; Anderson, 1971.

Schizophyllum

Species studied: *S. commune*; saprophyte.

Similar to *Coprinus*, particularly valuable for studies on structure and recombination of mating-type factors.

Reference: Raper, 1966.

References to Appendix

Anderson, G. E. (1971). *The life history and genetics of* Coprinus lagopus, Harris Biol. Supplies.

Bachmann, B. J. and Strikland, W. N. (1965). *Neurospora bibliography and index*, Yale University Press.

Barratt, R. W., Newmeyer, D., Perkins, D. D. and Garnjobst, L. (1954). *Adv. Gen.*, **6**, 1–93.

Barratt, R. W., Johnson, G. B. and Ogata, W. N. (1965). *Genetics*, **52**, 233–46.

Bergman, K., Burke, P. W., Čerdo-Olmédo, E., David, C. N., Delbrück, M., Foster, K. W., Goodell, E. W., Heisenberg, M., Meissner, G., Zalokar, M., Dennison, D. S. and Shropshire, W. (1969). *Bact. Rev.*, **33**, 99–157.

Coen, D., Deutsch, J., Netter, P., Petrochilo, E. and Slonimski, P. P. (1971). In *Control of Organelle Development* (Ed. P. L. Miller), pp. 449–96, Cambridge University Press.

Day, A. W. and Jones, J. K. (1968). *Genet. Res.*, **11**, 63–81.

Day, A. W. and Jones, J. K. (1969). *Genet. Res.*, **14**, 195–221.

Day, P. R. (1960). *Genetics*, **45**, 641–50.

Day, P. R. and Anderson, G. E. (1961). *Genet. Res.*, **2**, 414–23.

Elliott, C. G., Hendrie, M. R. and Knights, B. A. (1966). *J. Gen. Microbiol.*, **42**, 425–35.

Elliott, C. G. and MacIntyre, D. (1973). *Trans. Br. Mycol. Soc.*, **60**, 311–16.

Emerson, S. (1955). In *Handbuch der physiologisch—und pathologisch—chemischen Analyse* (Eds. K. Lang and E. Lehnartz), Vol. 2, Pars 2, (10th Edn.), pp. 443–537. Springer, Berlin.

Ephrussi, B. (1953). *Nucleo-cytoplasmic relations in micro-organisms*. Clarendon Press. Oxford.

Esser, K. (1969). *Neurospora News Letter*, **15**, 27–30.

Fowell, R. R. (1969). In *The Yeasts* (Eds. A. H. Rose and J. S. Harrison), Vol. 1, pp. 303–83, Academic Press, London and New York.

Holliday, R. (1961). *Genet. Res.*, **2**, 204–30.

Khaki, I. A. and Shaw, D. S. (1974). *Genet Res.*, **23**, 75–86.

Lissouba, P., Mousseau, J., Rizet, G. and Rossignol, J. L. (1962). *Adv. Genet.*, **11**, 343–80.

Mortimer, R. K. and Hawthorne, D. C. (1969). In *The Yeasts* (Eds. A. H. Rose and J. S. Harrison), Vol. 1, pp. 385–460, Academic Press, London and New York.

Olive, L. S. (1956). *Amer. J. Bot.*, **43**, 97–107.

Pontecorvo, G. C. (1953). *Adv. Genet.*, **5**, 141–238.

Raper, J. R. (1966). *The Genetics of Sexuality in Higher Fungi*, Ronald Press, New York.

Shaw, D. S. and Elliott, C. G. (1968). *J. Gen. Microbiol.*, **51**, 75–84.

Shaw, D. S. and Khaki, I. A. (1971). *Genet. Res.*, **17**, 165–7.

Singleton, J. R. (1953). *Amer. J. Botany*, **40**, 124–44.

CLASSIFICATION

C: indicates cytological studies   G: indicates genetical studies

Phycomycetes
Chytridiomycetes	*Allomyces arbuscula* and spp., C.
Hyphochytridiomycetes	—
Oomycetes	*Achlya* spp., G; *Phytophthora infestans*, CG; *P. cactorum*, CG; *Pythium* spp., G.
Zygomycetes	*Mucor mucedo*, G; *Phycomyces blakesleeanus*, CG.

Ascomycetes
Hemiascomycetes	*Saccharomyces cerevisiae* and spp., CG; *Schizosaccharomyces pombe*, G.
Plectomycetes	*Aspergillus nidulans*, CG; *A. niger*, *A. oryzeae*, *A. amstelodami*, G; *A. heterothallicus*, G.
Pyrenomycetes	*Cochliobolus* spp., G; *Gibberella fujikuroi*, G; *Glomerella cingulata*, CG; *Hypomyces solani*, CG; *Neurospora crassa*, *N. sitophila*, *N. tetrasperma*, CG; and spp., G; *Podospora anserina*, CG; *Sordaria fimicola*, CG; *S. brevicollis*, G; *Venturia inaequalis*, CG.
Discomycetes	*Ascobolus immersus*, G; *Humaria rutilans*, C; *Sclerotinia* spp., G.
Loculoascomycetes	—
Laboulbeniales	—

Basidiomycetes
Teliomycetes
Uredinales	*Melampsora lini*, G; *Puccinia graminis*, CG.
Ustilaginales	*Ustilago maydis*, *U. violacea*, *U. hordei*, G; *Tilletia* spp., G.
Tremellales	*Tremella mesenterica*, G; *Auricularia* spp., G.

Hymenomycetes
Agaricales	*Agaricus bisporus*, CG; *Collybia velutipes*, G; *Coprinus cinereus* (= *C. lagopus*) CG; *C. radiatus*, CG., *Pleurotus* spp., G.
Aphyllophorales	*Polyporus* spp., G; *Schizophyllum commune*, CG.
Gasteromycetes	*Crucibulum vulgare*, G; *Cyathus* spp., CG; *Mycocalia* spp., CG.

Deuteromycetes (Fungi Imperfecti)
Cladosporium fulvum, G; *Fusarium oxysporum*, G; *Helminthosporium* spp., CG; *Penicillium chrysogenum*, *P. expansum*, G; *Verticillium albo-atrum* and spp., G.

References

Abraham, E. P., Chain, E., Fletcher, C. M., Gardner, A. D., Heatley, N. G., Jennings, M. A. and Florey, H. W. (1941). *Lancet*, **2**, 177–89.

Ahmad, K. A. and Woods, R. A. (1967). *Genet. Res.*, **9**, 179–93.

Ahmad, M. (1953). *Ann. Bot.*, **17**, 329–42.

Ahmad, M. (1965). In *Incompatibility in Fungi* (Eds. K. Esser and J. R. Raper), pp. 13–23, Springer, Berlin.

Ahmad, M., Khalil, M. D., Khan, N. A. and Mozmadar, A. (1964). *Genetics*, **49**, 925–33.

Ahmed, S. I. and Giles, N. H. (1969). *J. Bacteriol.*, **99**, 231–37.

Ainsworth, G. C. (1971). *Ainsworth and Bisby's Dictionary of Fungi*, 6th edn., Commonwealth Mycological Institute, London.

Alikhanian, S. I. (1962). *Adv. Appl. Microbiol.*, **4**, 1–50.

Alikhanian, S. I. (1965). In *Biogenesis of Antibiotic Substances* (Eds. Z. Vaněk and Z. Hošťálek), pp. 33–41, Academic Press, New York.

Alikhanian, S. I. (1973). In *Genetics of Industrial Microorganisms* (Eds. Z. Vaněk, Z. Hošťálek & J. Cudlín), Vol. I, pp. 9–18, Elsevier, Amsterdam.

Alikhanian, S. I. and Borisova, L. N. (1956). *Isvestia Akad. Nauk. S.S.R.*, **2**, 74.

Allard, R. W., Jain, S. K. and Workman, P. L. (1968). *Adv. Genetics*, **14**, 55–131.

Allen, N. E. and McQuillan, A. M. (1969). *J. Bacteriol.*, **7**, 1142–48.

Anderson, N. A., Stretton, H. M., Groth, J. V. and Flentje, N. T. (1972). *Phytopathology*, **62**, 1057–65.

Arlett, C. F. (1957). *Nature*, **179**, 1250–1.

Arlett, C. F. (1960). *Heredity*, **15**, 377–88.

Arlett, C. F., Grindle, M. and Jinks, J. L. (1962). *Heredity*, **17**, 197–209.

Arst, H. N. and Cove, D. J. (1969). *J. Bacteriol.*, **98**, 1284–93.

Arst, H. N., McDonald, D. W. and Cove, D. J. (1970). *Molec. Gen. Genetics*, **108**, 129–45.

Ashida, J. (1965). *A. r. Phytopathology*, **3**, 153–74.

Atwood, K. F. and Mukai, F. (1953). *Proc. Natn. Acad. Sci. U.S.A.*, **39**, 1027–35.

Atwood, K. F. and Mukai, F. (1955). *Genetics*, **40**, 438–43.

Auerbach, C. (1959). *Z. Vererbungslehre*, **90**, 335–46.

Auerbach, C. (1967). In *Heritage from Mendel* (Eds. R. A. Brink and E. D. Styles), pp. 67–80, University of Wisconsin Press, Madison.

Auerbach, C. and Ramsey, D. (1968). *Molec. Gen. Genetics*, **103**, 72–104.

Avers, C. J., Rancourt, M. W., Lin, F. H. and Pfeffer, C. R. (1964). *J. Cell Biol.*, **23**, 7A.

Azevedo, J. L. and Roper, J. A. (1967). *J. Gen. Microbiol.*, **49**, 149–55.

Azevedo, J. L. and Roper, J. A. (1970). *Genet. Res.* **16**, 79–93.

Backus, M. P. and Stauffer, J. F. (1955). *Mycologia*, **47**, 429–63.

Bailey, D. L. (1950). *Canad. J. Res.*, **C28**, 535–65.

Baker, H. G. (1953). *Symp. Soc. Exptl. Biology*, **7**, 114–45.

Ball, C. (1971). *J. Gen. Microbiol.*, **66**, 63–9.

Ball, C. (1973a). In *Genetics of Industrial Microorganisms* (Eds. Z. Vaněk, Z. Hoštálek and J. Cudlín), Vol. II, pp. 227–37, Elsevier, Amsterdam.

Ball, C. (1973b). *Prog. Indust. Microbiol.*, **12**, 47–72.

Barnes, B. (1928). *Ann. Bot.*, **168**, 783–872.

Barnett, H. L. (1964). *Mycologia*, **56**, 1–19.

Barr, R., Caldwell, R. M. and Amacher, R. H. (1964). *Phytopathology*, **54**, 104–9.

Barratt, R. W. (1954). *Microbiol. Genet. Bull.*, **11**, 5–6.

Barratt, R. W. and Garnjobst, L. (1949). *Genetics*, **34**, 351–69.

Barratt, R. W., Newmeyer, D., Perkins, D. D. and Garnjobst, L. (1954). *Adv. Genetics*, **6**, 1–93.

Barron, G. L. (1962). *Canad. J. Bot.*, **40**, 1603–13.

Barrus, M. F. (1911). *Phytopathology*, **1**, 190–5.

Bartels-Schooley, J. and McNeil, B. H. (1971). *Phytopathology*, **61**, 816–9.

Bartos, P., Fleischmann, G., Samborski, D. J. and Shipton, W. A. (1969). *Canad. J. Bot.*, **47**, 1383–7.

Bateson, W. and Saunders, E. R. (1902). *Rep. Evol. Comm. R. Soc.*, **1**, 1–160.

Bauch, R. (1941). *Naturwissenchaften*, **29**, 503–4.

Beadle, G. W. (1945). *Physiol. Rev.*, **25**, 643–63.

Beadle, G. W. (1973). *Neurospora News Letter*, **20**, 13.

Beadle, G. W. and Coonradt, V. L. (1944). *Genetics*, **29**, 291–308.

Beadle, G. W. and Tatum, E. L. (1941). *Proc. Natn. Acad. Sci. U.S.A.*, **27**, 499–506.

Bean, J., Brian, P. W., and Brooks, F. T. (1954). *Ann. Bot.*, **18**, 129–42.

Beckett, A. and Wilson, I. M. (1968). *J. Gen. Microbiol.*, **53**, 81–7.

Bell, M. K. and Burnett, J. H. (1966). *Ann. Appl. Biol.*, **58**, 123–30.

Belling, J. (1933). *Genetics*, **18**, 388–413.

Bent, K. J., Cole, A. M., Turner, J. A. W. and Woolner, M. (1971). *Proc. 6th Br. Insectic. Fungic. Conf.*, **1**, 274–82.

Bernet, J. (1963). *C. r. hebd. Séanc. Acad. Sci. Paris*, **256**, 771–3.

Bernet, J. (1965). *Ann. Sci. Nat.: Bot. Biol. Veg.*, **6**, 611–768.

Bernet, J., Esser, K., Marcou, D. and Schecroun, J. (1960). *C. r. hebd. Séanc. Acad. Sci. Paris*, **250**, 2053–5.

Bertrand, H. and Pittenger, T. H. (1972). *Genetics*, **71**, 521–33.

Biggs, R. (1937). *Mycologia*, **29**, 686–706.

Black, W. (1952). *Proc. R. Soc. Edinb.*, **65**, 36–51.

Black, W. (1957). *Scottish Soc. Res. Pl. Breeding Rep.*, 43–9.

Blakeslee, A. F., Welch, D. S. and Bergner, A. D. (1927). *Bot. Gaz.*, **84**, 27–50.

Bodmer, W. F. (1970). In *Organization and Control in Prokaryotic and Eukaryotic Cells* (Eds. H. P. Charles and B. C. J. G. Knight) pp. 279–94, Cambridge University Press.

Bodmer, W. F. and Ganeson, A. T. (1964). *Genetics*, **50**, 717–38.

Boidin, J. (1958). *Rev. Mycol., Mem-Hors. Sér,* **6**, 1–387.
Bole-Gowda, B. N., Perkins, D. D. and Strickland, W. N. (1962). *Genetics,* **47**, 1243–52.
Bollen, G. J. (1971). *Neth. J. Pl. Path.,* **77**, 187–193.
Bollen, G. J. and Scholten, G. (1971). *Neth. J. Pl. Path.,* **76**, 299–312.
Boone, D. M. (1971). *A. r. Phytopathology,* **9**, 297–318.
Boone, D. M. and Keitt, G. W. (1956). *Amer. J. Bot.,* **43**, 226–33.
Boone, D. M. and Keitt, G. W. (1957). *Phytopathology,* **47**, 403–9.
Boone, D. M., Stauffer, J. M., Stahmann, M. A. and Keitt, G. W. (1956). *Amer. J. Bot.,* **43**, 199–204.
Bourchier, R. J. (1957). *Mycologia,* **49**, 20–8.
Brasier, C. M. (1970). *Amer. Nat.,* **104**, 191–204.
Braun, W. (1965). *Bacterial Genetics,* W. B. Saunders, Philadelphia.
Bresch, C., Muller, G. and Egel, R. (1968). *Molec. Gen. Genetics,* **102**, 301–6.
Brian, P. W. (1972). *Trans. Br. Mycol. Soc.,* **58**, 359–75.
Brierley, W. B. (1929). *Proc. Int. Congr. Pl. Sci.,* **2**, 1629–54.
Brock, T. D. (1961). *J. Gen. Microbiol.,* **26**, 487–97.
Brodie, H. J. (1951). *Canad. J. Bot.,* **29**, 224–34.
Brody, S. and Tatum, E. L. (1967). *Proc. Natn. Acad. Sci. U.S.A.,* **58**, 923–30.
Brown, J. F. and Sharp, E. L. (1970). *Phytopathology,* **60**, 529–33.
Brown, W. (1926). *Ann. Bot.,* **40**, 28–43.
Brown, W. F. and Elander, P. P. (1966). In *Development in Industrial Microbiology,* pp. 114–20, American Inst. Biol. Sci., Washington.
Buck, K. W. and Kempson-Jones, G. F. (1970). *Nature,* **225**, 945–6.
Buller, A. H. R. (1941). *Bot. Rev.,* **7**, 335–431.
Burgeff, H. (1912). *Ber. Deut. Bot. Ges.,* **30**, 679–85.
Burgeff, H. (1914). *Flora N. F.,* **107**, 259–316.
Burgeff, H. (1924). *Bot. abh. K. Goebel,* **4**, 1–135.
Burgoyne, L., Case, M. E. and Giles, N. H. (1969). *Genetics,* **61**, 789–800.
Burnett, J. H. (1956). *New Phytol.,* **55**, 50–90.
Burnett, J. H. (1968). *Fundamentals of Mycology,* Arnold, London.
Burnett, J. H. and Boulter, M. E. (1963). *New Phytol.,* **62**, 217–36.
Burnett, J. H. and Evans, E. J. (1966). *Nature,* **210**, 1368–9.
Burnett, J. H. and Partington, M. (1957). *Proc. R. Phys. Soc. Edinb.,* **26**, 61–8.
Butcher, A. C. (1969). *Heredity,* **24**, 621–31.
Butcher, A. C., Croft, J. H. and Grindle, M. (1972). *Heredity,* **29**, 263–83.
Buxton, E. W. (1954). *J. Gen. Microbiol.,* **10**, 71–84.
Buxton, E. W. (1956). *J. Gen. Microbiol.,* **15**, 133–9.
Buxton, E. W. (1959). In *Plant Pathology: Problems and Progress 1908–1958* (Ed. C. S. Holton), pp. 183–91, University of Wisconsin Press, Madison.
Buxton, E. W. (1960). In *Plant Pathology: An Advanced Treatise* (Eds. J. G. Horsfall and A. E. Dimond), Vol. II, pp. 359–407, Academic Press, New York.
Caglioti, M. T. and Sermonti, G. (1956). *J. Gen. Microbiol.,* **14**, 38–46.
Calvo, J. M. and Fink, G. R. (1971). *A.r. Biochemistry,* **40**, 943–68.
Carr, A. J. H. and Olive, L. S. (1959). *Amer. J. Bot.,* **46**, 81–91.
Case, M. E. and Giles, N. H. (1958). *Proc. Natn. Acad. Sci. U.S.A.,* **44**, 378–90.
Case, M. E. and Giles, N. H. (1964). *Genetics,* **49**, 529–40.
Case, M. E. and Giles, N. H. (1971). *Proc. Natn. Acad. Sci. U.S.A.,* **68**, 58–62.
Casselton, L. A. (1965). *Genet. Res.,* **6**, 190–208.

Casselton, L. A. and Condit, A. (1972). *J. Gen. Microbiol.*, **72**, 521–7.

Casselton, L. A. and Lewis, D. (1966). *Genet. Res.*, **8**, 61–72.

Casselton, L. A. and Lewis, D. (1967). *Genet. Res.*, **9**, 63–71.

Catcheside, D. G. (1966). *Australian J. Biol. Sci.*, **19**, 1039–46.

Catcheside, D. G. (1968). In *Replication and Recombination of Genetic Material* (Eds. W. J. Peacock and R. D. Brock), pp. 216–26, Australian Academy of Sciences, Canberra.

Catcheside, D. G., Jessop, A. P. and Smith, B. R. (1964). *Nature*, **202**, 1242–3.

Caten, C. E. (1971). *Heredity*, **26**, 299–312.

Caten, C. E. (1972). *J. Gen. Microbiol.*, **72**, 221–9.

Cayley, D. M. (1923). *J. Genet.*, **13**, 353–70.

Chain, E., Florey, H. W., Gardner, A. D., Heatley, N. G., Jennings, M. A., Orr-Ewing, J. and Sanders, A. G. (1940). *Lancet*, **239**, 226–8.

Chase, M. and Doermann, A. H. (1958). *Genetics*, **43**, 332–53.

Cherest, H. and de Robichon-Szulmajster, H. (1973). In *Genetics of Industrial Microorganisms* (Eds. Z. Vaněk, Z. Hošťálek and J. Cudlín), Vol. II, pp. 165–78, Elsevier, Amsterdam.

Cherewick, W. J. (1958). *Canad. J. Pl. Sci.*, **38**, 481–9.

Chester, V. E. (1963). *Proc. Roy. Soc.*, **B157**, 223–33.

Chovnick, A. and Fox, A. S. (1953). *Amer. Nat.*, **87**, 263–7.

Christensen, C. M. (1967). *The Moulds and Man*, University of Minnesota Press, Minneapolis.

Christensen, J. J. (1940). In *The Genetics of Pathogenic Organism* (Ed. F. R. Moulton), pp. 77–82, A.A.A.S., Science Press, Lancaster, Pa.

Clements, L. L., Day, A. W. and Jones, J. K. (1969). *Nature*, **223**, 961–3.

Clutterbuck, A. J. (1969). *Genetics*, **63**, 317–27.

Clutterbuck, P. W., Lovell, R. and Raistrick, H. (1932). *Biochem. J.*, **26**, 1907–8.

Coddington, A. and Fincham, J. R. S. (1965). *J. Mol. Biol.*, **12**, 152–61.

Coen, D., Deutsch, J., Netter, P., Petrochilo, E. and Slonimski, P. P. (1971). In *Control of Organelle Development* (Ed. P. L. Miller), pp. 449–96, Cambridge University Press.

Colson, B. (1935). *Ann. Bot.*, **49**, 1–18.

Connolly, V. and Simchen, G. (1968). *Heredity*, **23**, 387–402.

Cook, L. M. (1971). *Coefficients of Natural Selection*, Hutchinson, London.

Cooke, P., Roper, J. A. and Watmough, W. (1970). *Nature*, **226**, 276–7.

Costantin, J. and Matruchot, L. (1894). Brevet No. 236, 349. 17/2. *Bull. Pror. Industr.* Paris.

Cove, D. J. (1966). *Biochim. Biophys. Acta*, **113**, 51–6.

Cove, D. J. (1969). *Nature*, **224**, 272–4.

Cove, D. J. and Coddington, A. (1965). *Biochim. Biophys. Acta*, **110**, 312–8.

Cove, D. J. and Pateman, J. A. (1963). *Nature*, **198**, 262–3.

Cove, D. J. and Pateman, J. A. (1969). *J. Bacteriol.*, **97**, 1374–8.

Cox, B. S. (1965). *Heredity*, **20**, 505–22.

Cox, B. S. (1971). *Heredity*, **26**, 211–32.

Cox, B. S. and Bevan, E. A. (1961). *Trans. Br. Mycol. Soc.*, **44**, 239–42.

Cox, B. S. and Bevan, E. A. (1962). *New Phytol.*, **61**, 342–55.

Cox, B. S. and Gill, J. J. B. (1967). *New Phytol.*, **66**, 653–64.

Cox, B. S. and Parry, E. M. (1967). *New Phytol.*, **66**, 643–52.

Croft, J. H. cited in Jinks, J. L. (1966). See p. 338.

Croft, J. H. and Simchen, G. (1965). *Amer. Nat.*, **99**, 451–62.

Curren, T. (1967). *Canad. J. Bot.*, **45**, 2125–35.

Cutter, V. M. (1942). *Bull. Torrey Bot. Club*, **69**, 592–616.

Czarnecki, H. T. and van Engel, E. L. (1959). *Brewer's Digest*, **34**, 52–6.

Darlington, C. D. (1934). *Z. Vererbungslehre*, **67**, 96–114.

Darlington, C. D. (1935). *J. Genet.*, **31**, 185–212.

Darlington, C. D. (1937). *Recent Advances in Cytology*, 2nd edn., Churchill, London.

Darlington, C. D. (1958). *Evolution of Genetic Systems*, Oliver and Boyd, Edinburgh.

Darlington, C. D. and Mather, K. (1949). *Elements of Genetics*, Allen and Unwin, London.

Das Gupta, S. N. (1936). *Lucknow Univ. Studies*, **5**, 1–83.

Davies, O. L. (1964). *Biometrics*, **20**, 576–91.

Davis, R. H. (1959). *Genetics*, **44**, 1291–308.

Davis, R. H. (1960a). *Amer. J. Bot.*, **47**, 351–7.

Davis, R. H. (1960b). *Amer. J. Bot.*, **47**, 648–54.

Davis, R. H. (1966). In *The Fungi* (Eds. G. C. Ainsworth and A. S. Sussman), Vol. II, pp. 567–88, Academic Press, London and New York.

Day, A. W. (1972). *Canad. J. Bot.*, **50**, 1337–47.

Day, A. W. and Jones, J. K. (1968). *Genet. Res.*, **11**, 63–81.

Day, A. W. and Jones, J. K. (1969). *Genet. Res.*, **14**, 195–221.

Day, A. W. and Jones, J. K. (1971). *Genet. Res.*, **18**, 299–309.

Day, P. R. (1956). *Tomato Genet. Coop. Rep.*, **6**, 13.

Day, P. R. (1957). *Nature*, **179**, 1141–2.

Day, P. R. (1958). *Rep. John Innes Hort. Instn.*, **49**, 16–18.

Day, P. R. (1960). *A. r. Microbiol.*, **14**, 1–16.

Day, P. R. (1966). *A. r. Phytopathology*, **4**, 245–68.

Day, P. R. (1970). In *Root Diseases and Soil-borne Pathogens* (Eds. T. A. Toussoun R. V. Bega and P. E. Nelson), pp. 69–74, University of California Press, Berkeley.

Day, P. R. (1974). *Genetics of Host-parasite Interaction*. Freeman, San Francisco.

Day, P. R. and Anderson, G. E. (1961). *Genet. Res.*, **2**, 414–23.

de Bruyn, H. L. G. (1951). *Phytopath. Z.*, **18**, 339–59.

de Deken, R. H. (1966). *J. Gen. Microbiol.*, **44**, 157–65.

Dekker, J. (1972). In *Systemic Fungicides* (Ed. R. W. Marsh), pp. 156–74, Longman, London.

de Leeuw, A. (1967). *Genetics*, **56**, 554 (Abstr.).

de Robichon-Szulmajster, H. and Surdin-Kerjan, Y. (1971). In *The Yeasts* (Eds. A. H. Rose and J. S. Harrison), Vol. II, pp. 335–418, Academic Press, London and New York.

de Robichon-Szulmajster, H., Surdin-Kerjan, Y. and Cherest, H. (1973). In *The Genetics of Industrial Microorganisms* (Eds. Z. Vaněk, Z. Hošťálek and J. Cudlín), Vol. II, pp. 149–64, Elsevier, Amsterdam.

de Serres, F. J. (1964). *Genetics*, **50**, 21–30.

de Serres, F. J. (1969). *Mutation Res.*, **8**, 43–50.

de Serres, F. J. and Kølmark, H. G. (1958). *Nature*, **182**, 1249–50.

de Serres, F. J. and Webber, B. B. (1967). *Genetics*, **57**, 449–54.

Deutsch, J. and others (1970). Cited in Coen, D. and others (1970), see p. 331.

Diacumakos, E. G., Garnjobst, L. and Tatum, E. L. (1965). *J. Cell. Biol.*, **26**, 427–43.

Dickson, H. (1936). *Ann. Bot.*, **50**, 719–33.

Dielemon-von Zaayen, A. (1972). *Agric. Res. Rep.*, **782**, Wageningen.

Dodge, B. O. (1927). *J. Agr. Res.*, **35**, 289–305.

Dodge, B. O. (1931). *Mycologia*, **23**, 1–50.

Dodge, B. O. (1942). *Bull. Torrey Bot. Club*, **69**, 75–91.

Dodge, B. O., Singleton, J. R. and Rolnick, A. (1950). *Proc. Amer. Phil. Soc.*, **94**, 38–52.

Douglas, H. C. and Hawthorne, D. C. (1964). *Genetics*, **49**, 837–44.

Douglas, H. C. and Hawthorne, D. C. (1966). *Genetics*, **54**, 911–16.

Douglas, H. C. and Pelroy, G. (1963). *Biochim. Biophys. Acta*, **68**, 155–6.

Dubos, R. J. (1947). *The Bacterial Cell*, Harvard University Press, Cambridge, Mass.

Duggar, B. M. (1905). *Bull. U.S. Bur. Pl. Ind.*, **55**, 1–60.

Duncan, E. G. and MacDonald, J. A. (1967). *Mycologia*, **59**, 803–18.

East, E. M. (1910). *Amer. Nat.*, **44**, 65–82.

Ebba, T. and Person, C. (1973). *2nd Int. Congr. Pl. Pathology*, Abstr. 0705.

Egel, R. (1973a). *Molec. Gen. Genet.*, **121**, 277–84.

Egel, R. (1973b). *Molec. Gen. Genet.*, **122**, 339–43.

Eggertson, E. (1953). *Canad. J. Bot.*, **31**, 710–9.

Elander, R. P. (1967). *Abh. Deut. Akad. Wiss. Berl.*, 403–23.

Elander, R. P., Espenshade, M. A., Parthak, S. G. and Parr, C. H. (1973). In *Genetics of Industrial Microorganisms* (Eds. Z. Vaněk, Z. Hošťálek and J. Cudlín), Vol. II, pp. 239–53, Elsevier, Amsterdam.

El-Ani, A. S. (1954). *Amer. J. Bot.*, **41**, 110–3.

El-Ani, A. S. and Olive, L. S. (1962). *Proc. Natn. Acad. Sci. U.S.A.*, **48**, 17–9.

Ellingboe, A. H. (1961). *Phytopathology*, **51**, 13–5.

Ellingboe, A. H. (1965). In *Incompatibility in Fungi* (Eds. K. Esser and J. R. Raper), pp. 36–48, Springer, Berlin.

Ellingboe, A. H. (1972). *Phytopathology*, **62**, 401–6.

Ellingboe, A. H. and Raper, J. R. (1962). *Genetics*, **47**, 85–98.

Elliott, T. J. (1972). *Mushroom Sci.*, **8**, 11–8.

Emara, Y. A. (1972). *Canad. J. Genet. Cytol.*, **14**, 919–24.

Emara, Y. A. and Sidhu, G. (1974). *Heredity*, **32**, 219–24.

Emeis, C. C. (1958). *Brauerei Wiss. Beil.*, **11**, 160–3.

Emeis, C. C. (1964). *Proc. Eur. Brew. Conv.*, *Brussels*, 362–9.

Emeis, C. C. (1965). *Proc. Eur. Brew. Conv.*, *Stockholm*, 156–163.

Emeis, C. C. (1966). *Z. Naturf.*, **21b**, 816–7.

Emerson, R. (1941). *Lloydia*, **4**, 77–144.

Emerson, R. and Wilson, C. M. (1954). *Mycologia*, **46**, 393–434.

Emerson, S. (1948). *Proc. Natn. Acad. Sci. U.S.A.*, **34**, 72–4.

Emerson, S. (1950). *A. r. Microbiology*, **4**, 169–200.

Emerson, S. (1952). In *Heterosis* (Ed. J. W. Gowen), pp. 199–217, Iowa State College Press, Ames.

Emerson, S. (1956). *C. r. Trav. Lab. Carlsberg, sér Physiol.*, **26**, 71–86.

Emerson, S. (1963). In *Methodology in Basic Genetics* (Ed. W. J. Burdette), pp. 167–208, Holden-Day, San Francisco.

Emerson, S. (1966). *Genetics*, **53**, 475–85.

Emerson, S. and Yu-Sun, C. C. C. (1967). *Genetics*, **55**, 39–47.

Emmons, C. W. and Hollaender, A. (1939). *Amer. J. Bot.*, **26**, 467–75.

334

Ephrussi, B. (1953). *Nucleo-cytoplasmic Relations in Micro-organisms*, Clarendon Press, Oxford.

Ephrussi, B. and Hottinguer, H. (1950). *Nature*, **166**, 956.

Ephrussi, B., Hottinguer, H. de M. and Tavlitski, J. (1949). *Ann. Inst. Pasteur*, **76**, 419–50.

Ephrussi, B., Hottinguer, H. de M. and Roman, H. (1955). *Proc. Natn. Acad. Sci. U.S.A.*, **41**, 1065–71.

Eriksson, J. (1894). *Ber. d. Deutsch. bot. Ges.*, **12**, 292–331.

Esser, K. (1956). *Z. Vererbungslehre*, **87**, 595–624.

Esser, K. (1959). *Z. Vererbungslehre*, **90**, 29–52; 445–56.

Esser, K. (1962). *Biol. Zbl.*, **81**, 161–72.

Esser, K. (1969). *Neurospora News Letter*, **15**, 27–30.

Esser, K. (1971). *Molec. Gen. Genet.*, **110**, 86–100.

Esser, K. (1973). In *Handbook of Genetics* (Ed. R. R. King), von Nostrand-Reinhold, New York.

Esser, K. and Blaich, R. (1973). *Adv. Genetics*, **17**, 107–52.

Esser, K. and Kuenen, R. (1965). *Genetik der Pilze*, Springer, Berlin.

Esser, K. and Straub, J. (1956). *Z. Vererbungslehre*, **89**, 729–46.

Evans, H. J. (1959). *Chromosoma*, **10**, 115–35.

Ferguson, M. (1902), *Bull. U.S. Bur. Pl. Indust.*, **16**.

Fincham, J. R. S. (1951). *J. Genet.*, **50**, 221–9.

Fincham, J. R. S. (1959). *J. Gen. Microbiol.*, **21**, 600–11.

Fincham, J. R. S. (1966). *Genetic Complementation*, W. A. Benjamin, New York.

Fincham, J. R. S. (1970). *A. r. Genetics*, **4**, 347–72.

Fincham, J. R. S. and Coddington, A. (1963). *J. Mol. Biol.*, **6**, 361–73.

Fincham, J. R. S. and Day, P. R. (1971). *Fungal Genetics*, 3rd edn., Blackwell Scientific, Oxford.

Fincham, J. R. S. and Pateman, J. A. (1957). *Nature*, **179**, 741–2.

Fischer, G. W. (1940). *Mycologia*, **32**, 275–89.

Fleming, A. (1929). *Br. J. Exp. Path.*, **10**, 226–36.

Flentje, N. T. and Stretton, H. M. (1964). *Australian J. Biol. Sci.*, **17**, 686–704.

Flentje, N. T., Stretton, H. M. and Hawn, E. J. (1963). *Australian J. Biol. Sci.*, **16**, 450–67.

Flentje, N. T., Stretton, H. M. and McKenzie, A. R. (1967). *Australian J. Biol. Sci.*, **20**, 1173–80.

Flentje, N. T., Stretton, H. M. and McKenzie, A. R. (1970). In *Rhizoctonia solani: biology and pathology* (Ed. J. R. Parmeter), pp. 52–65, University of California Press, Berkeley.

Flor, H. H. (1935). *J. Agr. Res.*, **51**, 819–37.

Flor, H. H. (1940). *J. Agr. Res.*, **60**, 575–92.

Flor, H. H. (1942). *Phytopathology*, **32**, 653–9.

Flor, H. H. (1946). *J. Agr. Res.*, **73**, 335–57.

Flor, H. H. (1947). *J. Agr. Res.*, **74**, 241–62.

Flor, H. H. (1953). *Phytopathology*, **43**, 624–8.

Flor, H. H. (1954). *U.S.D.A. Tech. Bull.*, **1087**.

Flor, H. H. (1955). *Phytopathology*, **45**, 680–5.

Flor, H. H. (1956). *Adv. Genetics*, **8**, 267–382.

Flor, H. H. (1960). *Phytopathology*, **50**, 603–5.

Flor, H. H. (1971). *A. r. Phytopathology*, **9**, 275–96.

Fogel, S. and Hurst, D. D. (1967). *Genetics*, **57**, 455–81.
Fogel, S. and Mortimer, R. K. (1969). *Proc. Natn. Acad. Sci. U.S.A.*, **62**, 96–103.
Ford, E. B. (1965). *Genetic Polymorphism*, Faber and Faber, London.
Fowell, R. R. (1951). *J. Inst. Brew.*, **57**, 180–95.
Fowell, R. R. (1956). *C. r. Trav. Lab. Carlsberg sér. Physiol.*, **26**, 117–38.
Fowell, R. R. (1966). *Process. Biochem.*, **1**, 25–8.
Fowell, R. R. (1969). In *The Yeasts* (Eds. A. H. Rose and J. S. Harrison), Vol. I, pp. 303–83, Academic Press, London.
Fox, A. S. and Allen, M. K. (1964). *Proc. Natn. Acad. Sci. U.S.A.*, **52**, 412–9.
Franke, G. (1962). *Z. Vererbungslehre*, **93**, 109–17.
Freeman, G. H. and Perkins, J. M. (1971). *Heredity*, **27**, 15–23.
Fries, N. (1936). *Bot. Notiser*, 567–74.
Fries, N. (1940). *Symb. Bot. Upsaliensis*, **4**, 5–39.
Fries, N. (1943). *Arch. Mikrobiol.*, **13**, 182–90.
Fries, N. (1947). *Nature*, **159**, 199.
Fries, N. (1948a). *Trans. Brit. Mycol. Soc.*, **30**, 118–34.
Fries, N. (1948b). *Physiologia Pl.*, **1**, 330–41.
Fripp, Y. (1972). *Heredity*, **28**, 223–38.
Fripp, Y. and Caten, C. E. (1971). *Heredity*, **27**, 393–407.
Fripp, Y. and Caten, C. E. (1973). *Heredity*, **30**, 341–9.
Fritsche, G. (1967). *Abh. Deut. Akad. Wiss. Berlin*, 441–7.
Frost, L. C. (1961). *Genet. Res.*, **2**, 43–62.
Fulton, I. (1950). *Proc. Natn. Acad. Sci. U.S.A.*, **36**, 306–12.
Gajewski, W., Paszewski, A., Davidowicz, A. and Dudzinskz, B. (1968). *Genet. Res.*, **11**, 311–7.
Gallegly, M. E. (1970). In *Root Diseases and Soil-borne Pathogens* (Eds. T. A. Toussoun, R. V. Begor and P. E. Nelson), pp. 30–54, University of California Press, Los Angeles.
Gallegly, M. E. and Eichenmuller, J. J. (1939). *Amer. Potato J.*, **36**, 45–51.
Gallegly, M. E. and Marvel, M. E. (1955). *Phytopathology*, **45**, 103–9.
Gallegly, M. E. and Niederhauser, J. S. (1959). In *Plant Pathology: Problems and Progress 1908–1958* (Ed. C. S. Holton), pp. 168–82, University of Wisconsin Press, Madison.
Garber, R. H. and Beraha, L. (1965). *Genetics*, **52**, 487–92.
Garnjobst, L. (1953). *Amer. J. Bot.*, **40**, 607–14.
Garnjobst, L. (1955). *Amer. J. Bot.*, **42**, 444–8.
Garnjobst, L. and Wilson, J. F. (1956). *Proc. Natn. Acad. Sci. U.S.A.*, **42**, 613–8.
Garza-Chapa, R. and Anderson, N. A. (1966). *Phytopathology*, **56**, 1260–8.
Gauger, W. L. (1965). *Mycologia*, **57**, 634–41.
Georgopoulos, S. G. and Panopoulos, N. J. (1966). *Canad. J. Genet. Cytol.*, **8**, 347–9.
Georgopoulos, S. G. and Sisler, H. D. (1970). *J. Bacteriol.*, **103**, 745–64.
Georgopoulos, S. G. and Zaracovitis, C. (1967). *A. r. Phytopathology*, **5**, 109–30.
Giles, N. H. (1952). *Genetics*, **37**, 584 (Abstr.).
Giles, N. H. (1955). *Brookhaven Symp. Biol.*, **8**, 103–25.
Giles, N. H., Case, M. E., Partridge, C. W. H. and Ahmed, S. I. (1967). *Proc. Natn. Acad. Sci. U.S.A.*, **58**, 1930–7.
Giles, N. H., de Serres, F. J. and Partridge C. W. H. (1955). *Ann. N.Y. Acad. Sci.*, **59**, 536–52.
Gillie, O. J. (1966). *Genet. Res.*, **8**, 9–31.

336

Gillies, C. B. (1972). *Chromosoma*, **36**, 119–30.

Gits, J. J. and Grenson, M. (1967). *Biochim. Biophys. Acta*, **135**, 507–17.

Gooday, G. W. (1973). In *Microbial Differentiation* (Eds. J. M. Ashworth and J. E. Smith), pp. 269–94, Cambridge University Press.

Graham, K. M., Dionne, L. A. and Hodgson, W. A. (1961). *Phytopathology*, **51**, 264–5.

Grant, V. (1958). *Cold Spring Harbor Symp. Quant. Biol.*, **23**, 337–63.

Grant, V. (1963). *The Origin of Adaptations*, Columbia University Press, New York.

Green, G. J. (1964). *Canad. J. Bot.*, **42**, 1653–64.

Green, G. J. (1965). *Canad. J. Genet. Cytol.*, **7**, 641–50.

Green, G. J. (1971). *Canad. J. Bot.*, **49**, 2089–95.

Greenaway, W. and Cowan, J. W. (1970). *Trans. Br. Mycol. Soc.*, **54**, 127–38.

Gregory, P. H. (1973). *Microbiology of the atmosphere*, 2nd edn., L. Hill, London.

Grenson, M. (1973). In *The Genetics of Industrial Microorganisms* (Eds. Z. Vaněk, Z. Hošťálek and J. Cudlín), Vol. II, pp. 179–93, Elsevier, Amsterdam.

Griffiths, D. J. (1958). *Trans. Br. Mycol. Soc.*, **41**, 373–84.

Griffiths, D. J. and Carr, A. J. H. (1961). *Trans. Br. Mycol. Soc.*, **44**, 601–7.

Grigg, G. W. (1952). *Nature*, **169**, 98–100.

Grigg, G. W. (1958). *Australian J. Biol. Sci.*, **11**, 69–84.

Grindle, M. (1964). *Heredity.* **19**, 75–95.

Gross, S. R. (1952). *Amer. J. Bot.*, **39**, 574–7.

Gross, S. R. (1969). *A. r. Genetics*, **3**, 395–424.

Gross, S. R. and Fein, A. (1960). *Genetics*, **45**, 885–904.

Grover, R. K. and Chopra, B. L. (1970). *Acta Phytopathologica (Hungary)*, **5**, 113–21.

Gutz, H. (1966). *J. Bacteriol.*, **92**, 1567–8.

Hadwiger, L. A. and Schwochau, M. E. (1969). *Phytopathology*, **59**, 223–7.

Halvorson, H. O., Winderman, S. and Gormon, J. (1963). *Biochim. Biophys. Acta*, **67**, 42–53.

Hanna, W. F. (1924). *Ann. Bot.*, **38**, 791–4.

Hanna, W. F. (1926). *Trans. Br. Mycol. Soc.*, **11**, 219–38.

Hansen, H. N. (1938). *Mycologia*, **30**, 442–55.

Hansen, H. N. and Smith, R. E. (1932). *Phytopathology*, **22**, 953–4.

Hansen, H. N. and Smith, R. E. (1935). *Zentr. Bakt. Parasitenk. Abt. II*, **92**, 272–9.

Hansen, H. N. and Snyder, W. C. (1944). *Science*, **99**, 264–5.

Hansen, H. N. and Snyder, W. C. (1946). *Proc. Natn. Acad. Sci. U.S.A.*, **32**, 272–3.

Harding, H. (1973). *Canad. J. Bot.*, **51**, 9–13.

Harmsen, L. (1960). *Friesia*, **6**, 233–77.

Hartsell, S. E. (1956). *Appl. Microbiol.*, **4**, 350–5.

Hartwell, L. H. (1967). *J. Bacteriol.*, **93**, 1662–70.

Hastie, A. C. (1962). *J. Gen. Microbiol.*, **27**, 373–82.

Hastie, A. C. (1964). *Genet. Res.*, **5**, 305–15.

Hastie, A. C. (1967). *Nature*, **214**, 249–52.

Hastie, A. C. (1968). *Molec. Gen. Genetics*, **102**, 232–40.

Hastie, A. C. (1970). In *Root Diseases and Soil-borne Pathogens* (Eds. T. A. Toussoun, R. V. Bega and P. E. Nelson), pp. 55–62, University of California Press, Los Angeles.

Hastie, A. C. (1973). *Trans. Br. Mycol. Soc.*, **60**, 511–23.

Hastie, A. C. and Georgopoulos, S. G. (1971). *J. Gen. Microbiol.*, **67**, 371–3.

Hawker, L. E. (1957). *The Physiology of Reproduction in Fungi*, Cambridge University Press.

Hawthorne, D. C. (1963a). *Genetics*, **48**, 1727–9.

Hawthorne, D. C. (1963b). *Proc. 11th Int. Congr. Genetics*, **1**, 34–5 (Abstr.).

Hawthorne, D. C. and Mortimer, R. K. (1960). *Genetics*, **45**, 1085–110.

Hawthorne, D. C. and Mortimer, R. K. (1963). *Genetics*, **48**, 617–20.

Hawthorne, D. C. and Mortimer, R. K. (1968). *Genetics*, **60**, 735–42.

Hemmons, L. M., Pontecorvo, G. C. and Bufton, A. W. J. (1953). *Adv. Genetics*, **5**, 194.

Henderson, W. R. and Winstead, N. N. (1961). *Plant Dis. Rep.*, **45**, 272–3.

Herring, A. J. and Bevan, E. A. (1974). *J. Gen. Virology*, **22**, 387–94.

Heslot, H. (1958). *Rev. Cytol. et Biol. végét.*, **19**, Suppl. 2, 1–235.

Heslot, H. (1962). *Abh. Deutsch. Akad. Wiss. Berlin Kl. Medizin*, **1**, 193–228.

Hesseltine, C. W. (1965). *Mycologia*, **57**, 149–97.

Hignett, R. C. and Kirkham, D. S. (1967). *J. Gen. Microbiol.*, **48**, 269–75.

Hilger, J. (1973). *J. Gen. Microbiol.*, **75**, 23–31.

Hocking, D. (1967). *Trans. Br. Mycol. Soc.*, **50**, 207–20.

Hoffer, A. and Osmond, H. (1967). *The Hallucinogens*, Academic Press, New York.

Hoffman, G. M. (1966). *Arch. Mikrobiol.*, **56**, 40–59.

Hollaender, A. and Emmons, C. W. (1951). *Cold Spring Harbor Symp. Quant. Biol.*, **9**, 179–86.

Holliday, R. (1960). *Nature*, **178**, 987.

Holliday, R. (1961a). *Genet. Res.*, **2**, 204–30.

Holliday, R. (1961b). *Genet. Res.*, **2**, 231–48.

Holliday, R. (1964a). *Genetics*, **50**, 282–304.

Holliday, R. (1964b). *Genetics*, **50**, 323–35.

Holliday, R. (1965). *Mutation Res.*, **2**, 557–9.

Holliday, R. (1967). *Mutation Res.*, **4**, 265–74.

Holliday, R. (1968). In *Replication and Recombination of Genetic Material* (Eds. W. J. Peacock and R. D. Brock), pp. 157–74, Australian Acad. of Sciences, Canberra.

Hollings, M. and Stone, O. M. (1971). *A. r. Phytopathology*, **9**, 93–118.

Holloway, B. W. (1955). *Genetics*, **40**, 117–29.

Holton, C. S. (1931). *Phytopathology*, **21**, 835–42.

Holton, C. S. (1932). *Minn. Agric. Expt. Sta. Tech. Bull.*, **87**, 1–34.

Holton, C. S. (1944). *Phytopathology*, **34**, 586–92.

Holton, C. S. (1954). *Phytopathology*, **44**, 493 (Abstr.).

Holton, C. S. (1967). *Plant. Dis. Rep.*, **50**, 62–3.

Holton, C. S. and Halisky, P. M. (1960). *Phytopathology*, **50**, 766–70.

Holton, C. S. and Hoffmann, J. A. (1968). *A. r. Phytopathology*, **5**, 163–82.

Hoppe, P. E. (1936). *J. Agr. Res.*, **53**, 671–80.

Hora, F. B. (1959). *Trans. Br. Mycol. Soc.*, **42**, 1–14.

Horn, P. and Wilkie, D. (1966). *Heredity*, **21**, 625–35.

Horne, A. S. and Das Gupta, S. N. (1929). *Ann. Bot.*, **43**, 417–35.

Horowitz, N. H., Fling, M., MacLeod, H. L. and Sueoka, G. N. (1960). *J. Mol. Biol.*, **2**, 96–104.

Hotchkiss, R. D. (1971). *Adv. Genetics*, **16**, 327–48.

Hotta, Y., Ito, M. and Stern, H. (1966). *Proc. Natn. Acad. Sci. U.S.A.*, **56**, 1184–91.

Hough, J. S. (1957). *J. Inst. Brew.*, **63**, 483–7.

Houlahan, M. B., Beadle, G. W. and Calhoun, H. G. (1949). *Genetics*, **34**, 493–507.

Howard, H. W. (1968). *Abstr. First Int. Cong. Pl. Pathology*, 92.

Howe, H. B. and Terry, C. E. (1962). *Canad. J. Genet. Cytol.*, **4**, 447–52.

Hurst, D. D., Fogel, S. and Mortimer, R. K. (1972). *Proc. Natn. Acad. Sci. U.S.A.*, **69**, 101–5.

Ingold, C. T. (1940). *Trans. Br. Mycol. Soc.*, **24**, 29–32.

Ingram, R. (1968). *Trans. Br. Mycol. Soc.*, **51**, 339–41.

Ishitani, C., Ikeda, Y. and Sakaguchi, K. (1956). *J. Gen. Appl. Microbiol. Tokyo*, **2**, 401–30.

Ishitani, C. and Sakaguchi, K. (1956). *J. Gen. Appl. Microbiol. Tokyo*, **2**, 345–400.

Jacob, F. and Monod, J. (1961). *J. Mol. Biol.*, **3**, 318–56.

Jacob, F. and Monod, J. (1962). *Cold Spring Harbor Symp. Quant. Biol.* **26**, 193–211.

Jacob, F., Perrin, D., Sanchez, C. and Monod, J. (1960). *C. r. hebd. Séanc. Acad. Sci. Paris*, **250**, 1727–9.

Jacob, F., Ullman, A. and Monod, J. (1964). *C. r. hebd. Séanc. Acad. Sci. Paris*, **258**, 3125–8.

Jakob, H. (1962). *C. r. hebd. Séanc. Acad. Sci. Paris*, **254**, 3909–11.

Jessop, A. P. and Catcheside, D. G. (1965). *Heredity*, **20**, 237–56.

Jinks, J. L. (1952). *Proc. Roy. Soc.*, **B140**, 105–45.

Jinks, J. L. (1954). *Nature*, **174**, 209.

Jinks, J. L. (1956). *C. r. Trav. Lab. Carlsberg, Sér. Physiol.*, **26**, 183–203.

Jinks, J. L. (1957). *Proc. Roy. Soc.*, **B146**, 517–40.

Jinks, J. L. (1958). *Proc. Roy. Soc.*, **B148**, 314–21.

Jinks, J. L. (1959a). *Heredity*, **13**, 525–8.

Jinks, J. L. (1959b). *J. Gen. Microbiol.*, **21**, 397–409.

Jinks, J. L. (1963). In *Methodology in Basic Genetics* (Ed. W. J. Burdette), pp. 325–43, Holden-Day, San Francisco.

Jinks, J. L. (1966). In *The Fungi* (Eds. G. C. Ainsworth and A. S. Sussman), Vol. II, pp. 619–60, Academic Press, London.

Jinks, J. L., Caten, C. E., Simchen, G. and Croft, J. H. (1966). *Heredity*, **21**, 227–39.

Jinks, J. L. and Simchen, G. (1966). *Nature*, **210**, 778–80.

Johannsen, W. (1909). *Elemente der exakten Erblichkeitslehre*, Fischer, Jena.

Johnson, T. (1954). *Canad. J. Bot.*, **35**, 506–22.

Johnson, T. (1961). *Science*, **133**, 357–62.

Johnson, T. and Newton, M. (1933). *Proc. World's Grain Exh. and Conf.*, II, 219–23.

Johnson, T. and Newton, M. (1940). *Canad. J. Res.*, **C18**, 54–67.

Johnson, T. and Newton, M. (1946). *Bot. Rev.*, **12**, 335–92.

Johnson, T., Newton, M. and Brown, A. M. (1932). *Sci. Agr.*, **13**, 141–53.

Johnson, T., Newton, M. and Brown, A. M. (1934). *Sci. Agr.*, **14**, 360–73.

Johnston, J. R. and Mortimer, R. K. (1959). *J. Bacteriol.*, **78**, 292.

Jurand, M. K. and Kemp, R. F. O. (1973). *Genet. Res.*, **22**, 125–34.

Käfer, E. (1958). *Adv. Genet.*, **9**, 105–45.

Käfer, E. (1961). *Genetics*, **46**, 1581–609.

Käfer, E. (1962). *Genetica*, **33**, 59–68.

Käfer, E. (1965). *Genetics*, **52**, 217–32.

Käfer, E. and Chen, T. L. (1964). *Canad. J. Genet. Cytol.*, **6**, 249–54.

Katsuya, K. and Green, G. J. (1967). *Canad. J. Bot.*, **45**, 1077–91.

Keene, M. L. (1914). *Ann. Bot.*, **28**, 455–70.

Keitt, G. W. and Boone, D. M. (1954). *Phytopathology*, **44**, 362–70.

Keitt, G. W., Boone, D. M. and Shay, J. R. (1959). In *Plant Pathology: Problems and Progress* (Ed. C. S. Holton), pp. 157–67, University of Wisconsin Press, Madison.

Keitt, G. W., Langford, M. H. and Shay, J. R. (1943). *Amer. J. Bot.*, **30**, 491–500.

Kemp, R. F. O. (1961). *Trans. Br. Mycol. Soc.*, **55**, 493–6.

Kemp, R. F. O. (1974). *Trans. Br. Mycol. Soc.*, **62**, 547–55.

Kemp, R. F. O. and Bevan, E. A. (1959). *Trans. Br. Mycol. Soc.*, **42**, 308–11.

Kihlman, B. A. (1966). *Actions of Chemicals on Dividing Cells*, Prentice-Hall, N.J.

Kiritani, K. (1959). *Z. Vererbungslehre*, **90**, 182–9.

Kirkham, D. S. and Hignett, R. C. (1966). *Nature*, **212**, 211–2.

Kishy, K. (1962). *Ann. Phytopath. Soc. Japan*, **27**, 172–9; 180–8.

Kitani, Y. and Olive, L. S. (1967). *Genetics*, **57**, 767–82.

Kitani, Y. and Olive, L. S. (1969). *Genetics*, **62**, 23–66.

Kitani, Y., Olive, L. S. and El-Ani, A. S. (1962). *Amer. J. Bot.*, **49**, 697–706.

Knott, D. R. (1968). *Canad. J. Genet. Cytol.*, **10**, 311–30.

Knott, D. R. (1972). *J. Environ. Quality*, **1**, 227–31.

Köhler, E. (1930). *Planta*, **10**, 495–522.

Köhler, F. (1935). *Z. Vererbungslehre*, **70**, 1–54.

Kølmark, G. and Westergaard, M. (1952). *Nature*, **169**, 626.

Koltin, V., Berick, R., Stamberg, J. and Ben-Shaul, Y. (1973). *Nature New Biology*, **241**, 108–9.

Koltin, V., Stamberg, J. and Lemke, P. A. (1972). *Bact. Rev.*, **36**, 156–71.

Küntzel, H. (1969). *Nature*, **222**, 142–8.

Küntzel, H. and Noll, H. (1967). *Nature*, **215**, 1340–5.

Kuo, M. S., Yoder, O. C. and Scheffer, R. P. (1970). *Phytopathology*, **60**, 365–8.

Kwon, K. J. and Raper, K. B. (1967a). *Amer. J. Bot.*, **54**, 36–48.

Kwon, K. J. and Raper, K. B. (1967b). *Amer. J. Bot.*, **54**, 49–60.

Lamb, I. M. (1935). *Ann. Bot.*, **49**, 403–38.

Lambert, E. B. (1938). *Bot. Rev.*, **4**, 397–426.

Landner, L. (1971). *Heredity*, **27**, 385–92.

Lange, M. (1952). *Dansk Bot. Arkiv.*, **14**, 1–164.

La Rue, C. D. (1922). *Genetics*, **7**, 142–83.

Lawrence, G. J. (1973). Cited in Person and Mayo (1974), p. 344.

Leblon, G. (1972). *Molec. Gen. Genet.*, **116**, 322–35.

Lederberg, J. (1955). *J. Cell. Comp. Physiol.*, **45**, Suppl. **2**, 75–107.

Lederberg, J. and Lederberg, E. M. (1952). *J. Bacteriol.*, **63**, 399–406.

Lee, B. T. O. (1962). *Australian J. Biol. Sci.*, **15**, 160–5.

Lee, B. T. O. and Pateman, J. A. (1959). *Nature*, **183**, 698–9.

Lee, B. T. O. and Pateman, J. A. (1961). *Australian J. Biol. Sci.*, **14**, 223–30.

Lemke, P. A. (1969). *Mycologia*, **61**, 57–76.

Lemke, P. A., Nash, C. H. and Pieper, S. W. (1973). *J. Gen. Microbiol.*, **76**, 265–75.

Leonard, K. J. (1969). *Phytopathology*, **59**, 1858–63.

Lescure, L. A. (1956). *M.Sc. Thesis*, University of California, Davis.

Lester, H. E. and Gross, S. R. (1959). *Science*, **129**, 572.

Leu, L. S. (1967). *Ph.D. Thesis*, University of Wisconsin, Madison (cited in Boone, 1971; see p. 330).

Leupold, U. (1950). *C. r. Trav. Lab. Carlsberg Sér. Physiol.*, **24**, 381–480.

Leupold, U. (1956a). *J. Genet.*, **54**, 411–26.

Leupold, U. (1956b). *J. Genet.*, **54**, 427–39.

Leupold, U. (1958). *Cold Spring Harbor Symp. Quant. Biol.*, **23**, 161–70.

Levine, M. N. and Cotter, R. U. (1931). *Phytopathology*, **21**, 107.

Lewis, C. W. and Johnston, J. R. (1973). *Proc. Soc. Gen. Microbiol.*, **1**, 73.

Lewis, D. (1963). *Nature*, **200**, 151.

Lewis, D. H. (1973). *Biol. Rev.*, **48**, 261–78.

Lhoas, P. (1961). *Nature*, **190**, 744.

Lhoas, P. (1967). *Genet. Res.*, **10**, 45–61.

Lhoas, P. (1968). *Genet. Res.*, **12**, 305–15.

Lhoas, P. (1971). *Nature*, **236**, 248–9.

Lindegren, C. C. (1932). *Bull. Torrey Bot. Club*, **59**, 119–38.

Lindegren, C. C. (1933). *Bull. Torrey Bot. Club*, **60**, 133–54.

Lindegren, C. C. (1936). *J. Genet.*, **32**, 243–56.

Lindegren, C. C. (1949). *The Yeast Cell. Its Genetics and Cytology*, Educational Publishers, St. Louis.

Lindegren, C. C. (1953). *J. Genet.*, **51**, 625–37.

Lindegren, C. C. and Lindegren, G. (1937). *J. Heredity*, **28**, 105–13.

Lindegren, C. C. and Lindegren, G. (1939). *Genetics*, **24**, 1–7.

Lindegren, C. C. and Lindegren, G. (1942). *Genetics*, **27**, 1–24.

Lindegren, C. C. and Lindegren, G. (1943). *Proc. Natn. Acad. Sci. U.S.A.*, **29**, 306–8.

Lindegren, C. C. and Lindegren, G. (1944). *Ann. Mo. Botan. Garden*, **31**, 203–16.

Lindegren, C. C., Spiegelman, S. and Lindegren, G. (1945). *Arch. Biochem.*, **6**, 185–98.

Linnane, A. W., Lamb, A. J., Christodoulou, C. and Lukins, H. B. (1968a). *Proc. Natn. Acad. Sci. U.S.A.*, **59**, 1288–93.

Linnane, A. W., Saunders, G. W., Gingold, E. B. and Lukins, H. B. (1968b). *Proc. Natn. Acad. Sci. U.S.A.*, **59**, 903–10.

Lissouba, P. and Rizet, G. (1960). *C. r. hebd. Séanc. Acad. Sci. Paris*, **250**, 3408–10.

Little, R. and Manners, J. G. (1969a). *Trans. Br. Mycol. Soc.*, **53**, 251–8.

Little, R. and Manners, J. G. (1969b). *Trans. Br. Mycol. Soc.*, **53**, 258–9.

Littlefield, L. J. (1969). *Phytopathology*, **59**, 1323–8.

Loegering, W. Q. (1951). *Phytopathology*, **41**, 56–65.

Loegering, W. Q. (1966). *Hereditas*, Suppl. II, 167–77.

Loegering, W. Q. (1971). In *Biology of rust resistance in Forest Trees* (Eds. R. T. Bingham, R. J. Hoff and G. I. Macdonald), pp. 681–9, U.S.D.A. Forest Service Misc. Publ. **1221**.

Loegering, W. Q. and Powers, H. R. (1962). *Phytopathology*, **52**, 547–54.

Lu, B. C. (1964). *Amer. J. Bot.*, **51**, 343–7.

Lu, B. C. (1970). *J. Cell. Sci.*, **6**, 669–78.

MacDonald, K. D. (1964). *Nature*, **204**, 404–5.

MacDonald, K. D. (1966). *Antonie von Leeuwenhoek*, **32**, 431–41.

MacDonald, K. D. (1968). *Genet. Res.*, **11**, 327–30.

MacDonald, K. D., Hutchinson, J. M. and Gillet, W. A. (1963). *J. Gen. Microbiol.*, **33**, 385–94.

MacDonald, K. D., Hutchinson, J. M. and Gillett, W. A. (1964). *Antonie von Leeuwenhoek*, **30**, 209–24.

MacDonald, K. D., Hutchinson, J. M. and Gillett, W. A. (1965). *Genetica*, **36**, 378–97.

MacKinnon, J. M. and Johnston, J. R. (1972). *Heredity*, **28**, 347–55.

Mackintosh, M. E. and Pritchard, R. H. (1963). *Genet. Res.*, **4**, 320–2.

Macrae, R. (1967). *Canad. J. Bot.*, **45**, 1371–98.

Mainwaring, H. R. and Wilson, I. M. (1968). *Trans. Br. Mycol. Soc.*, **51**, 663–77.

Malcolmson, J. F. (1970). *Nature*, **225**, 971–2.

Malcolmson, J. F. and Black, W. (1966). *Euphytica*, **15**, 199–203.

Manney, T. R. (1964). *Genetics*, **50**, 109–21.

Manney, D. and Mortimer, R. K. (1964). *Science*, **143**, 581–2.

Marcou, D. and Schecroun, J. (1959). *C. r. hebd. Séanc. Acad. Sci. Paris*, **248**, 280–3.

Markert, C. L. (1949). *Amer. Nat.*, **83**, 227–31.

Martens, J. W., McKenzie, R. H. I. and Green, G. W. (1970). *Canad. J. Bot.*, **48**, 969–75.

Martini, A. (1960). *Riv. Vitic. Enol.*, **13**, 263–73.

Mas, P., Risser, G. and Rode, G. C. (1969). *Ann. Phytopathol.*, **1**, 213–6.

Mason, E. W. (1937). *Mycol. Pap. C.M.I.*, **4**, 68–99.

Mather, K. (1933). *J. Genet.*, **27**, 243–59.

Mather, K. (1938). *Biol. Rev.*, **13**, 252–92.

Mather, K. (1941). *J. Genet.*, **41**, 159–93.

Mather, K. (1942). *Nature*, **149**, 54–6.

Mather, K. (1943). *Biol. Rev.*, **18**, 32–64.

Mather, K. (1949). *Biometrical Genetics*, Methuen, London.

Mather, K. (1953). *Symp. Soc. Exptl. Biol.*, **7**, 66–95.

Mather, K. (1967). *Heredity*, **22**, 97–103.

Mather, K. and Harrison, B. J. (1949). *Heredity*, **3**, 1–52; 131–62.

Mather, K. and Jinks, J. L. (1958). *Nature*, **182**, 1188–90.

Mather, K. and Jinks, J. L. (1971). *Biometrical Genetics*, 2nd edn., Chapman and Hall, London.

Matile, Ph., Moor, H. and Robinow, C. F. (1969). In *The Yeasts* (Eds. A. H. Rose and J. S. Harrison), Vol. I, pp. 219–302, Academic Press, London.

Mayr, E. (1954). *Animal Species and Evolution*, Harvard University Press, Cambridge, Mass.

McClintock, B. M. (1945). *Amer. J. Bot.*, **32**, 671–8.

McGinnis, R. C. (1953). *Canad. J. Bot.*, **31**, 522–6.

McGinnis, R. C. (1956). *J. Hered.*, **47**, 254–9.

McIntosh, R. A., Luig, N. H. and Baker, E. P. (1967). *Australian J. Biol. Sci.*, **20**, 1181–92.

McIntosh, R. A., Watson, I. A. and Luig, N. H. (1973a). *2nd Int. Congr. Pl. Pathology*, Abstr. 0423.

McIntosh, R. A., Watson, I. A. and Luig, N. H. (1973b). *2nd Int. Congr. Pl. Pathology*, Abstr. 0881.

McKean, C. V. (1952). *Canad. J. Bot.*, **30**, 764–87.

McKenzie, A. R., Flentje, N. T., Stretton, H. M. and Mayo, M. J. (1969). *Australian J. Biol. Sci.*, **22**, 1270–8.

Metzenberg, R. L. (1972). *A. r. Genetics*, **6**, 111–32.

Metzger, R. J. and Trione, E. J. (1962). *Phytopathology*, **52**, 363 (Abstr.).

Meuris, P., Lacroute, F. and Slonimski, P. P. (1967). *Genetics*, **56**, 149–61.

Middleton, R. B. (1964). *Genetics*, **50**, 701–10.

Miles, P. G., Takemaru, T. and Kimuro, K. (1966). *Bot. Mag. Tokyo*, **79**, 693–705.

Miller, J. J. (1945). *Canad. J. Res.*, **C23**, 16–43.
Miller, J. J. (1946a). *Canad. J. Res.*, **C24**, 188–212.
Miller, J. J. (1946b). *Canad. J. Res.*, **C24**, 213–23.
Miller, R. E. (1971). *Mushroom Sci.*, **8**, 713–8.
Mills, D. I. and Ellingboe, A. H. (1971). *Molec. Gen. Genet.*, **110**, 67–76.
Mills, W. R. (1940). *Phytopathology*, **30**, 830–9.
Ming, Y. N., Lin, P. C. and Yu, T. F. (1966). *Scientia Sinica*, **15**, 371–8.
Mishra, N. C. (1971). *Genetics*, **67**, 55–9.
Mishra, N. C. and Tatum, E. L. (1970). *Proc. Natn. Acad. Sci. U.S.A.*, **66**, 638–45.
Mitchell, D. J., Bevan, E. A. and Herring, A. J. (1973). *Heredity*, **31**, 133 (Abstr.).
Mitchell, M. B. (1955). *Proc. Natn. Acad. Sci. U.S.A.*, **41**, 215–20.
Mitchell, M. B. (1956). *C. r. Trav. Lab. Carlsberg Sér Physiol.*, **26**, 285–98.
Mitchell, M. B. and Mitchell, H. K. (1952). *Proc. Natn. Acad. Sci. U.S.A.*, **38**, 442–9.
Mitchell, M. B. and Mitchell, H. K. (1956). *J. Gen. Microbiol.*, **14**, 84–9.
Mode, C. J. (1957). In *Biometrical Genetics* (Ed. O. Kempthorne), pp. 84–97, Pergamon Press, Oxford.
Mode, C. J. (1958). *Evolution*, **12**, 158–65.
Moore, D. (1967). *Genet. Res.*, **9**, 331–42.
Moore, D. (1969). *J. Gen. Microbiol.*, **55**, 121–5.
Morgan, A. R., Wells, R. D. and Khoronha, H. G. (1966). *Proc. Natn. Acad. Sci. U.S.A.*, **56**, 1899–906.
Morpurgo, G. (1961). *Aspergillus News Letter*, **2**, 10.
Mortimer, R. K. and Hawthorne, D. C. (1966). *Genetics*, **53**, 165–73.
Mortimer, R. K. and Hawthorne, D. C. (1969). In *The Yeasts* (Eds. A. H. Rose and J. S. Harrison), Vol. I, pp. 386–460, Academic Press, London.
Moseman, J. G. (1957). *Phytopathology*, **47**, 453 (Abstr.).
Moseman, J. G. (1966). *A. r. Phytopathology*, **4**, 269–90.
Moses, M. J. (1956). *J. Biophys. Biochem. Cytol.*, **4**, 633–8.
Mounce, I. and Macrae, R. (1938). *Canad. J. Res.*, **C16**, 364–76.
Muller, H. J. (1916). *Amer. Nat.*, **50**, 193–221.
Muller, H. J. (1964). *Mutation Res.*, **1**, 2–9.
Mundkur, B. D. (1953). *Nature*, **171**, 793.
Murray, N. (1963). *Genetics*, **48**, 1163–83.
Murray, N. (1968). *Genetics*, **58**, 181–91.
Nakai, S. and Mortimer, R. K. (1969). *Molec. Gen. Genet.*, **103**, 329–38.
Nakamura, H. and Sakurai, L. (1962). *Ann. Phytopath. Soc. Japan*, **27**, 84 (Abstr.).
Nakamura, K. and Egashira, T. (1961). *Nature*, **190**, 1129–30.
Narita, K. C. and Titani, K. (1969). *J. Biochem. (Tokyo)*, **65**, 259–67.
Nelson, R. R. (1961). *Phytopathology*, **51**, 736–7.
Nelson, R. R. (1964a). *Phytopathology*, **54**, 867–77.
Nelson, R. R. (1964b). *Evolution*, **18**, 700–4.
Nelson, R. R. (1970). *Canad. J. Bot.*, **48**, 261–3.
Nelson, R. R. (1972). *J. Environ. Quality*, **1**, 220–7.
Nelson, R. R., Huisingh, D. and Webster, R. K. (1967). *Phytopathology*, **57**, 1081–5.
Nelson, R. R. and Kline, D. M. (1962). *Phytopathology*, **52**, 1045–9.
Nelson, R. R. and Kline, D. M. (1963). *Phytopathology*, **53**, 101–5.
Nelson, R. R. and Kline, D. M. (1964). *Phytopathology*, **54**, 1207–9.

Nelson, R. R. and Kline, D. M. (1969a). *Phytopathology*, **59**, 164–7.

Nelson, R. R. and Kline, D. M. (1969b). *Canad. J. Bot.*, **47**, 1311–4.

Nelson, R. R., Wilcoxson, R. D. and Christensen, J. J. (1955). *Phytopathology*, **45**, 639–43.

Netzer, D. and Dishon, L. (1970). *Plant Dis. Rep.*, **54**, 909.

Newmeyer, D., Howe, H. B. and Galeazzi, D. R. (1973). *Canad. J. Genet. Cytol.*, **15**, 577–85.

Newmeyer, D. and Taylor, C. W. (1967). *Genetics*, **56**, 771–9.

Newton, M. and Johnson, T. (1932). *Dom. Canad. Dept. Agr. Bull.*, **160** (new series).

Newton, M., Johnson, T. and Brown, A. M. (1930a). *Sci. Agr.*, **10**, 721–31.

Newton, M., Johnson, T. and Brown, A. M. (1930b). *Sci. Agr.*, **10**, 775–98.

Nga, B. H. and Roper, J. A. (1968). *Genetics*, **58**, 193–209.

Nga, B. H. and Roper, J. A. (1969). *Genet. Res.*, **14**, 63–70.

Nicholas, D. J. D. (1965). In *The Fungi* (Eds. G. C. Ainsworth and A. S. Sussman), Vol. I, pp. 349–76, Academic Press, London.

Niederhauser, J. S., Cervantes, J. and Servin, L. (1954). *Phytopathology*, **44**, 406–8.

Niederpruem, D. J. (1969). *Arch. Mikrobiol.*, **64**, 387–95.

Nielsen, J. (1968). *Canad. J. Bot.*, **46**, 197–202.

Nilsson-Ehle, H. (1910). *Acta Univ. Lund, sér. 2*, **5**, 1–22.

Nobles, M. K. (1956). *Canad. J. Bot.*, **34**, 104–30.

Noll, H. (1970). *Symp. Soc. Exptl. Biol.*, **24**, 419–47.

Noronha-Wagner, M. and Bettencourt, A. J. (1967). *Canad. J. Bot.*, **45**, 2021–31.

Oeser, H. (1962). *Arch. Mikrobiol.*, **44**, 47–74.

Ohmori, K. (1967). *J. Antibiotics (Tokyo)*, **A20**, 109–14.

Olive, L. S. (1954). *Bull. Torrey Bot. Club*, **81**, 95–7.

Olive, L. S. (1956). *Amer. J. Bot.*, **43**, 97–107.

Oort, A. J. P. (1963). *Neth. J. Pl. Path.*, **69**, 104–9.

Östergren, G. and Levan, A. (1943). *Hereditas*, **29**, 496–8.

Papa, K. E. (1971). *J. Heredity*, **62**, 87–9.

Papa, K. E., Srb, A. M. and Federer, W. T. (1966). *Heredity*, **21**, 595–613.

Papa, K. E., Srb, A. M. and Federer, W. T. (1967). *Heredity*, **22**, 285–96.

Papazian, H. P. (1950). *Bot. Gaz.*, **112**, 143–63.

Papazian, H. P. (1951). *Genetics*, **36**, 441–59.

Parker-Rhodes, A. F. (1949). *New Phytol.*, **48**, 382–9.

Parker-Rhodes, A. F. (1950). *New Phytol.*, **49**, 328–34.

Parker-Rhodes, A. F. (1951). *New Phytol.*, **50**, 227–43.

Pateman, J. A. (1955). *Nature*, **176**, 1274–5.

Pateman, J. A. (1959a). *Heredity*, **13**, 1–21.

Pateman, J. A. (1959b). In *The Evolution of Living Organisms*, pp. 203–12, Royal Society of Victoria, Australia.

Pateman, J. A. (1960). *Z. Vererbungslehre*, **91**, 380–90.

Pateman, J. A. and Cove, D. J. (1967). *Nature*, **215**, 1234–7.

Pateman, J. A., Cove, D. J., Rever, B. M. and Roberts, D. B. (1964). *Nature*, **201**, 58–60.

Pateman, J. A. and Fincham, J. R. S. (1958). *Heredity*, **12**, 317–32.

Pateman, J. A. and Lee, B. T. O. (1960). *Heredity*, **15**, 351–61.

Pateman, J. A., Rever, B. M. and Cove, D. J. (1967). *Biochem. J.*, **104**, 103–11.

Pellizzari, E. D., Kuć, J. and Williams, E. B. (1970). *Phytopathology*, **60**, 373–6.

344

Perkins, D. D. (1949). *Genetics,* **34,** 607–26.

Perkins, D. D. (1962). *Genetics,* **47,** 1253–74.

Perkins, D. D. (1969). *Genetics,* **61,** Supplement, Abstr. 847.

Perkins, D. D., Newmeyer, D., Taylor, C. W. and Bennett, D. C. (1969). *Genetica,* **40,** 247–78.

Person, C. (1959). *Canad. J. Bot.,* **37,** 1101–30.

Person, C. (1966). *Nature,* **212,** 266–7.

Person, C. (1967). *Canad. J. Bot.,* **45,** 1193–1204.

Person, C. (1968). In *The Fungi* (Eds. G. C. Ainsworth and A. S. Sussman), Vol. III, pp. 395–415, Academic Press, London.

Person, C. and Mayo, G. M. (1974). *Canad. J. Bot.,* **52,** 1339–47.

Person, C., Samborski, D. J. and Rohringer, R. (1962). *Nature,* **194,** 561–2.

Peterson, D. H. (1963). In *Biochemistry of Industrial Microorganisms* (Eds. C. Rainbow and A. H. Rose), pp. 538–606, Academic Press, London.

Pincheira, G. and Srb, A. M. (1969). *Amer. J. Bot.,* **56,** 846–52.

Pittenger, T. H. (1954). *Genetics,* **39,** 326–42.

Pittenger, T. H. (1956). *Proc. Natn. Acad. Sci. U.S.A.,* **42,** 747–52.

Pittenger, T. H. and Atwood, K. C. (1954). *Genetics,* **39,** 987–8 (Abstr.).

Pittenger, T. H. and Atwood, K. C. (1956). *Genetics,* **41,** 227–241.

Pittenger, T. H. and Brawner, T. G. (1961). *Genetics,* **46,** 1645–63.

Pittenger, T. H., Kimball, A. W. and Atwood, K. C. (1955). *Amer. J. Bot.,* **42,** 954–8.

Pomper, S. and Atwood, K. C. (1955). In *Radiation Biology* (Ed. A. Hollaender), pp. 431–53, McGraw Hill, New York.

Pontecorvo, G. C. (1946). *Cold Spring Harbor Symp. Quant. Biol.,* **11,** 193–201.

Pontecorvo, G. C. (1947). *Analyst,* **71,** 411–3.

Pontecorvo, G. C. (1949). *J. Gen. Microbiol.,* **3,** 122–6.

Pontecorvo, G. C. (1959). *Trends in Genetic Analysis,* Columbia University Press, New York.

Pontecorvo, G. C. and Gemmell, A. R. (1944). *Nature,* **154,** 532–3.

Pontecorvo, G. C. and Käfer, E. (1958). *Adv. Genetics,* **9,** 71–104.

Pontecorvo, G. C. and Roper, J. A. (1952). *J. Gen. Microbiol.,* **6,** vii (Abstr.).

Pontecorvo, G. C., Roper, J. A. and Forbes, E. (1953a). *J. Gen. Microbiol.,* **8,** 198–210.

Pontecorvo, G. C., Roper, J. A., Hemmons, L. M., MacDonald, K. D. and Bufton, A. W. J. (1953b). *Adv. Genetics,* **5,** 141–239.

Pourquié, J. (1968). *Biochim. Biophys. Acta,* **209,** 269–77.

Powers, H. R. and Sando, W. J. (1957). *Phytopathology,* **47,** 453 (Abstr.).

Prasad, I. (1969). *Experientia,* **25,** 428.

Pratt, R. G. and Green, R. J. (1973). *Canad. J. Bot.,* **51,** 429–36.

Pritchard, R. H. (1954). *Caryologia,* **6,** (suppl.), 1117.

Pritchard, R. H. (1955). *Heredity,* **9,** 343–71.

Pritchard, R. H. (1960). In *Microbial Genetics* (Eds. W. Hayes and R. C. Clowes), pp. 155–80, Cambridge University Press.

Prout, T., Huebschman, C., Levene, H. and Ryan, F. J. (1953). *Genetics,* **38,** 518–29.

Prud'homme, N. (1970). *Molec. Gen. Genet.,* **107,** 256–71.

Puhalla, J. E. (1970). *Genet. Res.,* **16,** 229–32.

Puhalla, J. E. and Mayfield, J. E. (1974). *Genetics,* **76,** 411–22.

Putrament, A. (1967). *Molec. Gen. Genet.*, **100**, 307–20; 321–36.

Raa, J. and Sijpesteijn, A. (1968). *Neth. J. Pl. Path.*, **74**, 229–31.

Radding, C. M. (1973). *A. r. Genetics*, **7**, 87–111.

Radhakrishnan, A. N., Wagner, R. P. and Snell, E. E. (1960). *J. Biol. Chem.*, **235**, 2322–31.

Raestad, R. (1941). *Nyt mag. Naturvid.*, **81**, 207–31.

Rainbow, C. (1970). In *The Yeasts* (Eds. A. H. Rose and J. S. Harrison), Vol. III, pp. 147–224, Academic Press, London.

Raper, J. R. (1950). *Bot. Gaz.*, **112**, 1–24.

Raper, J. R. (1954). In *Sex in Microorganisms* (Ed. D. H. Weinrich), pp. 42–87, A.A.A.S., Washington.

Raper, J. R. (1966). *Genetics of Sexuality in Higher Fungi*, Ronald Press, New York.

Raper, J. R., Boyd, D. H. and Raper, C. A. (1965). *Proc. Natn. Acad. Sci. U.S.A.*, **53**, 1324–32.

Raper, J. R. and Miles, P. G. (1958). *Genetics*, **43**, 530–46.

Raper, J. R. and Raper, C. A. (1972). *Mushroom Sci.*, **8**, 1–9.

Raper, J. R., Raper, C. A. and Miller, R. E. (1972). *Mycologia*, **64**, 1088–117.

Raper, J. R. and San Antonio, J. P. (1954). *Amer. J. Bot.*, **41**, 69–86.

Raper, K. B., Alexander, D. F. and Coghill, R. D. (1944). *J. Bacteriol.*, **48**, 639–59.

Raper, K. B. and Fennell, D. I. (1953). *J. Elisha Mitchell Sci. Soc.*, **69**, 1–29.

Raper, K. B. and Fennell, D. J. (1965). *The Genus Aspergillus*, Williams and Wilkins, Baltimore.

Raut, C. and Simpson, L. W. (1955). *Arch. Biochem. Biophys.*, **57**, 218–28.

Reaume, S. E. and Tatum, E. L. (1949). *Arch. Biochem.*, **22**, 331–8.

Reddick, D. and Mills, W. (1938). *Amer. Potato J.*, **15**, 29–34.

Rees, H. and Jinks, J. L. (1952). *Proc. Roy. Soc.*, **B140**, 100–6.

Reese, E., Sanderson, K., Woodward, R. and Eisenberg, G. M. (1949). *J. Bacteriol.*, **57**, 15–21.

Rizet, G. (1957). *C. r. hebd. Séanc. Acad. Sci.*, Paris, **244**, 663–5.

Rizet, G., Lissouba, P. and Mousseau, J. (1960). *C. r. Séanc. Soc. Biol.*, Paris, **11**, 1967–70.

Rogers, R. P. (1968). *Canad. J. Bot.*, **46**, 1337–40.

Rogers, R. P. (1973). *Evolution*, **27**, 153–60.

Roman, H. (1955). *C. r. Trav. Lab. Carlsberg Sér. Physiol.*, **26**, 299–314.

Roman, H. (1963). In *Methodology in Basic Genetics* (Ed. W. J. Burdette), pp. 209–27, Holden-Day, San Francisco.

Roman, H., Phillips, M. M. and Sands, S. M. (1955). *Genetics*, **40**, 546–61.

Roper, J. A. (1950). *Nature*, **166**, 956.

Roper, J. A. (1952). *Experientia*, **8**, 14–5.

Roper, J. A. (1966). In *The Fungi* (Eds. G. C. Ainsworth and A. S. Sussman), Vol. II, pp. 589–617, Academic Press, London.

Roper, J. A. (1973). In *The genetics of Industrial Microorganisms* (Eds. Z. Vaněk, Z. Hošťálek and J. Cudlín), Vol. II, pp. 81–8, Elsevier, Amsterdam.

Roper, J. A. and Nga, B. H. (1969). *Genet. Res.*, **14**, 127–36.

Roshal, J. Y. (1950). *Ph.D. Thesis*, University of Chicago.

Rossen, J. M. and Westergaard, M. (1966). *C. r. Trav. Lab. Carlsberg Sér. Physiol.*, **35**, 233–60.

Rossignol, J. L. (1967). Cited in Fincham, (1970). See p. 334

Rothschild, D. and Suskind, S. R. (1966). *Science*, **154**, 1256.

Rowell, J. B. (1954). *Phytopathology*, **44**, 504.

Rowell, J. B., Loegering, W. Q. and Powers, H. R. (1963). *Phytopathology*, **53**, 932–7.

Rudert, F. and Halvorson, H. O. (1963). *Bull. Res. Coun. Israel*, Sect., **A11**, 337–44.

Ryan, F. J. (1947). *Cold Spring Harbor Symp. Quant. Biol.*, **11**, 215–27.

Ryan, F. J. and Lederberg, J. (1946). *Proc. Natn. Acad. Sci. U.S.A.*, **32**, 163–73; 293.

Sackston, W. E. (1962). *Canad. J. Bot.*, **40**, 1449–58.

Samborski, D. J. and Dyck, P. L. (1968). *Canad. J. Genet. Cytol.*, **10**, 24–32.

Sanford, G. B. and Skoropad, W. P. (1955). *Canad. J. Microbiol.*, **1**, 412–5.

Sansome, E. R. (1959). *Nature*, **184**, 1820.

Sansome, E. R. and Bannan, L. (1946). *Lancet*, **250**, 828–9.

Sarazin, A. (1952). *M.G.A. Bull.*, **33**, 281–5.

Sass, J. E. (1929). *Amer. J. Bot.*, **16**, 663–701.

Saunders, M. M. (1956). *M.Sc. Thesis*, University of Liverpool.

Schiemann, E. (1912). *Z. Vererbungslehre*, **8**, 1–34.

Schippers, B. and Snyder, W. C. (1967). *Phytopathology*, **57**, 328 (Abstr.).

Schroeder, N. T. and Provvidenti, R. (1969). *Plant Dis. Rep.*, **53**, 271.

Schwinghamer, E. A. (1959a). *Phytopathology*, **49**, 260–9.

Schwinghamer, E. A. (1959b). In *Plant Pathology: Problems and Progress 1908–1958* (Ed. C. S. Holton), pp. 192–201, University of Wisconsin Press, Madison.

Scott, W. A. and Tatum, E. L. (1970). *Proc. Natn. Acad. Sci. U.S.A.*, **66**, 515–22.

Sermonti, G. (1954). *Rend. Inst. Sup. Sanità*, **17**, 1348.

Sermonti, G. (1957). *Genetics*, **42**, 433–43.

Sermonti, G. (1959). *Ann. N.Y. Acad. Sci.*, **81**, 950–66.

Sermonti, G. (1961). *Sci. Repts. Inst. Super. Sanità*, **1**, 449–54.

Sermonti, G. (1969). *Genetics of Antibiotic-producing Microorganisms*, Wiley, London.

Setlow, R. B. (1964). *J. Cell. Comp. Physiol.*, **64** (suppl.), 51–68.

Sherman, F., Stewart, J. W., Panker, J. H., Putterman, G. J., Agrawal, B. B. L. and Margoliash, E. (1970). *Symp. Soc. Exptl Biol.*, **24**, 85–107.

Shipton, P. J. (1973). *2nd. Int. Congr. Pl. Path.*, Abstr. 0836.

Shult, E. E. and Lindegren, C. C. (1959). *Canad. J. Genet. Cytol.*, **1**, 189–207.

Siddiqi, O. H. (1961). *Genet. Res.*, **3**, 68–89.

Sidhu, G. and Person, C. (1971). *Canad. J. Genet. Cytol.*, **13**, 173–8.

Sidhu, G. and Person, C. (1972). *Canad. J. Genet. Cytol.*, **14**, 209–13.

Silva, J. (1972). *Physiol. Pl. Path.*, **2**, 333–7.

Simchen, G. (1965). *Genetics*, **51**, 709–21.

Simchen, G. (1966a). *Heredity*, **21**, 241–63.

Simchen, G. (1966b). *Genetics*, **53**, 1151–65.

Simchen, G. (1967). *Genet. Res.*, **9**, 195–210.

Simchen, G. and Connolly, V. (1968). *Genetics*, **58**, 319–26.

Simchen, G. and Jinks, J. L. (1964). *Heredity*, **19**, 629–49.

Simchen, G. and Stamberg, J. (1969). *Nature*, **222**, 329–32.

Singer, R. (1961). *Mushrooms and Truffles, Botany, Cultivation and Utilization*, L. Hill, London.

Sisler, H. D. (1971). *Proc. 2nd Int. Congr. Pesticide Chem.*, IUPAC (Tel-Aviv), 323.

Slonimski, P. P., Perrodin, G. and Croft, J. H. (1968). *Biochem. Biophys. Res. Comm.*, **30**, 232–9.

Smith, A. H. (1934). *Mycologia*, **26**, 305–11.

Smith, B. R. (1965). *Heredity*, **20**, 257–76.

Smith, B. R. (1968). *Heredity*, **21**, 481–98.

Smith, B. R. (1966). *Proc. XII Int. Conf. Genetics (Tokyo)*, **1**, 36.

Smith, D. G., Marchant, R., Maroudes, N. G. and Wilkie, D. (1969). *J. Gen. Microbiol.*, **56**, 47–54.

Smith, G. (1969). *An introduction to industrial mycology*. 6th edn., Arnold, London.

Snider, P. J. (1963a). *Genetics*, **48**, 47–55.

Snider, P. J. (1963b). *Amer. J. Bot.*, **50**, 255–62.

Snider, P. J. (1965). In *Incompatibility in Fungi* (Eds. K. Esser and J. R. Raper), pp. 52–70, Springer, Berlin.

Snow, R. (1966). *Nature*, **211**, 206–7.

Snyder, W. C. (1961). *Recent Advances in Botany*, Vol. I, pp. 371–4, University of Toronto Press, Toronto.

Somers, J. M. and Bevan, E. A. (1969). *Genet. Res.*, **13**, 71–83.

Sorger, G. J. (1963). *Biochem. Biophys. Res. Comm.*, **12**, 395–401.

Sorger, G. J. (1965). *Biochim. Biophys. Acta*, **99**, 234–45.

Sorger, G. J. (1966). *Biochim. Biophys. Acta*, **118**, 484–94.

Sost, H. (1955). *Arch. Protistenk*, **100**, 541–64.

Srb, A. M. (1958). *Cold Spring Harbor Symp. Quant. Biol.*, **23**, 269–77.

Srb, A. M. and Horowitz, N. H. (1944). *J. Biol. Chem.*, **154**, 129–39.

Stadler, D. R. and Towe, A. M. (1962). *Genetics*, **47**, 839–46.

Stadler, D. R. and Towe, A. M. (1963). *Genetics*, **48**, 1323–44.

Stahmann, M. A. and Stauffer, J. F. (1947). *Science*, **106**, 35–6.

Stakman, E. C. and Harrar, J. G. (1957). *Principles of Plant Pathology*, Ronald Press, New York.

Stakman, E. C. and Levine, M. N. (1922). *Minn. Agr. Expt. Sta. Tech. Bull.*, **8**.

Stakman, E. C., Levine, M. N. and Cotter, R. U. (1930). *Sci. Agr.*, **10**, 707–10.

Stakman, E. C., Levine, M. N. and Leach, J. G. (1919). *J. Agr. Res.*, **16**, 103–5.

Stakman, E. C. and Piemeisel, F. J. (1917). *J. Agr. Res.*, **10**, 429–96.

Stakman, E. C., Stewart, D. M. and Loegering, W. Q. (1962). *U.S.D.A. Agric. Res. Serv.*, **E617**.

Stamberg, J. (1968). *Molec. Gen. Genet.*, **102**, 221–8.

Stamberg, J. (1969). *Heredity*, **24**, 361–8.

Stamberg, J. and Koltin, Y. (1973). *Genet. Res.*, **22**, 101–11.

Stamberg, J. and Simchen, G. (1970). *Heredity*, **25**, 41–52.

Stark, C. (1961). *Die Gartenbauwiss*, **26**, 493–528.

Stauffer, J. F. (1961). *Sci. Repts. Inst. Sup. Sanità*, **1**, 441–6.

Stebbins, G. L. (1950). *Variation and Evolution in Plants*, Columbia University Press, New York.

Steinberg, R. A. (1937). *J. Agr. Res.*, **57**, 569–74.

Steinberg, R. A. and Thom, C. (1940). *Proc. Natn. Acad. Sci. U.S.A.*, **26**, 363–6.

Steinberg, R. A. and Thom, C. (1942). *J. Agr. Res.*, **64**, 645–52.

Stern, C. (1936). *Genetics*, **21**, 625–730.

St Lawrence, P. (1956). *Proc. Natn. Acad. Sci. U.S.A.*, **42**, 189–94.

Strickland, W. N. (1958). *Proc. Roy. Soc.*, **B148**, 533–42.

Strickland, W. N. (1960). *J. Gen. Microbiol.*, **22**, 583–8.

Strømnaes, Ø. (1968). *Hereditas*, **59**, 197–220.

Subak-Sharpe, H. (1958). *Proc. Roy. Soc.*, **B148**, 355–9.

Suskind, S. R. and Kurek, L. I. (1959). *Proc. Natn. Acad. Sci. U.S.A.*, **45**, 1193–6.

Swann, M. M. (1962). *Nature*, **193**, 1222–7.

Swiezynski, K. M. (1962). *Acta Soc. Bot. Pol.*, **31**, 169–84.

Sybalski, W. (1952). *Science*, **116**, 46–8.

Taber, W. A. and Taber, R. A. (1967). *The Impact of Fungi on Man*, Rand McNally and Company, Chicago.

Takahashi, T. (1959). *Rep. Kihara Inst. Biol. Res.*, **10**, 57–9.

Takahashi, T. (1964). *Bull. Brev. Sci.*, **10**, 11–22.

Takano, I. and Oshima, Y. (1967). *Genetics*, **57**, 875–85.

Tatum, E. L., Barratt, R. W. and Cutter, V. M. (1949). *Science*, **109**, 509–11.

Tatum, E. L., Barratt, R. W., Fries, N. and Bonner, D. (1950). *Amer. J. Bot.*, **37**, 38–46.

Teas, H. J. (1947). *Ph.D. Thesis*, California Institute of Technology.

Teas, H. J., Horowitz, N. H. and Fling, M. (1948). *J. Biol. Chem.*, **172**, 651–8.

Tector, M. A. and Käfer, E. (1962). *Science*, **136**, 1056–7.

Terry, C. E., Kilbey, B. J. and Branch-Howe, H. (1967). *Radiat. Res.*, **30**, 739–47.

Thanassoulopoulos, C. C., Giannopolitis, C. N. and Kitsos, G. T. (1971). *Phyt. Zeitschr.*, **70**, 114–20.

Thomas, P. L. and Person, C. (1965). *Canad. J. Genet. Cytol.*, **7**, 583–8.

Thorne, R. S. W. (1951). *C. r. Trav. Lab. Carlsberg. Sér. Physiol.*, **25**, 101–40.

Thorne, R. S. W. (1962). In *Colloque sur les levures*, pp. 83–102, École de Brasserie de Nancy, Nancy.

Thornton, R. J. and Johnston, J. R. (1971). *Genet. Res.*, **18**, 147–51.

Thurston, H. D. (1961). *Phytopathology*, **51**, 748–55.

Thurston, H. D. and Eide, C. J. (1952). *Phytopathology*, **42**, 481–2 (Abstr.).

Tillman, R. W. and Sisler, H. D. (1971). *Phytopathology*, **61**, 914.

Toxopeus, H. J. (1956). *Euphytica*, **5**, 221–37.

Turian, G. (1969). *Différenciation Fongique*, Masson et Cie, Paris.

Turkensteen, L. J. (1973). *Agric. Res. Rep.*, **810**.

Turner, W. B. (1971). *Fungal Metabolites*, Academic Press, London.

Tuveson, R. W. and Coy, D. O. (1961). *Mycologia*, **53**, 244–53.

Tuveson, R. W. and Garber, E. D. (1961). *Genetics*, **46**, 485–92.

Ullrich, R. C. (1973). *Mycologia*, **65**, 1234–49.

Vakili, N. G. (1959). *Diss. Abstr.*, **19**, 3103–4.

Vakili, N. G. and Caldwell, R. M. (1957). *Phytopathology*, **47**, 536 (Abstr.).

van der Plank, J. E. (1968). *Disease Resistance in Plants*, Academic Press, London.

Van Dijkman, A. (1972). *Natural resistance of Tomato Plants to Cladosporium fulvum: A Biochemical Study*, Utrecht, The Netherlands.

Van Dijkman, A., Dieleman, S. J. and Kaars Sijpesteijn, A. (1973). *Neth. J. Pl. Path.*, **79**, 70–80.

Van Dijkman, A. and Kaars Sijpesteijn, A. (1971). *Neth. J. Pl. Path.*, **77**, 14–24.

Van Dijkman, A. and Kaars, Sijpesteijn, A. (1973). *Physiol. Pl. Path.*, **3**, 57–67.

Varns, J. C. and Kuć, J. (1972). In *Phytotoxins in Plant Diseases* (Eds. R. K. S. Wood, A. Ballio and A. Graniti), pp. 465–8, Academic Press, London.

Wagner, R. P., Radhakrishnan, A. N. and Snell, E. E. (1958). *Proc. Natn. Acad. Sci. U.S.A.*, **44**, 1047–53.

Ward, H. M. (1903). *Ann. Mycol.*, **1**, 132–47.

Waterhouse, W. L. (1952). *Proc. Linnean. Soc. N.S.W.*, **77**, 209–58.

Watson, I. A. (1957). *Phytopathology*, **47**, 507–9.

Watson, I. A. (1970). *A. r. Phytopathology*, **8**, 209–30.

Watson, I. A. and Luig, N. H. (1962). *Proc. Linnean Soc. N.S.W.*, **87**, 99–104.
Watson, I. A. and Luig, N. H. (1968). *Phytopathology*, **58**, 70–3.
Watson, I. A. and Singh, D. (1952). *J. Australian Inst. Agr. Sci.*, **18**, 190–7.
Webb, R. B., Malina, M. M. and Benson, D. F. (1967). *Genetics*, **56**, 594–5.
Webster, R. K. and Nelson, R. R. (1968). *Canad. J. Bot.*, **46**, 197–202.
Wellman, F. L. and Blaisdell, D. J. (1940). *U.S.D.A. Tech. Bull.*, **705**.
Weresub, L. K. and Gibson, S. (1960). *Canad. J. Bot.*, **38**, 833–67.
Westergaard, M. (1957). *Experientia*, **13**, 224–34.
Westergaard, M. and Mitchell, H. K. (1947). *Amer. J. Bot.*, **34**, 573–7.
Westergaard, M. and von Wettstein, D. (1968). In *Effects of Radiation on Meiotic Systems. Int. Atom. Energy Agency, Vienna, Panel Proc. Ser Sti/Pub.*, **173**, 113–21.
Westergaard, M. and von Wettstein, D. (1970). *C. r. Trav. Lab. Carlsberg*, **37**, 239–68.
Westergaard, M. and von Wettstein, D. (1972). *A. r. Genetics*, **6**, 71–110.
Wheeler, H. (1969). *Proc. Symp. Crop Protection, N.Y. State Agric. Exp. Sta., Geneva,* 9–13.
Wheeler, H. and McGahen, J. W. (1952). *Amer. J. Bot.*, **39**, 110–9.
Whiffen, A. J. and Savage, G. M. (1947). *J. Bacteriol.*, **53**, 231–40.
Whitehouse, H. L. K. (1949). *Biol. Revs.*, **24**, 411–47.
Whitehouse, H. L. K. (1950). *Nature*, **165**, 893.
Whitehouse, H. L. K. (1957). *Nature*, **179**, 162–3.
Whitehouse, H. L. K. (1963). *Nature*, **199**, 1034–40.
Whitehouse, H. L. K. (1965). *Proc. 11th Int. Congr. Genet. 1963*, **2**, 87–8, Pergamon Press, Oxford.
Whitehouse, H. L. K. (1966). *Nature*, **211**, 708–13.
Whitehouse, H. L. K. (1967). *Nature*, **215**, 1352–9.
Whitehouse, H. L. K. (1970). *Biol. Rev.*, **45**, 265–315.
Whitehouse, H. L. K. and Hastings, P. J. (1965). *Genet. Res.*, **6**, 27–92.
Wickerham, L. J. and Burton, K. A. (1954). *J. Bacteriol.*, **67**, 303–8.
Wilcox, M. S. (1928), *Mycologia*, **20**, 3–16.
Wilkie, D. (1963). *J. Mol. Biol.*, **7**, 527–33.
Wilson, C. M. (1952). *Bull. Torrey Bot. Club*, **79**, 139–59.
Wilson, J. B. and Gallegly, M. E. (1955). *Phytopathology*, **45**, 473–6.
Wilson, J. F. (1963). *Amer. J. Bot.*, **50**, 780–6.
Wilson, J. F. (1969). *Proc. XI Int. Bot. Congr. (Seattle)*, **1**, 240.
Wilson, J. F., Garnjobst, L. and Tatum, E. L. (1961). *Amer. J. Bot.*, **48**, 299–305.
Wimber, D. E. and Prensky, W. (1963). *Genetics*, **48**, 1731–8.
Winge, Ö. (1935). *C. r. Trav. Lab. Carlsberg Sér. Physiol.*, **21**, 77–111.
Winge, Ö. (1941). *Scientia Genetica*, **2**, 171–89.
Winge, Ö and Laustsen, O. (1938). *C. r. Trav. Lab. Carlsberg Sér. Physiol.*, **22**, 235–44.
Winge, Ö. and Roberts, C. (1948). *C. r. Trav. Lab. Carlsberg Sér. Physiol.*, **24**, 263–315.
Winge, Ö. and Roberts, C. (1949). *C. r. Trav. Lab. Carlsberg Sér. Physiol.*, **24**, 341–6.
Winge, Ö. and Roberts, C. (1950). *C. r. Trav. Lab. Carlsberg Sér. Physiol.*, **25**, 35–83.
Winge, Ö. and Roberts, C. (1952). *C. r. Trav. Lab. Carlsberg Sér. Physiol.*, **25**, 141–71.

Winkler, H. (1930). *Die Konversion der Gene*, Gustav Fischer, Jena.

Witkin, E. M. (1966). *Radiation Res.* (Suppl.), **6**, 30–53.

Wolfe, M. S. (1971). *Proc. 6th Insectic. Fungic. Conf.* (*Brighton*), **3**, 724.

Wolff, S. (1969). *Int. Rev. Cytol.*, **25**, 279–96.

Woodward, D. O. and Munkres, K. D. (1966). *Proc. Natn. Acad. Sci. U.S.A.*, **55**, 872–80.

Woodward, V. W., de Zeeuw, J. R. and Srb, A. M. (1954). *Proc. Natn. Acad. Sci. U.S.A.*, **40**, 192–200.

Wright, R. E. and Lederberg, J. (1957). *Proc. Natn. Acad. Sci. U.S.A.*, **43**, 919–23.

Wright, S. (1931). *Genetics*, **16**, 97–159.

Wright, S. (1940). In *The New Systematics* (Ed. J. S. Huxley), pp. 161–183. Oxford University Press.

Wright, S. (1948). *Evolution*, **2**, 279–94.

Wynants, J. (1962). *J. Inst. Brew.*, **68**, 350–4.

Yotsuyanagi, Y. (1962). *J. Ultrastruct. Res.*, **7**, 141–58.

Young, C. S. H. and Cox, B. S. (1971). *Heredity*, **26**, 413–22.

Young, C. S. H. and Cox, B. S. (1972). *Heredity*, **28**, 189–200.

Zadoks, J. C. (1961). *T. Pl.-Ziekten*, **67**, 69–256.

Zetterberg, G. (1964). In *Photophysiology* (Ed. A. C. Giese), Vol. II., pp. 247–81, Academic Press, New York.

Zetterberg, G., Jonsson, U. and Karlsson, M. A. (1969). *Hereditas*, **62**, 97–104.

Zimmer, D. E., Schafer, J. F. and Patterson, F. L. (1963). *Phytopathology*, **53**, 171–6.

Zimmerman, F. K., Schwater, R. and Loer, V. (1966). *Z. Vererbungslehre*, **98**, 230–46.

Zollinger, W. D. and Woodward, D. O. (1972). *J. Bacteriol.*, **109**, 1001–13.

Author Index

356

357

Subject Index

Abies, 223, 224, 225
Acenaphthene, 44
Acervulus, 7
Acetate, 118
Achlya, 327
Aconidial growth, 26
Acridine, 295
Acriflavine, 46, 101, 323
Adaptation, *see* Training
Additive effects, 124, 128, 134, 137, 138, 139, 144, 145
Adenine, 30, 35, 36, 41, 63, 74, 77, 81, 82, 83, 110, 114, 245, 247, 292, 293, 309, 310
Aecidiospore (aecidium), 18, 100, 101, 275, 280
Aegricorpus, *see* Host-parasite interaction
Agaricales, 7, 11, 19, 108, 145, 146, 158, 165, 173, 175, 212, 218
Agaricus bisporus, 38, 86, 118, 168, 169, 235, 256, 327
  albidus, 257
  avellanus, 257
  , improvement of 256–258
Aggressiveness, 194, 195, 196, 198, 202, 206, 207, 210, 276–279, 280, 282, 283, 286, 287
Agropyron, 260, 283
  repens, 281
  scabrum, 281
Agrostis, 260, 280
  palustris, 284, 285
Alanine, 41, 309
Alkylating agents, 36, 82
Allele (allelomorph, allelism), *see also* Heteroallele, Homoallele, 54, 64, 84, 89, 94, 96, 101, 122, 124, 129, 137,

154, 163, 164, 166, 174, 226
  , multiple, 210, 268
Allomyces, 4, 8, 231
  arbuscula, 44, 45, 229, 230, 327
  javanicus, 229
  macrogynus, 229, 230
Allopatry, 229
Allopolyploidy, 230, 231
Alopecurus arundinaceus, 284
Amino acid(s), 4, 29, 40, 41, 50, 126, 175, 235, 247, 303, 307, 308, 309, 310, 313, 319
p-amino-benzoic acid, 29, 41, 77, 181, 313
2-amino-purine (2-AP), 32 36
Amixis, 158, 170–171
Ammonia, 316, 317
Amphithallism, *see* Homoheteromixis
Antheridium, 8, 321
Anthranilic acid, 314, 315, 316, 319
Aneuploid, 44, 45, 74, 80
Apex, hyphal, 4, 45, 85, 87, 91, 92, 93, 113, 115, 147, 174, 179, 182, 183, 184, 185
Apomixis, *see* Amixis
Apothecium, 16, 295
Apple, *see Malus*
Arginine, 30, 36, 41, 50, 83, 97, 98, 161, 187, 304, 319
Ascobolus, 16, 72, 324
  immersus, 8, 53, 294, 295, 296, 297, 327
Ascocarp, 8
Ascogenous hyphae, 10, 14, 171
Ascomycetes, 6, 8, 9, 10, 11, 14, 15, 16, 20, 21, 45, 53, 72, 73, 86, 96, 118, 120, 145, 147, 158, 159, 168, 173, 218, 226, 323, 324, 327
Ascospore, 7, 9, 11, 14, 15, 16, 37, 38, 42, 51, 53, 64, 67, 68, 80, 96, 101, 102,

Coupling, 50, 75, 76, 78, 131, 134
Covariance, 137, 138, 139
Cross-over, double, 50, 54, 56, 57, 62, 64, 67, 68, 69, 289
, multiple, 48, 56, 67, 292, 298, 301
, non-reciprocal, 21, 53
, reciprocal, 21, 48, 302
Crossing-over, 21, 48, 52, 69, 72, 76, 82, 126, 288, 289, 290, 291, 292, 293, 298, 302
, mitotic, 21, 72, 74, 75, 76, 77, 78, 81, 82
Crucibulum vulgare, 327
Culture collection, *see* Wild-type stock
, shake, 243, 269
, soil, 238
, submerged, 237, 238
, surface, 237, 238
Cyathus, 19, 327
stercoreus, 61, 165, 231
Cycloheximide, 36
Cynodon dactylon, 284, 285
Cysteine, 41, 115
Cystine, 41
Cytochrome, 115, 118
a, 101, 116
b, 101, 116
c, 116, 307, 308, 309, 310
c-reductase, 316, 317, 318, 319
oxidase, 101
Cytology, nuclear, 6
Cytoplasm, 4, 5, 21, 100, 106, 108, 112, 113, 118, 120, 125, 146, 147
, male-sterile, 286
Cytoplasmic factor, *see* Extra-chromosomal element
inheritance, *see* Extra-chromosomal inheritance
Cytosine, 35
Cytotype, 231

Dactylis, 260
glomerata, 284, 285
Degeneration strain, *see* Stock culture
Deletion, 32, 33, 34, 42, 44, 272
Deuteromycetes, *see* Fungi Imperfecti
Dextrose, *see* Glucose
Diacetyl, 255
Dichlorozoline, 204
Diaphoromixis, 17, 19, 31, 90, 135, 136, 157, 158, 161–165, 166, 173, 211, 222, 223, 224, 226, 324, 325
Diaporthe, 121
perniciosa, 88

Diethyl sulphate (DES), 32, 82
Diepoxybutane (DEB), 32, 239, 240
Differential host, 2, 3, 260, 266, 275, 285
viability, 49, 50, 52
Digitaria sanguinalis, 284
Dikaryon, 4, 8, 10, 11, 14, 17, 19, 48, 80, 81, 83, 84, 86, 87, 91, 96, 97, 104, 125, 127, 132, 135, 136, 137, 138, 139, 141, 142, 145, 146, 165, 166, 168, 170, 172, 180, 201, 218, 219, 220, 221, 222, 223, 228, 229, 259, 264, 266, 272, 273, 274, 275, 276, 277, 279, 325
Dimethirimol, 203, 204
Dimixis, 12, 13, 15, 16, 17, 31, 136, 157, 158, 161–165, 166, 167, 170, 173, 226, 262, 277, 321, 322, 323, 324
Diplodia, 198
Diploid, 4, 9, 12, 15, 17, 20, 21, 44, 45, 47, 48, 49, 53, 73, 74, 75, 76, 78, 80, 81, 82, 97, 101, 102, 103, 111, 125, 127, 135, 145, 151, 155, 159, 186, 192, 195, 201, 202, 230, 245, 246, 247, 248, 249, 250, 253, 254, 255, 273, 288, 289, 293, 297, 312, 317, 322
Diploidization, 166
Diplotene, 291
Disomic, 44, 324
Dispersal, of, fungi, 8, 152, 153, 154, 157, 170, 171, 175, 203, 221
Dispersion phase, *see* Repulsion
DNA, 32, 35, 36, 44, 45, 46, 80, 82, 119, 238, 290, 295, 298, 299, 300, 301, 302, 310, 311, 320
hybrid, 299, 301, 302
, mitochondrial (M-DNA), 45, 46, 103, 114, 115, 116, 119, 323
Dominance, 88, 124, 125, 127, 135, 137, 138, 139, 145, 179, 288
Drift, 174, 175
Drosophila, 20, 289, 303
Dual-phenomenon, 120, 238
Duplication, 42, 43, 44, 248

Effective breeding population, *see* Population size
Effective factor, 124, 125, 127, 128, 129, 130, 131, 132, 140, 145, 192, 193
Effective pairing, 292
Eleusine indica, 284, 285
Elymus, 260
Enrichment techniques, 38–40
Enzyme, 126, 147, 235, 299, 304, 305, 320
, adaptive, 312, 316
, aggregate, 314, 315, 316, 318, 319

Host *cont.*–
  range, 196, 283, 285, 287
  specificity, 202, 223, 224, 280
Humaria granulata, 8
  rutilans, see Neotiella
Hybrid inviability, 214, 226, 228
Hybridization, 229–230, 255, 280
  , introgressive, 229, 230
  , somatic, 281
Hydroxylamine, 35
Hypersensitivity, 270, 276
Hypha, 4, 5, 8, 117, 321, 322, 323, 324, 325
  , apex of, *see* Apex
Hyphal fragments, 180, 247, 256
  fusion, 5, 9, 11, 20, 85, 88, 89, 91, 100, 108, 109, 114, 118, 159, 176, 183, 201, 218, 219, 220, 225, 273, 279, 321, 322, 323, 324, 325
Hypoxanthine, 35, 41, 96, 322

Imperfect sexual reactions, 220
Inbreeding, 154, 157, 162, 168, 170, 172, 216, 221
  potential, 163, 164, 165
Incompatibility, heterogenic, *see also* Isolating mechanisms, 214–221, 231
  , homogenic, *see also* Mating systems, 158, 214
  , reciprocal, 216
Independent assortment, 51, 58–59, 62, 64
Induction, 311, 316, 317
Infertility, 192
Infra-red radiation, *see also* Temperature, 35
Inhibition zone, 88, 90, 215, 216, 217, 219–220, 225
Inositol, 41, 195, 223, 292
Interaction, *see* Epistasis, Genotype-environment, Nucleo-cytoplasmic
Interference, 21, 68–69, 70
  , chiasma, 68, 69, 289
  , chromatid, 68, 69, 289, 290
  , chromosomal, *see* Interference, chiasma
  , negative, 292, 298
Invasive spread, *see* Extra-chromosomal element
Inversion, 33, 42, 68, 296
Ionizing radiations, 29, 30, 32, 33, 34, 40, 42, 43, 49, 82, 205, 239, 240, 249, 271, 272, 295, 299, 302
Isogamy, 159
Isogenic line, 263, 264, 266, 279
Isolating mechanisms, 213–231, 287

, post-zygotic, 226–229
, pre-zygotic, 214, 226, 231
Isolation, by distance, 221–223
  , habitat (ecological), 223–226
  , genetic, *see* Isolation, reproductive
  , reproductive, 149, 213, 216, 218, 219, 220, 221, 223, 225, 226, 228, 229, 230, 231
Isoleucine, 41, 304, 305, 309
Itaconic acid, 29

Juncus effusus, 220

Karyogamy, 8, 9
Kasugamycin, 204
Kitazin, 204

Laccaria laccata, 212
Layered-plate technique, 41
Leptomitus lacteus, 4
Leptotene, 6
Lethal, balanced, 250
Lentinus edodes, 235
Leucine, 30, 41, 181, 182, 310
Life-cycle, 9, 10–11, 12–19
  , asexual, 10, 323
  , diploid, 10, 11, 127
  , haploid (haploid-monokaryotic), 10, 127
  , haploid dikaryotic, 10, 127
  , haploid-diploid, 10, 11, 127
Linkage, 43, 49, 50, 51–52, 54, 56, 57, 58, 59–60, 62, 63, 64, 65, 69, 71, 72, 74, 77, 84, 114, 124, 125, 128, 133, 134, 140, 145, 152, 164, 192, 193, 210, 212, 254, 255, 268, 275, 285, 289, 293, 294, 304, 311, 318
  group, *see also* Chromosome map, 51, 67, 76, 115, 156, 246, 249, 297, 307, 312, 313, 314, 315, 316
Linum usitatissimum, 204, 262, 263, 268, 270
Lolium multiflorum, 284, 285
  perenne, 285
Lychnis, 324
Lycopersicon esculentum, 205, 268, 269, 276, 277, 279, 282, 283
  pimpinellifolium, 276, 282
Lyophilization, 238, 253
Lysine, 30, 41, 50, 103, 187, 235, 309, 310
Lysis, 201, 228

Macroconidium, 38, 324
Macrophomina phaseoli, 6
Maize, *see* Zea

Maltose, 122, 125, 254
Malus, 266, 268, 269
Mannan, 255
Mapping function, 70
, X-ray, 82, 295
Marasmius oreades, 167, 201
Maternal inheritance, 121
Mating ability, 253
Mating system, 9, 17, 20, 136, 154, 157–173, 172, 173, 214, 222, 223, 257, 258, 321, 322, 323, 324, 325
Mating-type, 12, 16, 25, 31, 38, 80, 86, 90, 92, 108, 170, 171, 176, 211, 216, 227, 254, 257, 275, 276, 321
    factor, 31, 48, 63, 71, 83, 87, 158, 162, 163, 165, 167, 168, 201, 212, 289, 324, 325
    , symmetry of, 163, 164
Mean, 122, 127, 129
, components of, 129
Medium, complete, 37, 42, 43, 50, 247, 321, 322, 323, 324, 325
    , minimal, 27, 29, 37, 38, 40, 41, 42, 43, 49, 51, 52, 80, 97, 106, 181, 184, 244, 247, 321, 322, 323, 324, 325
Meiosis, 6, 7, 8, 9, 10, 11, 12, 13, 14, 15, 16, 17, 18, 19, 20, 23, 47, 48, 53, 61, 80, 81, 84, 131, 165, 167, 168, 171, 256, 288, 289, 290, 292, 301, 302, 321, 322, 323
    , four-strand stage of, 290, 291
Meiospore, 7, 11
Melampsora lini, 204, 209, 210, 262, 263, 268, 270, 272, 273, 275, 327
Melandrium, 324
Melanoprotein, 270
Melibiose, 101
Mercury, 121, 179
Merulius himantiodes, 219
Methionine, 36, 41, 50, 81, 110, 181, 223, 245, 304, 309, 319
Methylene blue, 35
Metric characters, 20, 122, 124, 125, 128, 129, 147, 227, 258
Microconidium, 16, 38, 200, 214, 216, 324
Migration, *see* Dispersal
Mildew, downy, *see* Perenosporales
, powdery, *see* Erysiphales
Mitochondrion, 4, 101, 114, 115, 116, 118, 119, 120
Mitomycin C, 49, 82
Mitosis, *see also* Nucleus, 6, 9, 14, 21, 47, 48, 165, 166, 180, 185, 301
Mitospore, 7

Mitotic non-conformity, *see also* Duplication, 44, 248, 276
recombination, *see* Recombination
Modifier, specific, *see also* Polygene, 167, 220
Molybdenum, 316, 317, 318
Monokaryon, 4, 7, 8, 10, 11, 19, 80, 81, 83, 84, 86, 87, 91, 123, 127, 135, 136, 139, 140, 141, 142, 145, 147, 168, 176, 219, 220, 224, 226, 264, 279, 325
Morphological mutant, *see* Mutant
Mortierella, 5
Mosaic, genetical, 176, 279
Mucor, 6, 8, 13, 87
mucedo, 88, 327
Multiple-factor hypothesis, 122, 124
Multivalent, 61, 231
Mushroom, *see* Agaricus
Mutagen, 20, 32–36, 236, 240–241, 249, 250, 252, 254, 258, 312, 321, 323, 324, 325
Mutants, *see also* under individual species, 25, 86, 96, 245, 255
, biochemical, 20, 21, 25, 27–30, 51
, chromosomal, 42–45
, colour, 25, 26, 27, 88, 174, 195, 294, 323
, extra-chromosomal, 45–46
, fermentation, 29, 323
, gene, 272
, identification of, 41–42, 43
, induced, 114, 153, 204, 236, 246, 247, 258, 271
, isolation of, 36–42, 43
, lethal, 29, 31–32, 33, 186, 187, 188, 201, 241, 249
, morphological, 25, 26, 27, 255, 322, 325
, physiological, 30, 31, 322
, resistant to toxic agents etc., 29, 31, 37, 82, 121, 179, 194, 321
, selection of, 38–42, 188–190, 236, 238–240
, spontaneous, 20, 29, 31, 114, 154, 187, 196, 204, 237, 239, 241, 248, 252, 271, 283, 321, 322
, super-suppressor, 30, 110, 111, 119, 126
, suppressor, 30, 82
, temperature-sensitive, 31, 37, 40, 42
Mutation, 6, 85, 106, 149, 151–154, 200, 209, 211, 228, 238
, back, 29
rate, 33, 188, 189, 190, 238, 241, 242, 271, 294

Mycelia sterila, 6, 8, 11
Mycelium, 5, 11, 12, 20, 38, 85, 93, 105, 185, 256, 258
Mycocalia, 327
  denudata, 88, 173, 218, 220
  duriaeana, 172, 173
Mycovirus, *see* Virus

Nectria haematococca, 194
Neotiella rutilans, 290, 291
Neurospora, 8, 16, 31, 35, 38, 40, 44, 48, 68, 97, 157, 238, 252, 289, 290, 292, 295, 303, 312
  crassa, 3, 25, 27, 32, 37, 42, 46, 50, 53, 54, 58, 69, 81, 87, 88, 167, 168, 172, 173, 179, 180, 183, 184, 191, 193, 204, 218, 226, 230, 289, 290, 323, 327
    mut. *a/A*, 65, 66, 67, 90
    mut. *abn*, 115, 116
    mut. *ac*, 110
    mut. *ade*, 32, 34, 43, 67
    mut. *al*, 51, 52, 67, 71, 94, 95, 145, 181, 183, 185, 186, 187
    mut. *am*, 93, 156, 302, 305, 306, 307
    mut. *arg*, 67, 97, 98, 99, 183, 304
    mut. *arom*, 313, 314, 315
    mut. *asco*, 27
    mut. *bal*, 26
    mut. *bu*, 26
    mut. *C*, 89, 90
    mut. *col*, 26, 51, 52, 71, 293, 294
    mut. *cr*, 67
    mut. *cys*, 297
    mut. *D*, 89, 90
    mut. *diffuse*, 25
    mut. *exn*, 116
    mut. *f*, 110, 116
    mut. *frost*, 26
    mut. *gap*, 65, 66
    mut. *his*, 156
    mut. *I/i*, 93
    mut. *iv*, 304, 305
    mut. *leu*, 67, 156, 181, 182
    mut. *lys*, 103, 104, 183
    mut. *m*, 51
    mut. *me*, 296
    mut. *mi*, 101, 103, 104, 110, 115, 116, 118, 119
    mut. *nic*, 183, 185, 186, 187
    mut. *nit*, 319
    mut. *osm*, 31
    mut. *pab*, 181
    mut. *pale*, 65, 66

    mut. *pan/pan-m*, 51, 52, 93, 94, 181, 183, 185, 186
    mut. *pdx/pdx-p*, 51, 52, 72, 293, 294, 297
    mut. *pe*, 51, 52
    mut. *pi*, 26
    mut. *poky*, 101, 116
    mut. *pyr*, 156, 293, 294
    mut. *rec*, 156, 302
    mut. *ro*, 26
    mut. *rg*, 26, 27
    mut. *sc*, 26
    mut. *S/s*, 110
    mut. *sfo*, 181
    mut. *SG*, 116
    mut. *stp*, 116
    mut. *str*, 25
    mut. *su*, 116
    mut. *tan*, 27
    mut. *td*, 30
    mut. *thin*, 25
    mut. *trp*, 70, 98, 99, 315, 316, 319
    mut. *ts*, 27
    mut. *ws*, 27
    mut. *y*, 71
  galapogosgenesis, 173
  sitophila, 65, 86, 110, 168, 173, 226, 230, 289, 323, 327
    mut. *het*, 90
  tetrasperma, 87, 90, 96, 168, 169, 172, 173, 179, 323, 324, 327
    mut. *a/A*, 87, 90
Nicotinic acid, 41, 43, 83, 185, 251
Nidulariales, 170
Nitrate, 316, 318
  reductase, 316, 317, 318, 319
  reduction, 316–318
Nitrite reductase, 316, 317
Nitrogen-mustard, 32, 35, 36, 239, 240, 241, 272
Nitrosoguanidine (NG), 32, 33, 36, 239, 240, 249, 295, 321, 322
Nitrous acid (NA), 32, 35, 49, 82, 240
Non-disjunction, chromosomal, 45, 48, 74, 276
Nuclear association, 8, 10
Nuclear ratio, 92, 93, 97, 180, 181, 183, 184, 185, 256
Nucleic acid, *see also* DNA, RNA, 33
  hydrolysate, 41, 325
Nucleo-cytoplasmic interaction, 103, 109–114
  ratio, 4
Nucleotide, 300, 302, 307, 309

Nucleus, 4, 9, 45, 115, 119
, cytology of, 6
, degeneration of, 215
, division of, *see* Meiosis, Mitosis
, , rate of, 94, 181, 185
, synchronous, 10, 86, 92, 167
fusion of, 7, 8, 48, 53, 84, 104, 159, 171
, membrane of, 6
, migration of, 5, 8, 90, 91, 93, 95, 118,
165, 176, 323, 324, 325
Nutrition, 3–4
Nyctalis, 171
Nystatin, 36, 39, 43

Oidium, 7, 19, 81, 146, 325
Oligogene, *see* Gene, major
One gene–one enzyme concept, 303–310
Oogonium, 160, 321
Oomycetes, 11, 12, 125, 158, 159
Oospore, *see* Zygote
Operator, 310, 311, 312, 315, 318
Operon concept, 310–318, 319
Ophiobolus graminis, 202, 226
avenae, 226
Ophiostoma multiannulatum, 38, 39, 40
Ornithine, 41, 304
Oryza sativa, 201, 284, 285
Ostiole, 7
Outbreeding, 154, 157, 162, 172, 173, 223,
268
potential, 162, 163, 164, 165, 170
Oxathiins, 204

Pachytene, 6, 290
Panaeolus campanulatus, 221
papilionaceus, 221
Panicum virgatum, 284, 285
Pantothenic acid, 41, 52, 93, 94, 186, 187
Panus stypticus, 212
Parasexuality, 9, 20, 21, 47, 48, 49, 154,
155, 156, 230, 236, 246, 247, 251, 252,
273–274, 321, 322, 323, 324, 325
Parasite, 4, 321, 324
, obligate, 18, 31, 157, 194, 195, 259, 286
Parental genome segregation, 248
Paspalum dilatatum, 284
notatum, 284, 285
Passaging, *see* Training
Pathogenicity, *see also* Aggressiveness,
Virulence, 25, 31, 120, 121, 194, 195,
198, 201, 202, 204, 210, 229, 259, 261,
262, 263, 264, 266, 272, 273, 280, 283
Penicillin, 29, 235, 236, 237, 242, 244, 246,
247, 248, 250, 251

, genetics of, 245–247
, improvement of, 237–252
Penicillium, 8, 82
brevicompactum, 204
chrysogenum, 29, 39, 49, 118, 188, 190,
237, 238, 239, 242, 243, 244, 245,
249, 250, 251, 254, 327
mut. *ade*, 245, 246
mut. *met*, 245, 246
mut. *nic*, 246, 249
mut. *pen*, 246
mut. *pro*, 246
mut. t_1–t_5, 246, 251
mut. *w*, 245, 246, 249
mut. *y*, 245, 246
corymbiferum, 204
cyclopium, 93, 184
expansum, 81, 327
notatum, see also P. chrysogenum, 96, 237
stoloniferum, 118
Peniophora, 224, 225
heterocystidea, 225, 226
mutata, 223, 225, 226
pini, 221
pini-duplex, 221
pini-pini, 221
populnea, 225, 226
pseudo-pini, 221
Pennisetum glaucum, 285
spicatum, 284
Perenosporales, 4
Peridiolum, 170, 220
Peridium, 170
Perithecium, 16, 26, 27, 38, 101, 104, 105,
116, 117, 159, 160, 161, 170, 171, 178,
214, 215, 216, 217, 227, 228
Permease, 29, 30, 319
Pestalozzia guepinii, 190
Phalaris arundinacea, 284, 285
tuberosa, 284, 285
Phaseolus, 259
vulgaris, 260
Phenylacetamide, 243
Phenylacetic acid, 243
Phenylalanine, 41, 313, 314, 315
Phialide analysis, 78–79
Phleum pratense, 260, 285
Phoma terrestris, 120
Photoreactivation, 35
Phycomyces, 7, 13, 32
blakesleeanus, 3, 25, 322, 327
mut. *mad*, 31
mut. *piloboloides*, 25
Phycomycetes, 10, 11, 12, 13, 44, 45, 73, 86,

Xanthine, 35
  dehydrogenase, 316, 318
X-irradiation, *see* Ionizing radiation
Xylaria curta, 231

Yeast, *see also Rhodotorula, Saccharomyces,
  Schizosaccharomyces, Sporobolomyces*, 5,
  11, 29, 30, 32, 38, 72, 101, 114, 115,
  116, 119, 167, 213, 235, 236, 254, 255,
  297, 320
  , crossing technique for, 253–254
  extract, 29, 322, 323, 325
  , flavour of, 255
  , flocculation of, 252, 255

, improvement of, 252–255
, killer character in, 119
, sensitive, 119
, tetraploid, 61, 73–74, 255, 322

Zea mays, 207, 284, 285, 286, 324
Zinc, 30
  sulphate, 30
Zoospore, 195, 276, 282, 321
Zygorrhynchus heterogamus, 8
Zygote, 13, 48, 101, 102, 103, 116, 161,
  162, 214, 217, 220, 226, 254, 321, 322,
Zygotene, 6

2